Driving Water Reuse by New Technologies, Land Use and Infrastructure Planning, and Legislation Policies

Driving Water Reuse by New Technologies, Land Use and Infrastructure Planning, and Legislation Policies

Guest Editors

Martin Wagner
Sonja Bauer

MDPI

Basel • Beijing • Wuhan • Barcelona • Belgrade • Novi Sad • Cluj • Manchester

Guest Editors
Martin Wagner
Institut IWAR
Technische Universität
Darmstadt
Darmstadt
Germany

Sonja Bauer
Ostbayerische Technische
Hochschule (OTH)
Amberg-Weiden
Amberg
Germany

Editorial Office
MDPI AG
Grosspeteranlage 5
4052 Basel, Switzerland

This is a reprint of the Special Issue, published open access by the journal *Water* (ISSN 2073-4441), freely accessible at: https://www.mdpi.com/journal/water/special_issues/7O4HV1792H.

For citation purposes, cite each article independently as indicated on the article page online and as indicated below:

Lastname, A.A.; Lastname, B.B. Article Title. *Journal Name* **Year**, *Volume Number*, Page Range.

ISBN 978-3-7258-6075-3 (Hbk)
ISBN 978-3-7258-6076-0 (PDF)
https://doi.org/10.3390/books978-3-7258-6076-0

Contents

About the Editors

Martin Wagner

Martin Wagner graduated from the Technische Universität Darmstadt, Germany, in 1983 and holds a doctoral degree in Civil Engineering. Since 1996, he has been the General Manager of the Institute IWAR, Technical University Darmstadt. In 2006, he was nominated as Professor Honorarium at the Qingdao Technological University, China. Since 2009, he has been the Head of the China Section of the German Water Partnership. In 2014, he received a certificate for Honorary Professorship from Tongji University Shanghai. His expertise focuses on wastewater treatment, aeration and gas transfer, energy in wastewater treatment plants, and water infrastructure solutions for fast growing urban areas. Prof. Wagner was honored with the "White Magnolia" of the Government of Shanghai, the top award for foreigners. He also received the Qilu-laurate, the highest award for foreigners in Shangdong Province (China).

Sonja Bauer

Sonja Bauer studied spatial planning at the Technische Universität Dortmund. This was followed by activities in a private planning office for urban development as well as with a municipality in the field of urban planning and urban development. From 2012 to 2019, Ms. Bauer worked as a research assistant at the Department of Land Management at the Technische Universität Darmstadt, where she received her PhD in 2016. From September 2019 until the end of August 2022, she was a professor for Land Management at Hochschule für Technik Stuttgart. On September 01 2022, Ms. Bauer was appointed to the Ostbayerische Technische Hochschule Amberg-Weiden to represent the professorship of Land Management. Her research focus is on water reuse for sustainable spatial development.

Preface

Water scarcity and the increasing demand for sustainable water management have accelerated global efforts to develop, integrate, and regulate water reuse. This Special Issue brings together a collection of studies that address the technological, spatial, and policy challenges of water reuse, highlighting practical solutions and innovative approaches across diverse contexts. The contributions are grouped into four key thematic areas.

1. Basics and Planning Framework

Urban water management must increasingly rely on resilient and integrated planning approaches. Two contributions address this dimension: "Challenges and Innovations in Urban Drainage Systems: Sustainable Drainage Systems Focus" explores sustainable drainage systems (SUDS) as a paradigm shift in stormwater management, while "From District to City Scale: The Potential of Water-Sensitive Urban Design (WSUD)" emphasizes the transition from localized initiatives to city-scale strategies that align urban design with natural hydrological processes.

2. Wastewater Management and Treated Wastewater Reuse: Regional Case Studies

Case-based insights demonstrate the opportunities and complexities of wastewater reuse at the regional level. "Re-use of Treated Wastewater for Crop Irrigation" examines water suitability, fertilization potential, and the impacts on soil properties, while "Impacts of Wastewater Management and Enhancing the Landscape of the Mae Kha Canal" evaluates a quasi-experimental intervention in Thailand. Complementing these, "Geospatial Approaches to Improve Water Availability through Demand Assessment in Agriculture"presents a case study from Germany, illustrating how spatial analysis supports the efficient allocation of treated wastewater resources.

3. Technical Innovations and Processes for Water/Wastewater Treatment

Technological advancements play a critical role in expanding water reuse. Several contributions present novel materials, processes, and analytical tools. "Designing a High-Performance Oil–Water Filtration System" and "Improved Oil/Water Separation by Employing Packed-Bed Filtration of Modified Quartz Particles" focuses on enhanced quartz-based filtration technologies. "Biomimicry-Inspired Automated Machine Learning for Sustainable Water Reuse" introduces an adaptive system for wastewater treatment, inspired by swarm intelligence. Finally, "Colorimetric Quantification for Residual Poly-DADMAC in Water Treatment" provides a rapid analytical method to monitor treatment efficiency and mitigate the potential risks.

4. Legal Framework and Regulatory Aspects

Sound governance and regulatory frameworks are essential to support water reuse adoption. "Integrating Spatial Concerns into Water Reuse Regulations: Insights from the European Union and the Iberian Peninsula" analyzes how spatial considerations can be incorporated into policy, providing lessons that link legislation with practical planning needs.

Together, these contributions advance our understanding of the multifaceted challenges and opportunities in water reuse. By integrating technical innovation, spatial planning, case-based evidence, and regulatory perspectives, this Special Issue provides a comprehensive overview of current progress and future directions. We hope that it will serve as a valuable resource for researchers, practitioners, and policymakers working toward sustainable and resilient water management systems worldwide.

Martin Wagner and Sonja Bauer
Guest Editors

water

MDPI

Article

Geospatial Approaches to Improve Water Availability through Demand Assessment in Agriculture Based on Treated Wastewater: A Case Study of Weinstadt, Baden-Württemberg

Sourav Karmaker [1,*], Sanchalita Bandyopadhyay [2] and Sonja Bauer [3]

1 Faculty of Computer Science, Mathematics and Geoinformatics, Stuttgart University of Applied Sciences (HFT Stuttgart), 70174 Stuttgart, Germany

2 TELUS International Europe, 45127 Essen, Germany; sanchalita1994@gmail.com

3 Department of Electrical Engineering, Media and Computer Science, Ostbayerische Technische Hochschule Amberg-Weiden, 92224 Amberg, Germany; s.bauer@oth-aw.de

* Correspondence: souravkm021@gmail.com

Abstract: The sustainable management of water scarcity is a globally crucial issue. Germany has established efficient water management systems, but the agricultural sector still struggles with water scarcity as the demand surpasses the available water supply. In this work, the primary aim was to establish a framework for making water accessible for irrigation and additional use in households through the effective utilization of recycled water from wastewater treatment facilities. The research inquiries were focused on evaluating the changes in the CROPWAT agricultural irrigation model, determining the spatial distribution of zonal severity, estimating the capacity of urban roof catchments, and evaluating the economic value addition of retreated water from the existing wastewater treatment plant supply. According to the findings, the annual amount of water required for agriculture in the designated study location is approximately 2.9 million m^3. Although there is no initial need for irrigation water, the demand for irrigation water increases during the development, active growth, and mature stages of maize, winter wheat, and wine grapes, reaching around 189 mm, 223 mm, and 63 mm, respectively. According to our observations, the annual water supply in Weinstadt is around 4 million m^3. On the other hand, the compensated volume of water to the current water supply calculated from the urban roof rainfall is estimated to be 0.8 million m^3, which is considered valuable from an economic standpoint. This economically efficient volume of water would reduce the current treated water supply, which indicates an opportunity for enhanced agricultural irrigation.

Keywords: water scarcity; irrigation water; roof catchment; CROPWAT; Germany

1. Introduction

Water scarcity is growing as a concerning issue in many regions, which makes water a vital resource that must be managed sustainably. The United Nations Sustainable Development Goal 6 "Clean Water and Sanitation" is dedicated to water and aims to significantly increase global water recycling and safe reuse by 2030. Irrigation water reuse practices have been established in different countries, i.e., the USA, Israel, African nations such as Namibia, Windhoek very long years ago and some projects implemented among the EU nations as well. But, in 2012, the European Union recognized the potentiality of water reuse as a solution to the problems of water scarcity and drought [1].

Germany is one of the countries in Europe with a higher water availability compared to other European countries because it is surrounded by oceans and seas, with numerous rivers and lakes running through it. Agriculture still has greater effects on water bodies through substance emissions and alterations to their physical structure. The use of agricultural fertilizers leads to excessive nutrient discharges into water bodies and contributes significantly to nitrate pollution and over-enrichment of nutrients (eutrophication) in rivers, lakes, and seas [2]. Due to this reason, Germany and other EU countries have made outstanding advances in managing their hydraulic resources to meet the needs of the agriculture, industrial, and tourism sectors, as well as the needs of a rising population.

From June 2023 onwards, treated wastewater would be used for irrigation purposes and excess water would be supplied to households. In this regard, a new regulation on the minimum treated water requirements will be imposed to encourage water use in agricultural irrigation and reduce the water scarcity brought about by climate change in the European Union and enable the deployment of water reclamation while promoting the circular economy, protecting the safety of human health and the environment as well [3].

Germany has several large-scale water treatment plants that provide clean and safe drinking water to its residents. Undoubtedly, traditional wastewater treatment systems have improved the quality of life in urban areas and lowered environmental stress [4]. For better reusability of water, the most considerable concern is to know the potentiality of wastewater treatment plants and also consider the urban roof water harvesting for how it adds economic value to the water users [5]. Nevertheless, with the potential for recycling wastewater and the nutrients it contains, the goal of improving sustainable resource management in wastewater treatment has become increasingly important [4].

Different prior research provided a detailed study of water availability, highlighting the depiction of irrigation water requirements, urban water harvesting, and the potential of wastewater treatment plants in assessing water stress scenarios. For addressing the water scarcity issue, a Geographic Information System (GIS)-based study was conducted to determine the potential volume of treated wastewater and afterwards the water quality standardized for suitable irrigation [6]. To meet the rising water needs in the EU, Jodar-Abellan et al. [7] proposed a conceptual model that compared to the present situation in several municipalities where the primary source of increased water requirements is clearly agriculture and urban expansion. Jia et al. [8] focused on the application of GIS techniques to assess the potentiality of wastewater reclamation in different urban land use scenarios in Beijing. Ramirez et al. [9] employed the water–energy–food nexus approach to analyze the impact of capturing, treating, and repurposing wastewater for irrigation. Jaramillo and Restrepo [10] presented a discussion on the positive and negative impacts of using treated wastewater and proposed "end of ripe" conventional solutions so that the irrigation purposes can be improved. To address this concern, Moseki et al. [11] utilized the CROPWAT model to determine the reference evapotranspiration rate (ETo), actual evapotranspiration (ETc), irrigation water requirements (IWR), and the effect of irrigation on yield. According to the study of Bilibio and Hensel [12], water scarcity situations in Germany were analyzed using the FAO-CROPWAT model to calculate the effective precipitation, crop evapotranspiration, actual evapotranspiration, and water deficit from 2014 to 2016. Consequently, Surendran et al. [13] studied the crop water requirements of different paddy varieties by using the CROPWAT model to estimate the climatic water balance and proposed projected water future demand for irrigation, industrial and domestic purposes. The crucial factor in designing irrigation systems is usually determining the effective rainfall to assess the amount of water that should be supplied through irrigation. Bokke and Shoro [14] compared various rainfall models using small-scale weather data and found that the USDA-SC method resulted in the lowest net irrigation water requirements in water-scarce regions, whereas the dependable rain method resulted in the highest net irrigation water requirements in water-sufficient regions.

After the estimation of the IWR in CROPWAT, some of the research articles demonstrated the geo-spatial distribution and some research took remote-sensing techniques for analyzing the crop water demand. Feng et al. [15] applied spatial distribution techniques for planning and managing maize cultivation at various stages of development on irrigated farmlands. Al-Najar [16] demonstrated the Gaza Strip's spatial distribution results of irrigation water needs for citrus, almonds, date palm, and grapes based on data from eight weather stations. Bhardwaj et al. [17] observed the Land Use Land Cover (LULC) analysis for evapotranspiration differentiation from the district-level collected data. But Adamala et al. [18] assessed the Normalized Difference Vegetation Index (NDVI) and LULC for the calculation of crop evapotranspiration. The actual evapotranspiration (ETc) of the wheat crop was estimated using the crop coefficient (Kc) that has relationship with the NDVI outcome maps and the reference evapotranspiration (ETo). Machine learning and multicriteria analysis were used for the modeling and evaluation of reference evapotranspiration. The study by Kadkhodazadeh et al. [19] employed six machine learning techniques, namely multiple linear regression (MLR), multiple non-linear regression (MNLR), multivariate adaptive regression splines (MARS), model tree M5 (M5), random forest (RF) and least-squares boost (LSBoost). In urban areas, roofs are often the primary choice for collecting rainwater. Farreny et al. [20] established guidelines for selecting roofs that would optimize the amount and quality of the collected rainwater on four types of roofs: clay tiles, metal sheet, polycarbonate plastic, and flat gravel. Maqsoom et al. [21] examined a Building Information Modeling (BIM)-based approach considering the water scarcity that created a 3D building model based on the average roof area and population. The water demand in the nearby city of Ludwigsburg of this proposed study area was evaluated using a method that can precisely predict the pressure on local water resources in Germany on a single building level and can also be scaled to a regional level while maintaining detail [22].

Therefore, the primary goal of this research is to create a context for the availability of water for irrigation and household use, estimating the potential of water recycling from wastewater treatment plants in Weinstadt Municipality of Baden Württemberg, Germany. Thus, it is necessary to gather the required factors, environmental derivers, and workflow for building a wastewater reuse framework to fulfill the goals. The main research questions are as follows:

- What is the extent of the agriculture irrigation water requirement (IWR) varied according to agro-climate variable in the CROPWAT model at different crop life stages?
- How can the definition of the urban roof catchment surface yield the volume of the water-retaining capacity for each roof type?
- How to economically compensate the agriculture irrigation and current water supply from the WWTP (wastewater treatment plant) by the urban roof catchment harvesting capacity?

The following section provides a detailed description of the water requirements, including the collection and preparation of data, the use of the CROPWAT model to determine irrigation water requirements, the calculation of urban water requirements, and the estimation of economic value.

2. Materials and Methods

This research consisted of several phases of water requirement calculations (Figure A1). The 1st phase described the datasets collection, preparation, and data types. The 2nd phase described the data processing in CROPWAT (CropWat 8.0 is a decision support tool that has a semantic model for agriculture cropping pattern authentication and irrigation scheduling calculation based on soil, climate, and crop datasets) and spatial integration for agricultural water requirements. The 3rd phase presented the urban water requirement analysis. The 4th phase showed the estimation of economic value and visualizations.

2.1. Study Area

The study area is Weinstadt, a town in Baden Württemberg, Germany. Weinstadt means "wine city", which is located in the Rems-Murr region, with an area of 31.7 km^2. As suggested by its name, Weinstadt is well renowned for its wineries and vineyards. By focusing on the biggest wine grape production area in the southern region and considering the other agriculture crop plantation (maize, winter wheat, etc.), the quantity of irrigation water usage has soared already as well as indeed required more. Nevertheless, the goal of this study is to utilize additional harvested water for irrigation in the form of recycled water from the WWTP, for which the local wastewater treatment facility should also be considered as well. That is why Weinstadt was chosen as the research area (Figure 1).

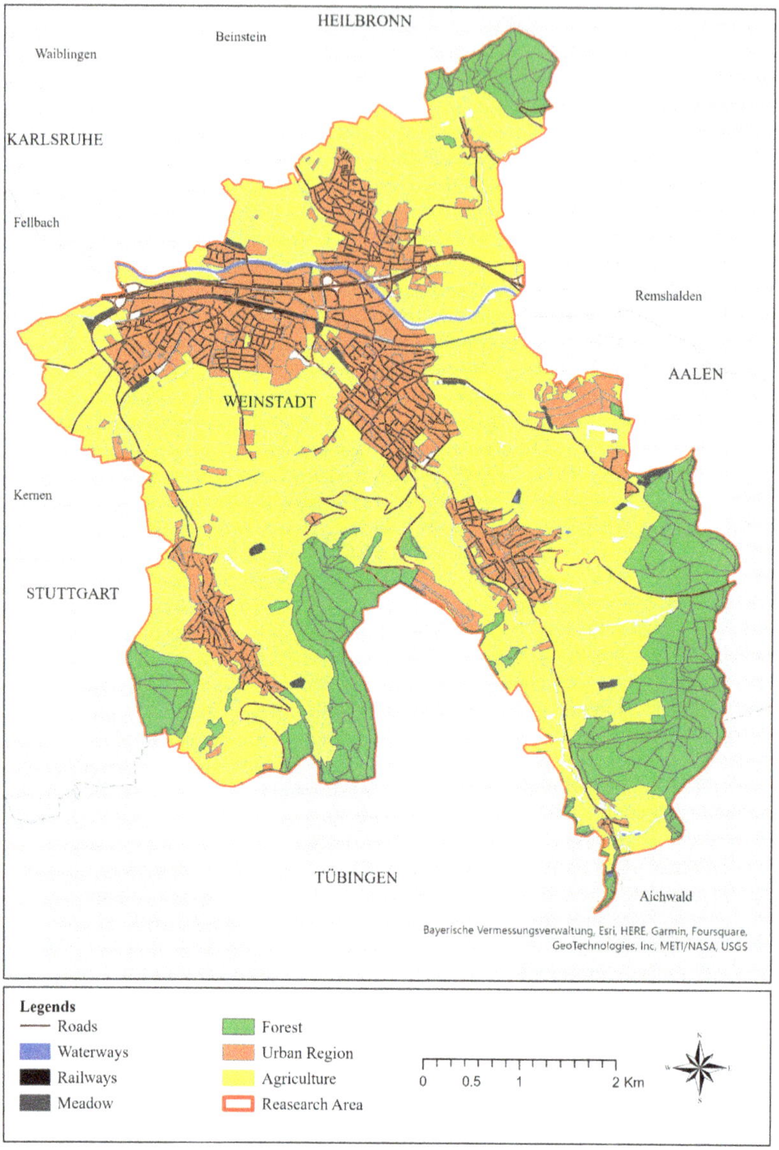

Figure 1. Research area: Weinstadt, Baden-Württemberg.

2.2. Datasets

The most important complexity of this research is the collection and processing of different kinds of datasets. As this research is fully focused on the water requirement issue of Weinstadt Municipality in Baden-Württemberg, the authority of the city planning office has provided the datasets containing the urban point clouds, digital elevation model, digital land use model, digital terrain model, true orthophotos, building level of details (LOD2)—Citygml 2.0 (a file format for virtual 3D city model), ALKIS 2500 land parcel group datasets (ALKIS stands for Liegenschaftskatasterinformationssystem, which contains information about the real estate cadastre), WWTP yearly water yield and topographic maps. It is possible to set up predefined weather stations of Germany by using ClimWat 2.0

weather station datasets. But these stations are very far away from each other. However, the availability of the historic 30-year dataset and nearby distance of 5 cities are selected for manual weather stations, for which the spatial analysis has being carried off. The main climate indicators are considered the minimum and maximum temperatures, precipitations, humidity, average sunshine hours and wind speed for the cities of Karlsruhe, Fellbach, Heilbronn, Aalen and Tübingen. The temperature, humidity, rainfall, and sunshine hours are collected from an international climate database [23]. Here, all the datasets are rendered from the European Centre for Medium-Range Weather Forecasts (ECMWF) data. These historic monthly climatic variables, such as temperature, humidity, precipitation, are collected between 1991 and 2021. Yet, the sunshine hours are recorded from 1999 to 2019. Though the historic monthly wind speed was not available in the previous site, this dataset is collected from an international weather database for the same previous cities [24]. The wastewater treatment plant (Klärwerk Weinstadt) is one of the biggest water treatment plants in the Stuttgart region. This WWTP has a yearly average volume of water 4000,000 m^3 purified. In Weinstadt, three agricultural crops, i.e., winter wheat, wine grapes and maize, are frequently practiced but another crop plantation is less practiced. According to the soil map BGR geoportal of Germany, Stuttgart lies in the region of brown soil with sandy clay rocks [25]. Rinaldi et al. [26] proposed that red wine production is better on loam soils. So, medium soil texture like loam soil is considered for the agriculture water requirement calculation for this research.

2.3. *Determination of Agriculture Water Requirements*

2.3.1. Reference Evapotranspiration

The amount of solar radiation that reaches the topsoil surface is the key factor affecting how much moisture evaporates from cropped soil. When the crop is very young, soil evaporation accounts for the majority of the water loss, but after the crop is mature and completely covers the soil, transpiration takes over as the primary mechanism [27]. The CROPWAT model calculates the crop water requirements and develops the irrigation schedule, which will help with the proper simulation of the water supply. So, CROPWAT windows used the FAO Penman–Monteith method for reference evapotranspiration and radiation with the help of measured weather data (Equation (1)) [15,28].

$$\mathrm{ETo} = \frac{0.408\Delta(\mathrm{Rn} - \mathrm{G}) + \frac{900}{\mathrm{T}+273}\mathrm{u}_2(\mathrm{e_s} - \mathrm{e_a})}{\Delta + \gamma(1 + 0.3442)} \tag{1}$$

where ETo = Reference evapotranspiration; Rn = Net radiation ($MJ/m^2/day$); G = Soil heat flux density ($MJ/m^2/day$); T = Mean daily air temperature at 2 m height (°C); u_2 = Wind speed at 2 m height (m/s); $e_s - e_a$ = Vapor pressure deficit of the air (kPa); Δ = Slope of the vapor pressure (kPa $°C^{-1}$); γ = Psychometric constant (kPa $°C^{-1}$).

2.3.2. Effective Rainfall

The crops may receive this water from irrigation, rainfall, or a mix of irrigation and precipitation. Though there is some rainfall, it is not enough to meet the crop's water demands. Effective rainfall is the portion of precipitation that is retained in the soil profile and aids in crop development [28]. In this study, the dependable rain (FAO/AGLW Formula) method is employed (Equation (2)) because Bokke and Shoro [14] proposed that the dependable rain method is preferable for the water sufficient area and implemented for small-scale irrigation as well.

$$\begin{aligned} \mathrm{Peff} &= 0.6 \times \mathrm{P} - 10 \text{ for Pmonth} \leq 70 \text{ mm} \\ \mathrm{Peff} &= 0.8 \times \mathrm{P} - 24 \text{ for Pmont} > 70 \text{ mm} \end{aligned} \tag{2}$$

where P = Precipitation; Peff = Effective Precipitation in in mm/month.

2.3.3. Irrigation Water Requirement (IWR)

The irrigation water requirement (IWR) is the quantity of water required to meet the crop's water needs following any significant rainfall, for a disease-free crop growing in substantial fields with distortive soil and water conditions and with sufficient nutrients [29]. So, the final irrigation water requirement is determined by calculating the reference evapotranspiration (ETo) and actual evapotranspiration (ETc). Afterwards, the ETc is combined with the decade of crop lifecycle for the final estimation of the irrigation water requirement of every life stage of the crop. This process is a sequential workflow in CROPWAT. Equation (3) is used to determine the actual evapotranspiration, and it is represented by the rate of ETc in mm/day.

$$\mathrm{ETc} = \mathrm{Kc} \times \mathrm{ETo} \quad (3)$$

where ETc = Crop water requirements; Kc = Crop coefficient; ETo = Reference evapotranspiration.

2.3.4. Spatial Distribution of Agricultural Water Requirements

The spatial distributions of winter wheat, wine grapes and maize are performed in two stages. The most used methods that build and maintain a digital elevation model (DEM—replicates the terrain surface of earth landscape [30]) use the Inverse Distance Weighted (IDW) approach, which is demonstrated in this research for the spatial distribution of crop irrigation water requirements (IWRs). The IDW settings for this study are output cell size 5 m, power 2, and search radius "variable", and the cell size projection method is also selected as the preserve resolution. The whole process is performed in the ArcGIS Pro 3.1 mapping environment. In the first stage, the DEM and GIS-assisted method is deployed to calculate the spatial distribution of ETo, ETc, and effective rainfall [15]. Again, with the same interpolation settings, the IDW approach is performed based on the monthly water requirement for the crop cultivation stage like initial, development, middle, late and whole crop growth water. Standard symbology followed the highest values in red color; medium values identified in yellow color and lowest values are in deep sky-blue, with the 10 classes considered for proper IWR visualization. In the second stage, it is very certain that knowing of the specific agriculture area is for each crop. The test ALKIS dataset of Baden-Württemberg clearly defined the ground space and cultivable lands are considered as agricultural lands. So, the agriculture areas for winter wheat and maize are considered green open spaces and current cultivation lands. But separately, there is the indication of wine garden areas. In line with the cultivation practices, it is expected that 70% of areas are defined as the winter wheats area because this crop has the mass production and one of the major crops in Germany than maize. On the contrary, 30% of the area is covered by maize. This estimated area defines the spatial distribution of the irrigation water demand with the calculated IWR from the CROPWAT model.

2.3.5. Agriculture Water Requirements Zonal Model

The zonal model is made to describe which area has more water required and accordingly less water required areas for agriculture irrigation. The ArcGIS model builder tool is used for this water variability processing in different zones of Weinstadt (Figure A2). The primary variable is the land parcel of the agriculture area of cultivation land and wine grounds. From the entire irrigation water requirements (IWRs), interpolation obtained the crop water requirements values, which would be necessary to extract for analysis. Afterwards, data normalization was performed for obtaining the dataset value of the water requirement in a simplified form of 0 to 5 scale (Equation (4)). Moreover, 0 is the lower water required area and, continuously, 5 indicated the higher water required area. So, a weighted parameter index is utilized between the winter wheat area and maize area because same area chosen for the both crops so that all the corps' water requirement lies in a specific percentile in between 100% [31]. Finally, merged all the outputs corresponding to land parcels with the purpose of effectively symbolizing the vulnerable zonal area of Weinstadt according to the zonal severity of the water required situations.

$$Zi = \frac{X - X_{min}}{X_{max} - X_{min}} \times SR \quad (4)$$

where, Zi = Normalized attribute; X = Value attribute; X_{min} = Minimum of value attribute; X_{max} = Maximum of value attribute; SR = Severity scale ranges (0 to 5); Weighted index (cultivable lands) = 70% × [Winter Wheat] + 30% × [Maize].

2.4. *Urban Roof Water Harvesting*

2.4.1. Roof Type Definition

For this analysis, monthly average rainfall datasets were used as the primary water source. The urban roof water harvesting technique is taken under the building information management (BIM) process. In the building level of details dataset, ten types of roofs are identified. The roof types are basically defined by the roof code (standardized roof ID by Open Geospatial Consortium) that each roof has (Figure 2) [32]. These datasets are processed in the ArcGIS pro workspace. Actually, the data interoperability plugin is used by using the quick import tools, which can read the 3D building surface and can be also worked in an attribute table.

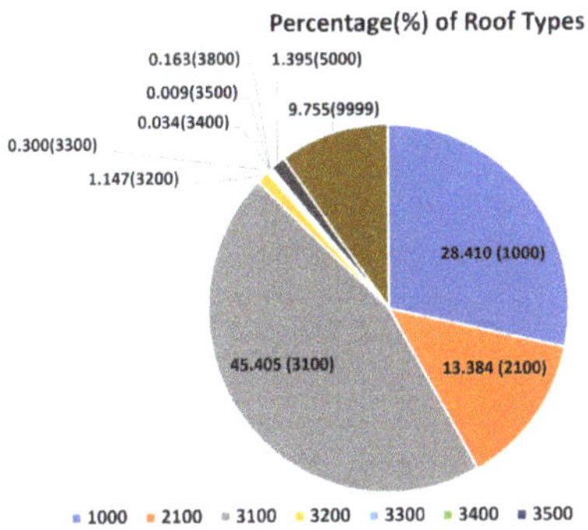

Roof Code	Roof Description
1000	Flat
2100	Pent
2200	Offset pent
3100	Pitched
3200	Hip
3300	Half-hipped
3400	Mansard
3500	Pyramid Hip
3600	Cone
3700	Dome
3800	Shed
3900	Arch
4000	Tower/Hexagonal
5000	Mixed form
9999	Miscellaneous

Figure 2. Percentage of building roof types.

2.4.2. Urban Water Catchment Area

The roof area is the major element of roof water harvesting for every roof type. However, some buildings have flat roofs, and some have sloped roofs. To calculate this inclined area, this study used the sloped area calculation by area calculator tool in the FME (Feature Manipulation Engine) Workbench and wrote it as the .csv format to be joined with the CityGML 2.0 building ID. Therefore, the following Figure 3 shows the FME workbench working schema.

Figure 3. FME (Feature Manipulation Engine) Workbench workflow for sloped area calculation.

2.4.3. Roof Runoff Coefficient

The roof runoff coefficient (RC) is the vital factor used to calculate the roof water harvesting volume. The runoff coefficient is a dimensionless figure that calculates the amount of rainfall that actually becomes runoff after accounting for leakage, spills, soaking of the catchment surface, and evaporation losses. To ensure that the water collection is as effective as possible, the RC must be considered while choosing roof types that have only been evaluated from the CityGML datasets and additional roof types, i.e., green roofs are not taken under analysis. The runoff coefficient is divided into two types. One, sloping roofs, which can be concrete/asphalt, metal and aluminum. Two, flat roofs with bituminous, gravel and level cement. In the Weinstadt area, most of the flat roofs are covered with gravel and the sloped roofs are made with concrete/asphalt. Table 1 has the description of the runoff coefficient.

Table 1. Runoff coefficient [33] (reproduced with permission).

Roof		RC	Reference
Sloping Roof	Concrete/Asphalt	0.9	[34]
	Metal	0.95	[34]
		0.81–0.84	[35]
	Aluminum	0.7	[36]
Flat Roof	Bituminous	0.7	[36]
	Gravel	0.8–0.85	[34]
	Level Cement	0.81	[35]

2.4.4. Catchment Water Volume

The urban roof catchment water volume calculation is the method that will serve the water shortage situations and have economic benefits as well [32]. For this progression, the monthly average rainfall, surface catchment area and the run runoff coefficient area had the relationship to operate for main water volume calculation (Equation (5)).

$$\text{Roof Water Volume} = \text{P} \times \text{A} \times \text{RC} \tag{5}$$

where P = Average precipitation (m); A = Roof catchment area (m^2); RC = Runoff co-efficient.

2.5. Economic Value Assessment

To summarize the final analysis, an economic value assessment has been made, which is the important socio-economic factor in the research. This value assessment is similar to the beneficial part for the municipality, which is performed on the basis of reduction for the current water supply so that the deducted amount of water can be used for future extreme situations. This process was performed by subtracting between the current volume of water supply from the wastewater treatment plant and the calculated harvested water from different roof types. The following Equation (6) shows the economic value estimation.

$$\text{EV} = \text{Ws} - \text{Whp} \tag{6}$$

where EV = Economic value; Ws = Water supply from WWTP; Whp = Harvested water potential.

2.6. Visualization Workflow

Visualization is very vital to show the outcomes of the agriculture and urban water demand. All the layers used in the analysis are in 2D and 3D formats. So, the ArcGIS visualization API is an appropriate platform where all the 2D and 3D layers are deployed as contents. ESRI shapefiles are generally uploaded. But CityGML 3D building objects are converted into scene layer packages (.slpk) file formats with the purpose of being readable by the ArcGIS online application. An application dashboard is customized with viewer controls, layer appearance and legends. All the parcels of the 2D layers of the agriculture water demand and 3D buildings of the rainfall harvesting showed the pop-up tables constructed on their attribute values (Figures A2 and 4).

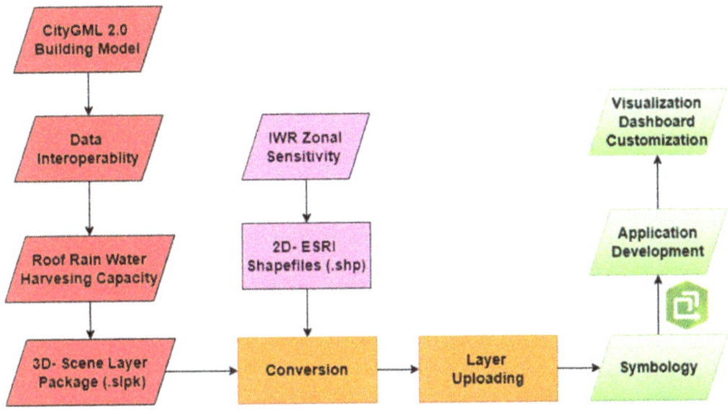

Figure 4. Visualization workflow for all the geospatial layers.

3. Results

Initially, the described results showed the agricultural IWR analysis for the CROPWAT model development outputs and several stages of crop development in various stages. The following stage is the urban roof water harvesting from the perspective of the monthly rainfall. Afterwards, there was included a description of how economic values were added by considering the present water supply.

3.1. Agriculture Water Requirements

3.1.1. Spatial Distribution of Calculated CROPWAT Model Parameters

The spatial distribution in Weinstadt for the water deficit parameters, i.e., ETo (reference evapotranspiration), effective rainfall and ETc (actual evapotranspiration) by which the IWR is defined. The spatial distribution considered the empirical distributions of the calculated CROPWAT model parameters. After performing the Digital Elevation Model (DEM)-based calculations of the spatial distribution procession, the ranges of the reference evapotranspiration values lie between 1.780 and 1.788 (Figure 5a). The highest value for the crop water requirements (actual evapotranspiration) is 1109 mm and the lowest value is 1104 mm (Figure 5b). The effective rainfall value that is calculated from the CROPWAT model for Weinstadt ranged between 419 mm and 429 mm yearly (Figure 5c).

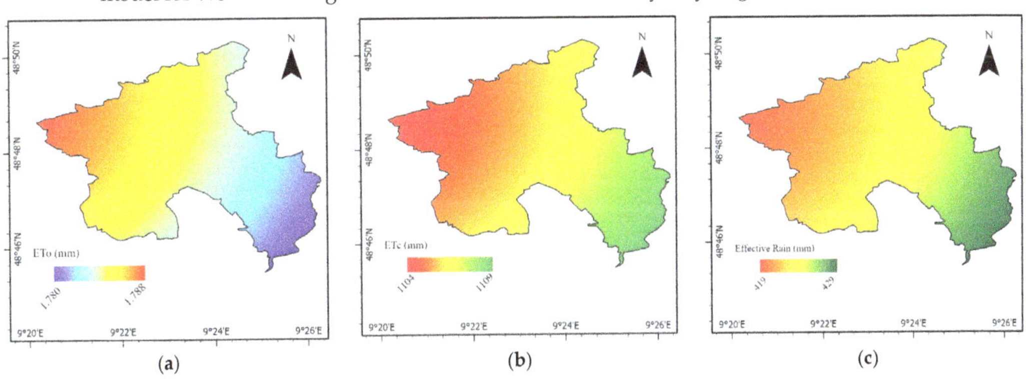

Figure 5. Estimated spatial distribution of (**a**) reference evapotranspiration; (**b**) crop water requirements and (**c**) effective rainfall in Weinstadt Municipality. This figure is the different observations of the CROPWAT model parameters.

3.1.2. Irrigation Water Requirements (IWR) for Maize

Weinstadt's irrigation water demand for maize is estimated from the CROPWAT model. The result is the clipped IWR values of maize in the active growth stage, development stage, maturity stage, and whole stages after the spatially distributed from five cities. Initially, no irrigation water was required in the month of May. The variation in irrigation water requirements in the development stage is very less. The average IWR value is 28 mm in June. In the active growth stage of maize, the irrigation water variation is a very smaller amount in the month of July. Among the 10 classes of the maize active growth stage, higher and lower values range between 113 mm and 115 mm. In the mature stage (starting middle of August and ending in September), the total values ranged between 47 mm and 48 mm. Eventually, a total of 188 mm to 191 mm water is required for the maize in Weinstadt.

3.1.3. Irrigation Water Requirements (IWR) for Wine Grapes

The spatial distribution of wine grapes' IWR has three stages—namely, active growth stage, mature stage, and whole period (cultivation to harvesting). Primarily, no water is required in the initial and development stages from August to the first week of February, because enough water stayed in the ground surface. In the active growth stage, which starts from the mid-term of February and lasts to the late mid-term of June, in that period the volume of irrigation water requirement lies between 40 mm and 42 mm. In the mature stage (late June to July), the ranges of the IWR are 21 mm to 22 mm. So, the total irrigation required water for wine grapes in Weinstadt is about 59 mm to 63 mm.

3.1.4. Irrigation Water Requirements (IWR) for Winter Wheat

The total ranching period of winter wheat is an entire year of time, which starts in September and lasts up to August. The spatial distribution of the winter wheat irrigation water requirement is taken under the four stages, i.e., initial, development, active growth, and fully matured stage. In the initial stage (beginning of September) and early stage of development (till the end of February), no water is needed for irrigation because of the availability of enough precipitation. Therefore, the IWR in the late development stage ranged between 23 mm and 27 mm from March to the mid-term of April. The active growth started immediately after the development stage, which lasted until July and the value of the IWR ranges between 151 mm and 153 mm. The mature stage irrigation water requirement values lie between 49 mm and 50 mm in the months of July and August. Finally, in the whole period of winter wheat cultivation, the total value of the IWR ranges between 221 mm and 226 mm for Weinstadt.

3.1.5. Agriculture Irrigation Water Requirements

Figure 6 has portrayed the varied stages of the IWR for maize, wine grape and winter wheat. Though there was no initial IWR, only wine grapes had no IWR in the development stage and all three crops had higher water requirements in the mature stages. In fact, winter wheat and maize required more water compared to wine grapes. The final mature stage had a much less amount of water required for all the crops.

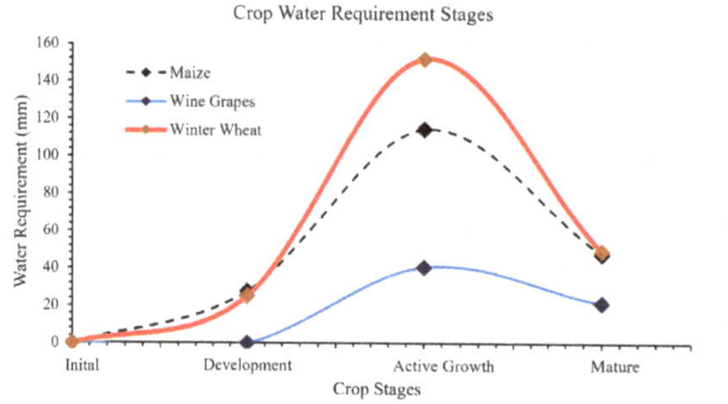

Figure 6. Agriculture water requirement from the CROPWAT model: variation of the irrigation water requirement in the crop lifecycle.

3.1.6. Zonal Severity for IWR

Figure 7a–c show the zonal severity, which generally described the required water of irrigation with spatial merging with the CROPWAT model outputs. The western part Endersbach of Weinstadt has extremely high irrigation water needed for crop cultivation. But the western parts of Grossherppach and Struempfelbach have moderate IWR severity. Although the eastern parts of Struempfelbach, Beutelsbach, and Schnait have medium and lower irrigation water requirements. Because the eastern parts have forest area, that is why the irrigation water requirement is very marginal.

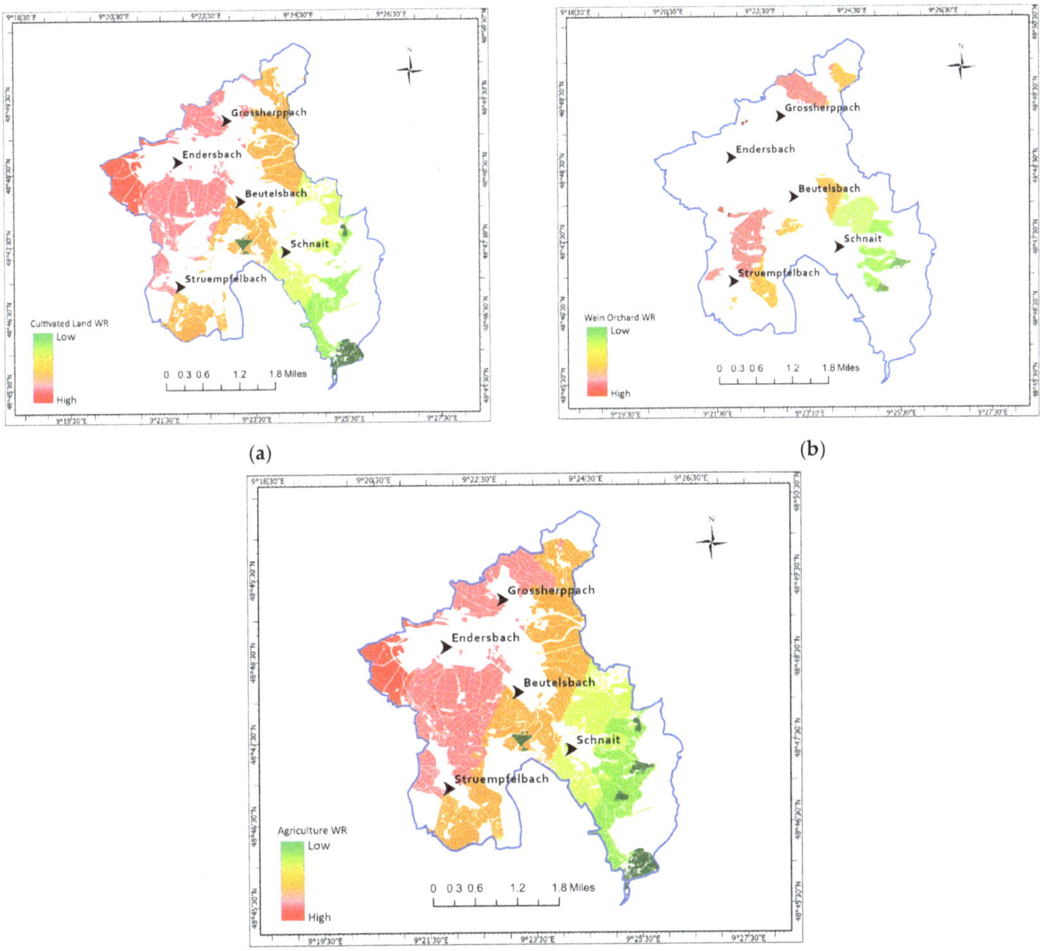

Figure 7. Zonal IWR severity—(**a**) cultivated lands for maize and winter wheat; (**b**) wine orchards and (**c**) final agriculture IWR zonal severity.

By focusing on the zonal severity area, the total volume of required water has been calculated (Table 2). So, the maize total area assumed here is 375×10^3 m^2 and the calculated IWR is 70.8×10^3 m^3. Similarly, winter wheat is having an area of 875×10^3 m^2 with a total required yearly water of 195.3×10^3 m^3. Wine grapes have 28.6×10^3 m^3 of required water for the area 470×10^3 m^2. Definitively, the total volume of agriculture water required is 294.7×10^3 m^3 in Weinstadt Municipality.

Table 2. Volume of total required water in the agricultural fields of Weinstadt.

Crop Name	Area (10^3 m^2)	Potential Irrigation Water Requirement (mm)	Total Water Requirement (10^3 m^3)
Maize	375	189.0	70.8
Winter Wheat	875	223.3	195.3
Wine Grapes	470	60.92	28.6
Total			**294.7**

3.2. Urban Rainfall Water Harvesting

The urban water harvesting from rainfall is mainly carried out by calculation of the roof area. There are other factors, i.e., runoff coefficient, average rainfall. Figure 8 shows the monthly chart and detailed harvested water volume for different roof types. In flat roofs, the yearly rainwater harvested is one of highest roof types, where the value is about 25.1 × 10^3 m^3. The pent roof has a harvested water volume of 5.7 × 10^3 m^3. But the pitched roof type collected the highest volume of 38.9 × 10^3 m^3 water. The roof types like hipped, half-hipped, shed, mixed and miscellaneous have retained the volume of water of 1.2 × 10^3 m^3; 0.3 × 10^3 m^3; 0.6 × 10^3 m^3; 2.2 × 10^3 m^3 and 12.1 × 10^3 m^3. In contrast, the mansard and pyramid roof types have a very little amount of rainwater harvested (0.02 × 10^3 m^3 and 0.005 × 10^3 m^3). Eventually, the total amount of 86 × 10^3 m^3 is harvested in different roof types in Weinstadt.

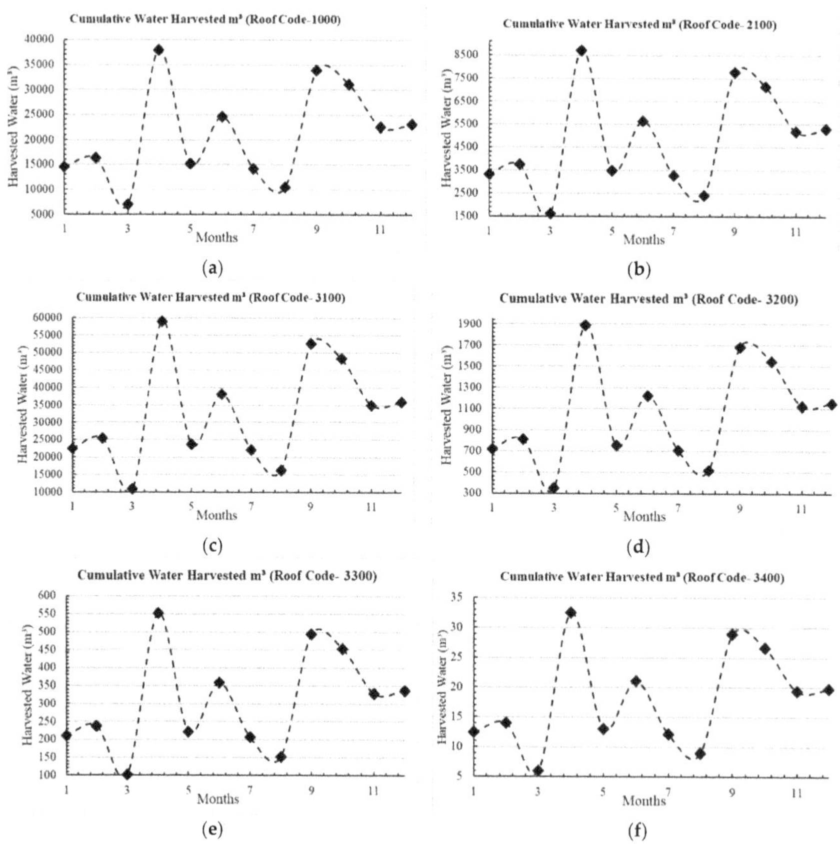

Figure 8. *Cont.*

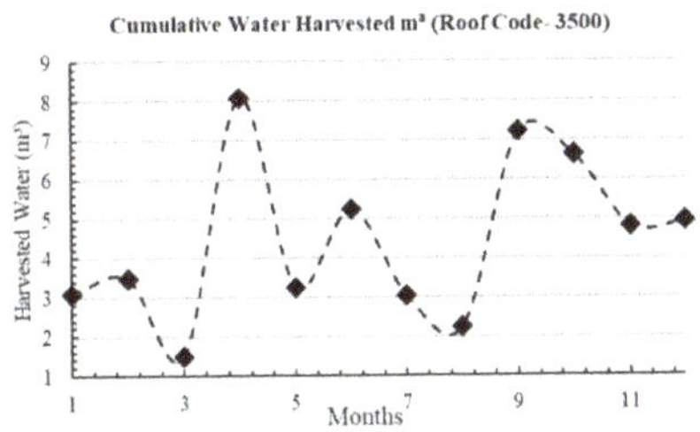

(g)

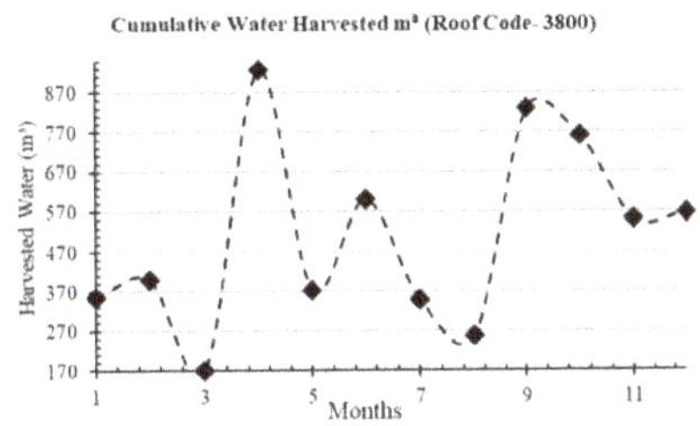

(h)

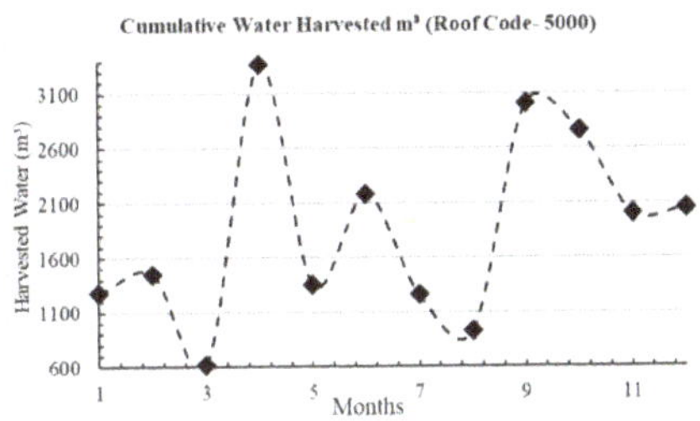

(i)

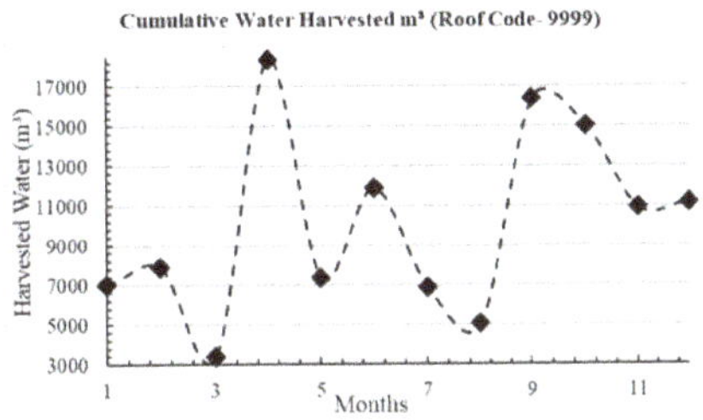

(j)

Figure 8. Volume of harvested water in different roof types. (**a**) flat roof; (**b**) pent roof; (**c**) pitched roof; (**d**) hip roof; (**e**) half-hipped roof; (**f**) mansard roof; (**g**) pyramid roof; (**h**) shed roof; (**i**) mixed form roof; and (**j**) miscellaneous roof.

Figure 9 shows the different building groups based on the higher to lower water harvesting capacity. The roof categories half hipper, mansard, pyramid and shed (3300, 3400, 3500 and 3800) have very lower rainwater harvesting capacity because they are very less in numbers and in areas as well (Figure 9a). But the roof types like pent, hipped, mixed and other types (2100, 3200, 5000, 9999) have more water harvesting category (Figure 9b,c). But the highest group of rainwater harvesting roof types are the pitched roofs (3100) (Figure 9e). These buildings have pitched roofs in residential areas as well as being greater in areas. However, the flat roof types (1000) also have more capacity compared to the other roofs but not other than pitched roofs (Figure 9d).

(a)

(b)

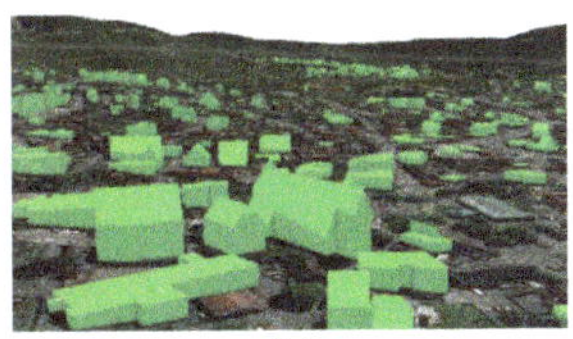

(c)

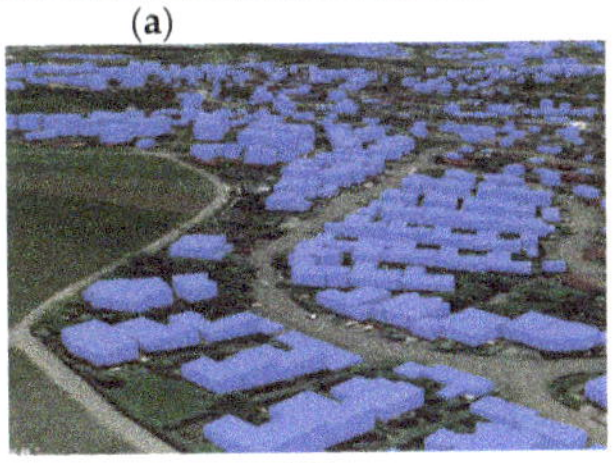

(d)

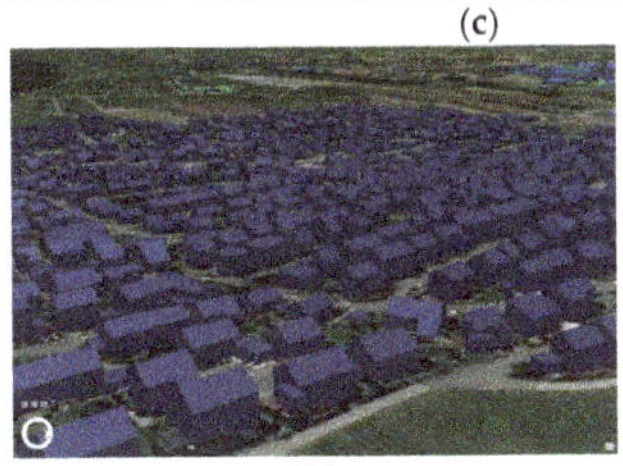

(e)

Figure 9. Symbology of roof types with their rainfall water harvesting capacity. (**a**) very less water harvesting group; (**b**) less water harvesting group; (**c**) moderate water harvesting group; (**d**) high water harvesting group; and (**e**) extremely high-water harvesting group (source of basemap: Esri Web Map).

3.3. Economic Impact of Water Efficiency

In this research, economic value addition is taken to estimate the potential reduced water that would be saved or used for future use purposes. Though this measured amount of water is not possible to calculate in the factual currency value because some factors, including water quality, ecological value and sustainable management, are also interrelated. However, the output of the total yearly rainfall water harvested from different roof types is about 86 × 10^3 m^3. But the total supply chain volume is about 380 × 10^3 m^3 in the year 2022. After comparison between the datasets, the reduced usage amount of potential water is about 294 × 10^3 m^3. So, from this perspective, the rainfall water harvested amount can be the potential water amount of economic values calculated for Weinstadt Municipality. Figure 10 and Table A1 show the total monthly variability of different economic value-induced parameters, i.e., urban catchment water volume, WWTP water volume and economic valued water. All three parameters are correlated to each other. However, the month of April has the highest rainfall as well as the urban roof harvested water quantity is higher in 2022, and the precipitation and water supply from the wastewater treatment plant are also greater. That is why the potential reduced water volume is also higher. Simultaneously, the other months, September, October, November, have a higher water supply from the wastewater treatment plant, but September to February has a higher declining amount of water that would be saved because of urban rainfall water harvesting.

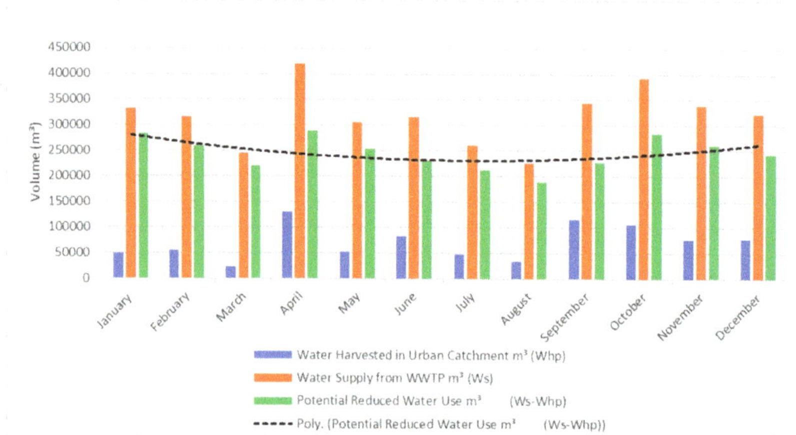

Figure 10. Potential economic irrigation water value.

4. Discussion

The agricultural irrigation water demand is very important to partially estimate the current water scarcity situation. To calculate the exact irrigation water requirement in the selected study area of Weinstadt, the empirical CROPWAT model is employed for different crop types. In the processing of stages, the reference evapotranspiration, effective rainfall, crop water requirements and final irrigation water requirements are measured for every specific crop (Figure 5). Afterwards, the results of the IWR for specific crops like winter wheat, maize and wine grapes have been spatially distributed by implementing a DEM-based approach (IDW) with the closest weather station datasets. In this way, the agricultural water volume is calculated, which is spatially modeled for zonal suitability by defining the multicriteria evaluation (Figure 7). In this research, there is another objective describing the assessments of roofs' ability to retain water depending on the kinds of roofs achieved regarding the urban rainfall water harvesting. The most valuable element for this stage is to understand the CityGML roof categories and to calculate the roof exact area, whether it is sloped or flat. By knowing the various roof runoff coefficients, a very effective volume of harvested water is calculated according to the different roof groups. Among all the roof categories, the residential areas have enormous quantities of runoff catchments and the larger areas calculated as well as the higher harvested volume of water identified (Figure 9). Water scarcity can be eradicated by the reliability of the economic analysis using urban rainfall harvesting. In addition, an economic water economic value estimation is performed by differentiating between the yearly current water supply to the municipality and the measured harvested rainwater. The is the best way to analyze the water supply assessment potentiality of the wastewater treatment plant.

4.1. Comparative Overview with Related Research

Numerous researchers have obtained the agricultural water requirement, urban rainwater availability and economic values in a visualization platform separately. But it is very unusual to find related research that is combined with all the above-mentioned three outputs. So, this study was carried out to overcome the real-life urban agriculture problem by doing similar, which helps to increase the potentiality of water use in an efficient way. The following Table 3 is the optimal description of the water availability analysis with the other methodologies and outputs.

Table 3. Parameters and outputs comparison with previous research.

References	Datasets and Time Frame	Methodological Philosophy	Research Methods	Findings	Study Regions
[37]	Wastewater treatment plant nominal flow rate, soil textures and depth, land use, DEM; Time period: 2009/2010 (Landsat TM imagery), 2000 (Google Earth data and land use map)	Analytical hierarchical process for geospatial integration	Study area characterization by classification, standardizing the sub criteria, sensitivity analysis and cross validation	31% of the aquifer is fitting for irrigation, GIS sensitivity ranking cases 1–5.	Tunisia
[28]	Agro climatic data, crop data showing and harvesting; Time period: 2017–2021 (agroclimatic data)	Irrigation water requirement (IWR) and irrigation scheduling for cultivated crops	CROPWAT model for calculation of Eto and effective rainfalls, calculation of evapotranspiration	IWR: 3108.0 mm-sugarcane, 1768.5 mm—banana, 1655.7 mm—cotton, 402.5 mm—wheat	Pakistan
[13]	Ago-ecological datasets, crop data; Time periods: Not defined	Total water requirement in various agro-ecological zone in order to estimate ground water balance	CROPWAT model 8.0 used for calculation of evapotranspiration	Net irrigation requirement: Paddy: 442 to 1483 mm; water demand: 1146 mm^3	India
[16]	CROPWAT station dataset and crop data; Time period: Not defined	Irrigation water requirement estimation for spatial modeling	CROPWAT model for crop water requirement and water qualitative measurements	IWR: 763 mm/year-citrus, 722 mm/year-almonds, 1083 mm/year-date palm, 591 mm/year-grapes.	Palestine
[15]	Meteorological data; Time period: 1961 to 2001	Spatial distribution of crop water requirement	CROPWAT model for irrigation water requirement and irrigation scheduling, DEM based methods	Spatial distribution of ETc of spring maize 324.57–500.55 mm; water deficit ratio up to 40%	China

Table 3. *Cont.*

References	Datasets and Time Frame	Methodological Philosophy	Research Methods	Findings	Study Regions
[21]	Daily rainfall, 2D and 3D model of building; Time period: January 2014–December 2018 (daily rainfall)	Rainwater harvesting assessment through Building Information Modeling (BIM)	Calculation of potential roofing catchment size, rainwater harvesting potential and fixing of tank capacity	Collected harvested rainfall water volume: 8190 L/yr to 103,300 L/yr	Pakistan
[22]	CityGML building models; Time period: Not defined	Urban water demand assessment	Implementation of water analysis workflow of SimStadt, log-log model for water demand assessment	Industrial water demand: 397 to 579 m^3 and Predicted precipitation: 248 mm by 2030	Germany
Our Proposed Methodology	Climate dataset, wastewater treatment Plant water supply volume, ALKIS maps Time period: 1991–2021 (temperature, humidity, rainfall), 1991–2019 (sunshine hours), 2022 (wind speed), 2021–2022 (WWTP supply), 2021 (ALKIS maps)	Agriculture water demand assessment by using the potentiality of urban rainwater harvesting and wastewater treatment plant supply	Employment of CROPWAT model for IWR, zonal severity analysis, urban roof catchment area measurement, economic value.	IWR estimation: 189 mm-maize, 223.3 mm-winter wheat, 60.92 mm for wine grapes, spatial volume of IWR 294.7×10^3 m^3/yearly. Sensitivity phases 0–5. Rainfall water harvested volume: 86×10^3 m^3/yearly	Germany

4.2. *Influence on Agriculture and Urban Water Demand*

From the modeling outcomes of CROPWAT and the spatial distribution, it is understood that a huge amount of cubic meter water is needed to irrigate in Weinstadt's agriculture area. Most of the agricultural lands are in medium high to severe water required regions. Extremely severe and other classes of lands are very less in percentage. But still every land parcel has water demand. So, Figure 11 shows the consequences of the IWR zonal severity. Most of the irrigation water sources in Germany are natural sources. The main source of irrigation water is ground water. In Germany, approximately 77% of the irrigation water utilized is blue water (the proportion of irrigation water that is attributed to the imported agricultural crops) [2], while 11% comes from surface water and the public or private supply networks, spring water makes up the majority [38]. The amount of ground water or water from rivers and lakes utilized in the production of a product is referred to as blue water. In the production of agricultural crops, this refers to the amount of extra irrigation that is used [2]. Between 80 and 150 mm (mm) or 425 and 800 million cubic meters (m^3) of irrigation water are used annually [39]. According to the federal statistics office of Germany, 364 billion m^3 water was supplied, including irrigation water, from 1961 to 2020. Beyond this huge amount of water supply nationally, the present demand for irrigation water is estimated in this study. If the percentage of the blue water supply is assumed static confronting the irrigation demand, the harvested urban water can repay the rest of the amount or be used for future use when the water supply would be very low. Hence, with the calculated reduced potential, the amount of water will be distributed in urban areas. The rest of the economically valued water will be helped to recharge the ground water in the summer seasons. The outcomes of this study not only improve the subsequent water resources but also explicate the climate change impacts on water resource management.

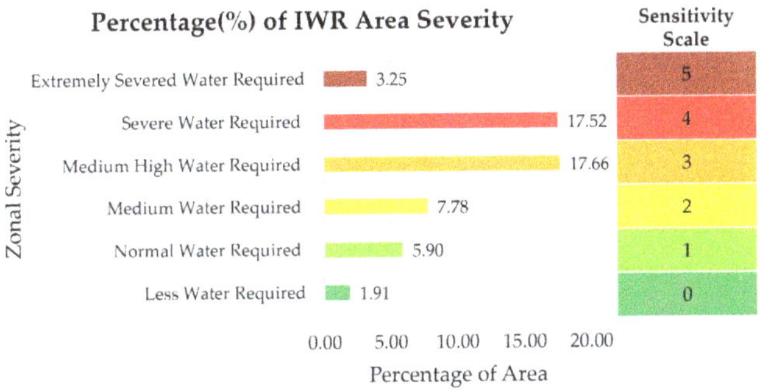

Figure 11. Percentile of zonal agriculture land sensitivity.

5. Conclusions

To conclude, this research on agriculture and urban water availability is very significant for the whole regional system. All three objectives are analyzed in a very scientific way to evaluate the ultimate benefit of encountering future water stress situations as a pilot study for all over Germany.

The agriculture water demand simulation in the CROPWAT model evaluation showed that the irrigation water demand is very high in Weinstadt Municipality all the year round 2022. The comparative aspects of the results, such as the irrigation water requirements, are much higher in the south–western region surrounding the study areas. In addition, the results from the spatial distribution between the crop studies are varied considerably based on their showing to harvesting periods. Wine grapes needed a very lower amount of irrigation water because of the elevated terrain cultivation and substantial amounts of water owing to the grounds. In this regard, winter wheat is the highest irrigation water required crop and maize is the medium conditions in both of their active growth stages. The zonal severity of the IWR is also very effective for this region because this analysis distinguishes the environmental effects. The western part of the study area is more like the urban regions, that is why the water demands in agricultural lands are very severe. Sequentially, when the IWR severity is directed to eastern parts of Weinstadt, lower sensitive areas are observed due to having the green areas.

Another objective signifies the urban water potentiality calculation so that it can be compared with present sources of water supply reductions. Basically, there are 10 types of LOD 2 building roof types observed. In this study, the primary source of urban water harvesting is considered only rainfall, which varied in different roof types. The pitched and flat roofs have the most harvested capacity linked to other types of roofs that have medium or less harvesting capacity. This harvested volume of water is potential water, which reduced the present supply volume from the WWTPs. This study referred to it as the economic value for the IWR. In addition to the outcomes of the analysis, water framework development has been proposed for the municipality and users to be followed. To do so, a visualization application is developed so that the total idea of protecting future scarcity-induced water availability is disseminated easily to the user levels in a very interactive manner. Apart from these above-mentioned analyses, there are many crucial factors to be taken care of in the future, such as database management and decision clarification. These are as follows:

- A comparative analysis can be performed using various methods in the CROPWAT model. In the depth of effective rainfall, other methods would be checked in CROPWAT models for the water sufficient region like Germany.
- The snowfall as a water harvesting source could add the more reduced amount of water, which will improve the more proficient amount of potential economic water values.

- The proper survey data on the degree of sloping, roof materials and roof covering can have huge impacts on the urban water availability potentials.

Author Contributions: S.K.: Conceptualization, methodology, visualizations, writing—original draft preparation; S.B. (Sanchalita Bandyopadhyay): Conceptualization, methodology, analysis—article reviewed; S.B. (Sonja Bauer): Conceptualization, methodology, analysis—article reviewed and supervision. All authors have read and agreed to the published version of the manuscript.

Funding: This research received no external funding.

Data Availability Statement: The data presented in this study are available on request from the corresponding author.

Acknowledgments: The work presented in this paper was developed in the master's thesis project by Sourav Karmaker at Stuttgart University of Applied Sciences (HFT Stuttgart), Germany. The authors express their sincere gratitude to Weinstadt Town Planning Department and Klärwerk Weinstadt authority for providing the geodatasets and supporting this project.

Conflicts of Interest: Sanchalita Bandyopadhyay was employed by TELUS International Europe. The remaining authors declare that the research was conducted in the absence of any commercial or financial relationships that could be construed as a potential conflict of interest.

Appendix A

Table A1. Potential economic impact estimation.

Months	Water Harvested in Urban Catchment (Whp) 10^3 m^3	Water Supply from WWTP (Ws) 10^3 m^3	Potential Reduced Water Use (Ws − Whp) 10^3 m^3
January	4.9	33.3	28.3
February	5.6	31.7	26.1
March	2.4	24.5	22.1
April	13	42	28.9
May	5.2	30.6	25.4
June	8.4	31.6	23.1
July	4.8	26.2	21.4
August	3.6	22.6	19
September	11.6	34.4	22.8
October	10.7	36.6	25.8
November	7.7	34	26.2
December	7.9	32.4	24.4
Total	**86**	**380**	**294**

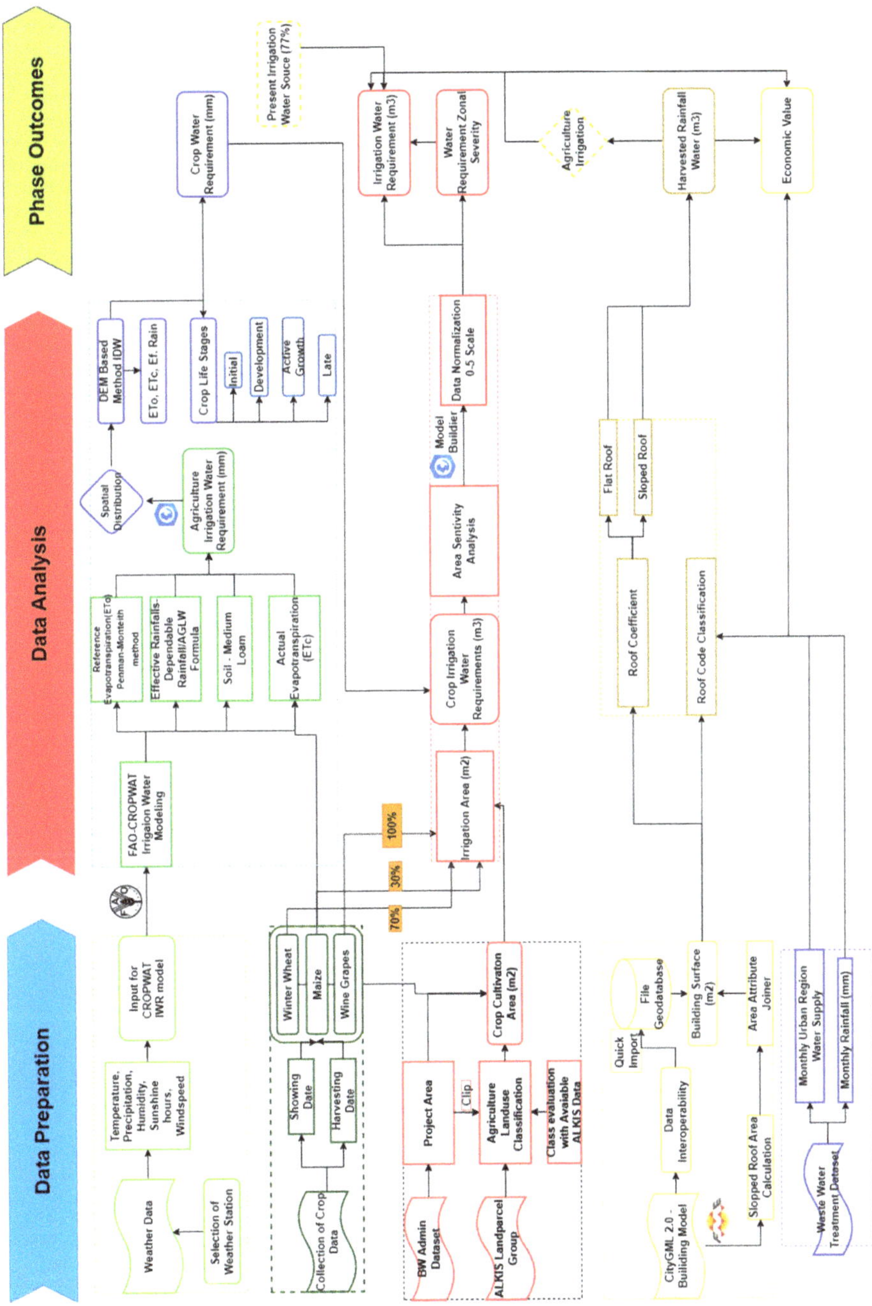

Figure A1. Proposed conceptual IWR framework.

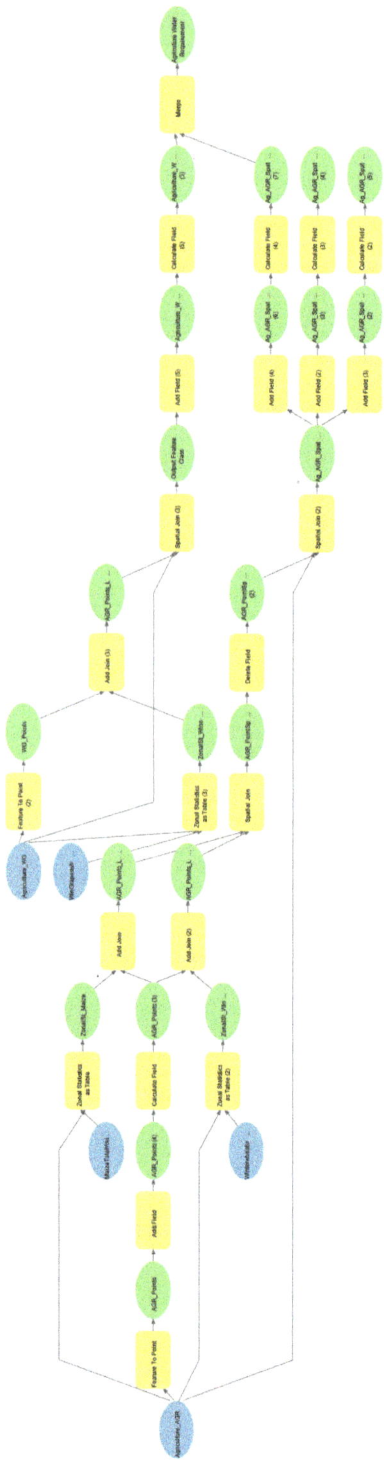

Figure A2. Agriculture water requirement zonal severity model.

Figure A3. User accessible web application.

References

1. Helmecke, M.; Fries, E.; Schulte, C. Regulating water reuse for agricultural irrigation: Risks related to organic micro-contaminants. *Environ. Sci. Eur.* **2020**, *32*, 4. [CrossRef]
2. Stoll, J. Water Resource Management in Germany. Umweltbundesamt. 2018. Available online: https://www.umweltbundesamt.de/publikationen/water-resource-management-in-germany (accessed on 3 February 2023).
3. Schiller, A. Umweltbundesamt. New EU Regulation on Minimum Requirements for Water Reuse. 2021. Available online: https://www.umweltbundesamt.de/en/topics/water/water-resource-management/water-reuse/new-eu-regulation-on-minimum-requirements-for-water (accessed on 26 September 2022).
4. Maaß, O.; Grundmann, P. Governing Transactions and Interdependences between Linked Value Chains in a Circular Economy: The Case of Wastewater Reuse in Braunschweig (Germany). *Sustainability* **2018**, *10*, 1125.
5. Inter3. Wasserwiederverwendung in Deutschland Ermöglichen. Available online: https://www.inter3.de/aktuelles/meldungen (accessed on 6 September 2022).
6. Barbagallo, S.; Cirelli, G.L.; Consoli, S.; Licciardello, F.; Marzo, A.; Toscano, A. Analysis of treated wastewater reuse potential for irrigation in Sicily. *Water Sci. Technol.* **2012**, *65*, 2024–2033. [CrossRef] [PubMed]
7. Jodar-Abellan, A.; Fernández-Aracil, P.; Melgarejo-Moreno, J. Assessing Water Shortage through a Balance Model among Transfers, Groundwater, Desalination, Wastewater Reuse, and Water Demands (SE Spain). *Water* **2019**, *11*, 1009. [CrossRef]
8. Jia, H.; Guo, R.; Xin, K.; Wang, J. Research on wastewater reuse planning in Beijing central region. *Water Sci. Technol.* **2005**, *51*, 195–202. [CrossRef]
9. Ramirez, C.; Almulla, Y.; Fuso Nerini, F. Reusing wastewater for agricultural irrigation: A water-energy-food Nexus assessment in the North Western Sahara Aquifer System. *Environ. Res. Lett.* **2021**, *16*, 044052. [CrossRef]
10. Jaramillo, M.; Restrepo, I. Wastewater Reuse in Agriculture: A Review about Its Limitations and Benefits. *Sustainability* **2017**, *9*, 1734. [CrossRef]
11. Moseki, O.; Murray-Hudson, M.; Kashe, K. Crop water and irrigation requirements of *Jatropha curcas* L. in semi-arid conditions of Botswana: Applying the CROPWAT model. *Agric. Water Manag.* **2019**, *225*, 105754. [CrossRef]
12. Bilibio, C.; Hensel, O. The Water Deficit of Evapotranspiration Covers on Potash Tailing Piles Using CropWat. *Agric. Eng. Int. CIGR J.* **2021**, *23*, 75–91.
13. Surendran, U.; Sushanth, C.M.; Mammen, G.; Joseph, E.J. Modelling the Crop Water Requirement Using FAO-CROPWAT and Assessment of Water Resources for Sustainable Water Resource Management: A Case Study in Palakkad District of Humid Tropical Kerala, India. *Aquat. Procedia* **2015**, *4*, 1211–1219. [CrossRef]
14. Bokke, A.S.; Shoro, K.E. Impact of effective rainfall on net irrigation water requirement: The case of Ethiopia. *Water Sci.* **2020**, *34*, 155–163. [CrossRef]
15. Feng, Z.; Liu, D.; Zhang, Y. Water requirements and irrigation scheduling of spring maize using GIS and CropWat model in Beijing-Tianjin-Hebei region. *Chin. Geogr. Sci.* **2007**, *17*, 56–63. [CrossRef]
16. Al-Najar, H. The integration of FAO-CropWat Model and GIS Techniques for Estimating Irrigation Water Requirement and Its Application in the Gaza Strip. *Nat. Resour.* **2011**, *2*, 146–154. [CrossRef]
17. Bhardwaj, R.K.; Sharma, S.; Kumar, D. Impact of LULC Dynamics on Evapotranspiration using GIS: Case Study of Uttarakhand. *J. Mater. Environ. Sci.* **2022**, *13*, 631–639.
18. Adamala, S.; Rajwade, Y.A.; Krishna Reddy, Y.V. Estimation of wheat crop evapotranspiration using NDVI vegetation index. *J. Appl. Nat. Sci.* **2016**, *8*, 159–166. [CrossRef]
19. Kadkhodazadeh, M.; Valikhan Anaraki, M.; Morshed-Bozorgdel, A.; Farzin, S. A New Methodology for Reference Evapotranspiration Prediction and Uncertainty Analysis under Climate Change Conditions Based on Machine Learning, Multi Criteria Decision Making and Monte Carlo Methods. *Sustainability* **2022**, *14*, 2601. [CrossRef]
20. Farreny, R.; Morales-Pinzón, T.; Guisasola, A.; Tayà, C.; Rieradevall, J.; Gabarrell, X. Roof selection for rainwater harvesting: Quantity and quality assessments in Spain. *Water Res.* **2011**, *45*, 3245–3254. [CrossRef]
21. Maqsoom, A.; Aslam, B.; Ismail, S.; Thaheem, M.J.; Ullah, F.; Zahoor, H.; Musarat, M.A.; Vatin, N.I. Assessing Rainwater Harvesting Potential in Urban Areas: A Building Information Modelling (BIM) Approach. *Sustainability* **2021**, *13*, 12583. [CrossRef]
22. Bao, K.; Padsala, R.; Thrän, D.; Schröter, B. Urban Water Demand Simulation in Residential and Non-Residential Buildings Based on a CityGML Data Model. *ISPRS Int. J. Geo-Inf.* **2020**, *9*, 642. [CrossRef]
23. Climate-Data. Available online: https://en.climate-data.org/ (accessed on 5 October 2022).
24. Weater-Data. Available online: https://weatherspark.com/ (accessed on 7 October 2022).
25. BGR Geoportal. Available online: https://www.bgr.bund.de/ (accessed on 10 October 2022).
26. Rinaldi, S.; Bonamente, E.; Scrucca, F.; Merico, M.; Asdrubali, F.; Cotana, F. Water and Carbon Footprint of Wine: Methodology Review and Application to a Case Study. *Sustainability* **2016**, *8*, 621. [CrossRef]
27. Choudhary, D. Methods of Evapotranspiration. 2018. Available online: http://rgdoi.net/10.13140/RG.2.2.14533.76007 (accessed on 23 January 2023).
28. Solangi, G.S.; Shah, S.A.; Alharbi, R.S.; Panhwar, S.; Keerio, H.A.; Kim, T.W.; Memon, J.A.; Bughio, A.D. Investigation of Irrigation Water Requirements for Major Crops Using CROPWAT Model Based on Climate Data. *Water* **2022**, *14*, 2578. [CrossRef]

29. Alemayehu, Y.A.; Steyn, J.M.; Annandale, J.G. FAO-type crop factor determination for irrigation scheduling of hot pepper (*Capsicum annuum* L.) cultivars. *S. Afr. J. Plant Soil.* **2009**, *26*, 186–194. [CrossRef]
30. Lakshmi, S.E.; Yarrakula, K. Review and critical analysis on digital elevation models. *Geofizika* **2019**, *35*, 129–157. [CrossRef]
31. Rabbi, S.E.; Shant, R.; Karmakar, S.; Habib, A.; Kropp, J.P. Regional mapping of climate variability index and identifying socio-economic factors influencing farmer's perception in Bangladesh. *Environ. Dev. Sustain.* **2021**, *23*, 11050–11066. [CrossRef]
32. Repository.De. Index of /Schemas/Adv/Citygml/Codelisten/. Available online: https://repository.gdi-de.org/schemas/adv/citygml/Codelisten/ (accessed on 25 January 2023).
33. Ugai, T. Evaluation of Sustainable Roof from Various Aspects and Benefits of Agriculture Roofing in Urban Core. *Procedia Soc. Behav. Sci.* **2016**, *216*, 850–860. [CrossRef]
34. Lancaster, B. *Rainwater Harvesting for Drylands and Beyond*; Rainwater Harvesting for Drylands and Beyond; Rainsource Press: Tucson, Arizona, 2006; Volume 1.
35. Liaw, C.; Tsai, Y. Optimum Storage Volume of Rooftop Rain Water Harvesting Systems for Domestic Use1. *J. Am. Water Resour. Assoc.* **2004**, *40*, 901–912. [CrossRef]
36. Ward, S.; Memon, F.A.; Butler, D. Harvested rainwater quality: The importance of appropriate design. *Water Sci. Technol.* **2010**, *61*, 1707–1714. [CrossRef] [PubMed]
37. Anane, M.; Bouziri, L.; Limam, A.; Jellali, S. Ranking suitable sites for irrigation with reclaimed water in the Nabeul-Hammamet region (Tunisia) using GIS and AHP-multicriteria decision analysis. *Resour. Conserv. Recycl.* **2012**, *65*, 36–46. [CrossRef]
38. Schoenwiese, G. Umweltbundesamt. WW-R-1: Water Use Index. 2021. Available online: https://www.umweltbundesamt.de/en/topics/climate-energy/climate-impacts-adaptation/impacts-of-climate-change/monitoring-report-2019/indicators-of-climate-change-impacts-adaptation/cluster-water-regime-water-managementmarine-coastal/ww-r-1-water-use-index (accessed on 20 September 2022).
39. Germany, Federal Republic of. Available online: https://www.icid.org/i_d_germany.pdf (accessed on 3 February 2023).

water

MDPI

Article

Colorimetric Quantification for Residual Poly-DADMAC in Water Treatment

Ilil Levakov [1], Ido Maor [1,2,†], Chen Barak [1], Yael Kirshenbaum [2] and Giora Rytwo [1,2,*]

[1] Environmental Physical Chemistry Laboratory, MIGAL-Galilee Research Institute, Kiryat Shmona 1101602, Israel; ilill@migal.org.il (I.L.)
[2] Environmental Sciences & Water Sciences Department, Tel Hai College, Upper Galilee 1220800, Israel
* Correspondence: rytwo@telhai.ac.il or giorarytwo@gmail.com; Tel.: +972-4-7700516
† Deceased.

Abstract: Poly-DADMAC (PD) is a commonly used organic polymer in water treatment, known for its effectiveness as a coagulant. However, its presence as a residue in water raises concerns related to membrane fouling and the potential formation of carcinogenic compounds. Therefore, fast and simple quantification is necessary to efficiently control and monitor the optimal dose of poly-DADMAC with minimal negative effects. This study introduces a new colorimetric quantification method for poly-DADMAC, based on complexation with a cationic dye (fast green-FG). The method was examined through varying conditions, which included different analytical and commercial poly-DADMAC formulations and concentrations. These experiments confirm its effectiveness in quantifying poly-DADMAC with a detection limit of 3.22 μg L^{-1} (0.02 μM based on monomers' molecular weight), which is one order of magnitude lower than regulatory requirements (50 μg L^{-1}). To validate the method, the effect of pH was examined, and implementation demonstrations were conducted on cyanobacteria and cowshed-polluted water samples. This research introduces a fast, cost-effective innovative method to accurately quantify poly-DADMAC, enhancing water treatment strategies for high-quality purification and water reuse

Keywords: cationic polyelectrolyte coagulant; fast green; spectrophotometric quantification; water treatment

1. Introduction

Coagulation and flocculation are important preliminary steps in wastewater and potable water treatment facilities. The processes include the removal of non-dissolved and particulate matter, including algae, pathogens, organic materials, minerals, and others. This is achieved by the addition of materials, such as ferric chloride, alum (aluminium sulfate), or organic polymers, which destabilize the colloids resulting in small suspended particles agglomerating into larger settleable flocs [1]. Poly-diallyl-dimethylammonium chloride (poly-DADMAC, PD, CAS 26062-79-3) is a homopolymer also known as polyquarternium-6, with a chemical formula of $(C_8H_{16}NCl)_n$ (Figure 1). In commercial formulations, its molecular weight typically ranges from a few hundred to several thousand kDa, depending on the number of monomers (each with a molecular weight of 161.7 Da) [2]. PD is the most common polymer used in various coagulation processes [3], including in the pulp and paper industry, either alone [4] or in combination with other polymers [5]. It was the first cationic polyelectrolyte approved by the Food and Drug Administration of the U.S.A. (FDA) for use in potable water treatment [6]. Recent studies have also tested its influence when adsorbed on cellulose-based products [7] or for the preparation of specific adsorbing matrices for per- and polyfluoroalkyl substances [8].

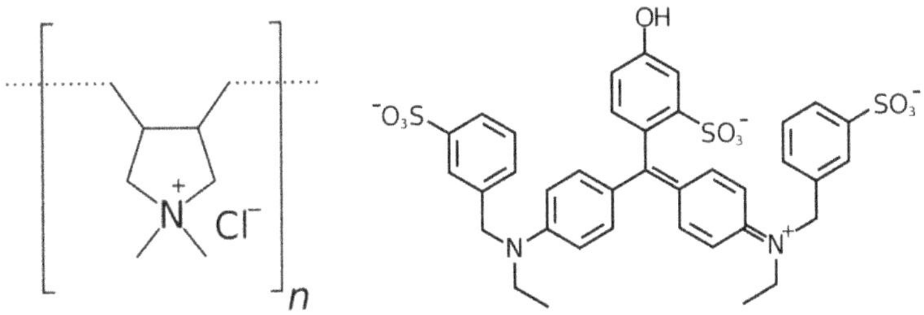

Figure 1. Chemical structure of polyDADMAC (PD, polyquaternium-6, **left**) and fast green (FG, E143, **right**).

The wide utilization of PD and other organic polymers can be attributed to several advantages. These include their remarkable cationic properties, high-charge density that leads to the formation of stronger and easily separable flocs, effective bridging capabilities allowing for higher solid content in the sludge phase, and minimal impact on pH levels, necessitating only a single pH adjustment during the process. However, a few disadvantages are associated with the use of PD as a coagulant, mainly regarding the residuals remaining after the process. While PD can serve as an effective coagulant, its presence in the subsequent treatment process might cause severe fouling to the filtration membranes. Although novel membranes with better anti-fouling and anti-microbial properties have been developed [9,10], residual PD molecules of even low concentrations can desorb onto the membrane surface, potentially leading to pore blocking [11]. Fouling depends on the type, charge, size, and concentration of the polymer, and higher molecule-weight polymers have larger fouling potential [12,13]. Membrane fouling, particularly in processes like reverse osmosis, microfiltration, and nanofiltration, can lead to reduced water flux and increased energy consumption. It necessitates more frequent membrane cleaning or replacement, subsequently elevating operational costs in water treatment plants [14]. An additional disadvantage is the reaction of residual PD with disinfection by-products, such as chloramine and ozonation, forming carcinogenic nitrosamines [15–19]. Therefore, several American and European standardization organizations have limited the residual amount of PD in drinking water at $\leq$50 μg/L~0.310 μM, which is considered to be also toxic to aquatic organisms [20].

For these reasons, fast and simple quantification of residual PD is required at relatively low limits of detection (LOD), to control and monitor the optimal coagulant dose with minimal negative effects. Chromatography-based analytical methods for the quantification of polymers are usually very sensitive to interferences by other organic molecules [21]. Indeed, a recent study presented a method based on ion-pair chromatography, which accurately measured residual DADMAC monomers [22] without measuring the amounts of the polymer. Other accurate methods, such as gel permeation chromatography, epoxidation, fluorescent tagging, and co-precipitation, require advanced, relatively expensive, and long procedures [2,3,23–26]. A highly sensitive method was developed based on the use of gold nanoparticles [27], applied in South African treated water [28] and recently developed to Lovibond color filters [29], but the preparation of the nanoparticles requires a relatively complicated procedure.

Cationic polyelectrolytes, such as PDs, can bind to anionic molecules, such as tannic acid or anionic dyes like rose Bengal, methyl-orange [23], Ponceau S [30], and acid orange [31], resulting in an insoluble polymer-anion complex that can be separated by centrifugation. This separation method allows the quantification of the polymer concentration by measuring the remaining dye. A similar method was used for quantification of the cationic biopolymer chitosan, based on complexation with Cibacron Brilliant Red 3B-A [32]. However, in all those studies, LODs were two to three orders of magnitude higher than the requirement.

In this study, we present a simple colorimetric quantification method based on the addition of fast green (Figure 1, FG, ethyl-[4-[[4-[ethyl-[(3-sulfophenyl) methyl] aminophenyl]-(4-hydroxy-2-sulfophenyl)methylidene]-1-cyclohexa-2,5-dienylidene]-[(3-sulfophenyl)methyl] azanium), and its complexation with PD. FG is an anionic tri-aryl methane food dye known also as "food green 3", "green 1724", and "E143". It is stable in a wide range of pH and has a brilliant blue-green color with an absorption maximum of 624 nm, and two additional smaller absorption bands at 420 and 304 nm [33]. The very high molar absorptivity (ε_{624} > 100,000 $M^{-1}cm^{-1}$) enables high sensitivity and yields a procedure suitable for trace quantification of PD with LOD < 3 μg L^{-1} (0.018 μM), one order of magnitude lower than regulation requirements. The primary objective of this research was to establish the proof of concept for the proposed quantification method in a simple, fast, and cost-effective manner. This approach serves as the first step in mitigating PD's impact on membrane fouling and potential environmental risks. Additionally, we conducted preliminary implementations of the method under different conditions.

2. Materials and Methods

2.1. Materials

Analytical polydiallyl dimethylammonium chloride (PD; medium molecular weight, 200,000 to 350,000) and fast green (FG) were purchased from Sigma-Aldrich (Jerusalem, Israel). A commercial-grade coagulant with 40% PD as an active ingredient manufactured by SNF Ltd. (FLOQUAT® FL-45) was purchased from Amgal-Depotchem Ltd. (Beer Tuvia, Israel), SNF representatives in Israel. NC24 clay polymer nanocomposites were prepared as described in the literature [34–36] using 10 g/L sepiolite S9 provided by Tolsa S.A. (Madrid, Spain), and 45 g/L FL-45. DKG kaolinite clay polymer nanocomposites [37] were prepared in a similar way, using 10 g/L MPO kaolinite provided by Agat Minerals and Yehu Clays Ltd. (Yeruham, Israel) and 45 g/L FL-45. A total of 1 mM HCl was prepared from 37% stock (Merck, Rehovot, Israel) and used for pH adjustment.

2.2. Complexation/Calibration Experiments Description

In order to prepare calibration curves and demonstrate the method's applicability, several complexation experiments were conducted, including a wide range of analytic poly-DADMAC with fast green dye. Three separate experiments were conducted to examine different analytic PD concentrations. These concentrations included low, medium, and high PD ranges: 0.02–1 μM (0.0322–0.1617 mg L^{-1}), 1–30 μM (0.1617–4.851 mg L^{-1}), and 10–200 μM (1.617–32.34 mg L^{-1}) PD solutions, all of which were dissolved in double-distilled water. To each PD range, 0.8, 20, and 100 μM of FG were added to the solutions in the low, medium, and high series, respectively. After the addition of FG, pH values were adjusted to 3–3.5 by adding 20 μL of 1 mM HCl (see Section 2.4). Subsequently, all samples were kept at room temperature (23 ± 1 °C) on an orbital shaker (200 rpm) for 1 h. FG spectra were measured using a 1 cm cuvette in a Cary 60 UV–Vis spectrophotometer (Agilent Technologies, Santa Clara, CA, USA). Absorbance was determined using the absorption bands at 624 and 420 nm with a molar absorptivity of 103,500 and 11,600 $M^{-1}cm^{-1}$, respectively. To enhance sensitivity, absorbances of the low-range FG solutions were measured at 624 nm, utilizing a 5 cm cuvette. A visual representation of the preparation procedure and the experimental setup is available in Figure S1 of the Supplementary Material.

To confirm the formation of the PD-FG complexes, 50 mL samples containing 20 μM FG and a range between 2.5 and 50 μM PD solutions were prepared as described above, centrifuged, and the sediment was lyophilized (Christ Alpha 1-2 LD Plus, Germany) and measured for the attenuated total reflection-Fourier transform infrared (ATR-FTIR) spectra on a Nicolet iS10 FTIR (Nicolet Analytical Instruments, Madison, WI, USA), using a SMART ATR device with a diamond crystal plate, within a range of wavenumber comprised from 4000 to 525 cm^{-1} and analyzed with OMNIC 9.3.32 software (Thermo Fisher Scientific, Madison, WI, USA), including full spectra ATR correction.

2.3. Implementation in Water with Commercial PD

Commercial PD samples were examined in medium-range concentrations of 2–50 μM (0.323–8.085 mg L^{-1}) PD dissolved in double-distilled water with the addition of 20 μM FG. FG spectra were measured in the UV–Vis spectrophotometer using a 1 cm cuvette as described in Section 2.2.

2.4. Influence of pH

To assess the performance of the quantification method in actual water samples, two coagulation-flocculation experiments were conducted. These experiments were carried out using water contaminated with cyanobacteria and cowshed effluents as test samples. The evaluation involved testing both a commercial PD formulation and clay-PD nanocomposites developed within our research group.

2.5. Quantification of Poly-DADMAC in Coagulation Water Treatment Samples

To evaluate the efficiency of the quantification method in "real" water samples, two coago-flocculation experiments were performed in cyanobacteria-polluted water and cowshed effluents. Experiments included testing commercial PD formulation, and clay-PD nanocomposites developed in our research group [35–37]

The first coagulation experiment was performed on cyanobacteria culture suspension using concentrations of two poly-DADMAC-based coagulants: industrial PD and kaolinite-PD nanocomposites (DKG). The cyanobacteria suspension consisted of the Microcystis family, which was isolated from Lake Kinneret and cultivated in batch cultures using BG11 medium [38]. The experiments were carried out in 300 mL glass beakers and included two sets of PD and DKG coagulants. Each set comprised five distinct coagulant concentrations: 14, 28, 42, 53, and 71 μM PD, either in its free form or pre-adsorbed onto kaolinite clay particles (referred to as DKG coagulant). The coagulants were added to 250 mL cyanobacteria suspensions. After the addition, the suspensions were stirred thoroughly for 1 min and left to settle for 1 h.

The second coagulation experiment was performed on cowshed effluents. The wastewater was taken from the Kfar Blum industrial cowshed (Upper Galilee, Israel) in July 2023. To determine the required amount of coagulant needed for neutralization [39], the electrokinetic colloidal charge of the wastewater was measured in a particle charge detector (PCD) (BTG Mütek, PCD-05, Eclépens, Switzerland). Three poly-DADMAC-based coagulants were tested: commercial PD, kaolinite-PD nanocomposite (DKG), and sepiolite-PD nanocomposite (NC). The doses were equivalent to 100% of the requirement to neutralize the colloids (4.3 mM charges). Each experiment was conducted in 300 mL glass beakers, with each treatment having 3 replicates. After the addition, the suspensions were thoroughly stirred for 1 min and allowed to settle for 1 h

In both experiments, turbidity was measured with a LaMotte 2020i turbidimeter (Chesterton, MD, USA) before and after the treatment to estimate the efficiency of clarification and coagulation. The residual PD concentrations were quantified using the complexation method as described in Section 2.2. Specifically, 10 mL of the supernatant was sampled, and the pH values were adjusted to 3–3.5 by adding 20 μL of 1 mM HCl. Subsequently, suitable volumes of FG from a 1000 μM stock solution were added to each sample, yielding an FG concentration of 20 μM. Test tubes were placed on an orbital shaker (200 rpm) for 1 h, and samples were measured in the spectrophotometer as described in Section 2.2.

3. Results

3.1. Quantification of Analytical Poly-DADMAC Concentrations

As mentioned in Section 2.2, three separate calibration experiments were conducted to quantify PD at different concentration levels: low, medium, and high. Each range included the addition of different FG concentrations. The quantification method is based on the measurement of free fast-green molecules in the solution, according to its distinct absorption bands at wavelengths 420 and 624 nm. Figure 2 shows the measured absorbance values in specific wavelengths as a function of the known PD concentrations. To increase sensitivity at the medium and low PD range, the selected wavelength was 624 nm. To avoid the need for dilutions at the high range the 420 nm (with lower molar absorptivity) was selected. Linear regression calculations between absorbance and PD concentrations were performed for all three ranges (see Figure 2), showing excellent correlations ($R^2 > 0.982$) in the high and medium ranges (1–200 μM, 0.1617–32.4 mg L^{-1}). According to the results, as the PD concentration increases, absorbance values decrease, indicating the formation of a PD-fast green complex that reduces available fast green molecules in the solution. This inverse relationship between dye absorbance and PD concentration serves as the foundation for quantitatively analyzing PD through complexation with the FG dye, similar to the effect observed in the chitosan quantification method [32]. The relatively low R^2 ($R^2 = 0.866$) in the 0.02–1 μM PD range is attributed to increased measurement noise and experimental variability at these low concentrations. The lower PD concentration used for the calibration experiment was determined as the detection limit of the method (0.02 μM).

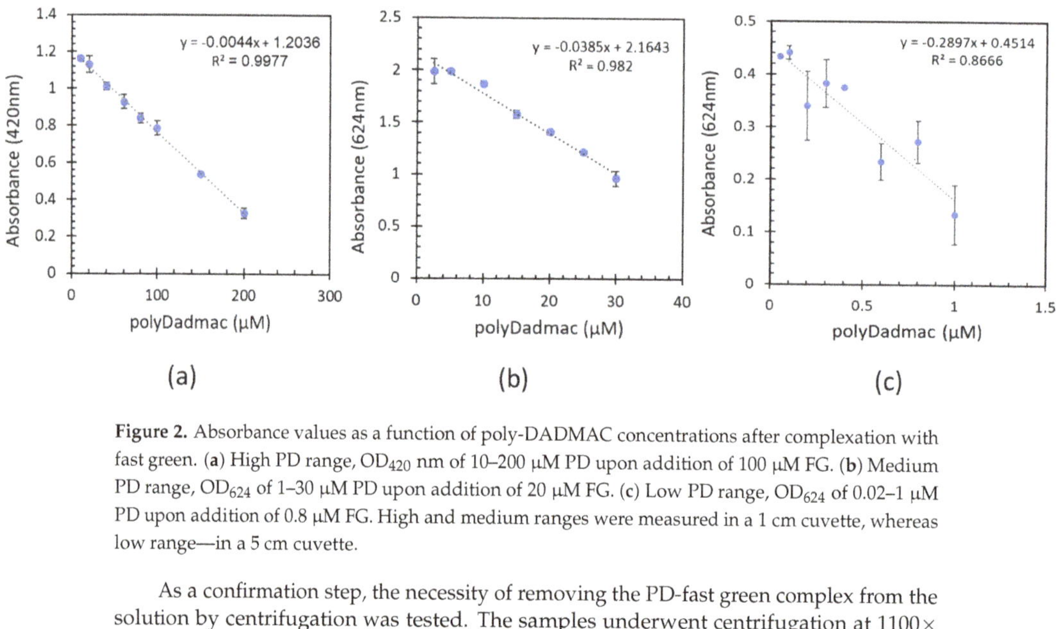

Figure 2. Absorbance values as a function of poly-DADMAC concentrations after complexation with fast green. (**a**) High PD range, OD_{420} nm of 10–200 μM PD upon addition of 100 μM FG. (**b**) Medium PD range, OD_{624} of 1–30 μM PD upon addition of 20 μM FG. (**c**) Low PD range, OD_{624} of 0.02–1 μM PD upon addition of 0.8 μM FG. High and medium ranges were measured in a 1 cm cuvette, whereas low range—in a 5 cm cuvette.

As a confirmation step, the necessity of removing the PD-fast green complex from the solution by centrifugation was tested. The samples underwent centrifugation at 1100× *g* for 20 min, followed by spectrophotometer measurements. According to the results, centrifugation of the mixture solution after complexation did not affect the absorbance values (data not shown).

3.2. FTIR Measurements of PD-FG Complexes

Figure 3 shows the FTIR spectra of dried PD, FG, and sediments of samples with 20 μM FG complexed with 2.5 and 50 μM PD. The raw PD spectrum shows a distinct absorption band at 1464 cm^{-1}, attributed to quaternary ammonium bound to chloride anions [40]. Additionally, a broad peak associated with aliphatic secondary amines at 3100–3600 cm^{-1} [26] and bands related to C-H stretching vibrations of methine, methylene, and methyl groups [41] at 3000–2800 cm^{-1} can be clearly observed. The raw FG dye spectrum exhibits a series of absorption bands related to aromatic structures and sulfonate vibrations [41] at 1340 and 1170 cm^{-1} In both the FG-2.5 μM PD and FG-50 μM PD complexes, the vibrations associated with quaternary ammonium bound to chloride anions completely disappear, suggesting that the complexation occurs through the exchange of chloride ions with the anionic dye. The spectra also display distinct bands corresponding to the C-H and secondary amine groups of PD, along with aromatic and sulfonate bands from FG. These observations confirm the formation of the dye-coagulant complex. Similar spectra were recorded for all measured FG-PD complexes, and shown in Figure S2 in the Supplementary Material.

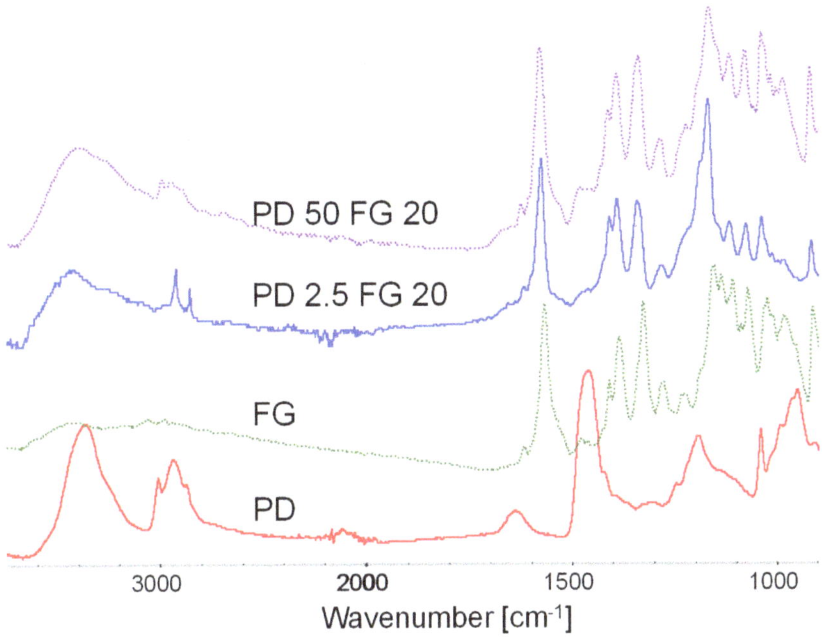

Figure 3. FTIR spectra of dried polyDADMAC (PD, red line), raw fast green (FG, green line), sedimented complexes formed by 2.5 (PD 2.5 FG 20, blue line), and 50 μM PD (PD 50 FG 20, purple line). Note that the x-axis is split at 2000 cm^{-1}.

3.3. *Influence of pH*

Given that the accuracy of spectroscopy measurements relies on the stability of the spectrum, we conducted experiments to assess the impact of pH on FG spectra. The absorbance spectrum of FG was measured at different pH values ranging from 2 to 9.1 (Figure 4). The results indicate that within the pH range of 2 to 7, the spectrum of FG remained constant, with no changes in the molar absorptivity at any wavelength. This suggests that there were no significant alterations in the chromophore within this acidic to neutral pH range. However, as the pH increased above 7, the spectrum deviated from the constant trend observed in the acidic to neutral pH range. The large absorption band at 624 nm shifted hypsochromically (to lower wavelengths), and molar absorptivity slightly increased, while the absorption band at 420 nm disappeared. These findings are crucial to determine the pH requirements and limitations relevant when applying the quantification method presented hereby and highlight the importance of pH control to a range of 2–7 during the analysis process. In our study, the pH of all samples was adjusted to a range of 3–3.5 by adding very small volumes of concentrated HCl before measuring the spectrum

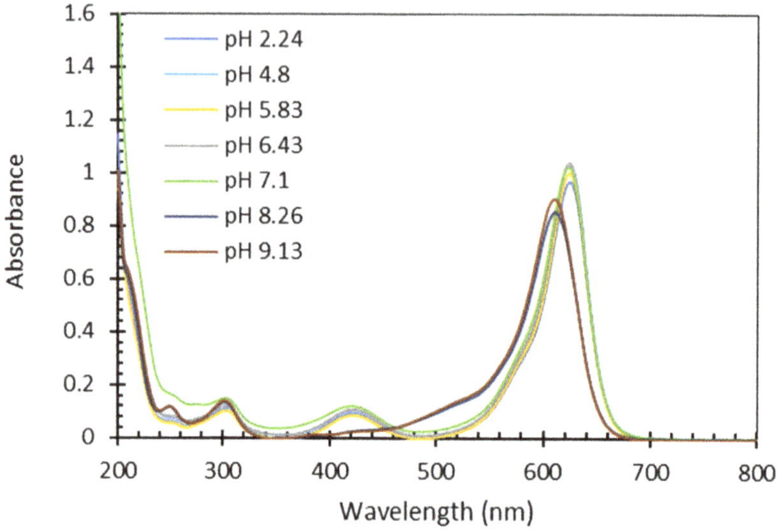

Figure 4. UV-Vis spectra of 10 μM FG solutions at different pH values (2.24–9.13).

3.4. *Quantification of Commercial Poly-DADMAC*

Complexation experiments were conducted with commercial PD in a range of 1–50 μM polymer concentrations, with the addition of 20 μM FG according to the medium range presented in Section 3.1. Figure 5 shows the absorbance of the remaining dye as a function of commercial PD concentrations, after the formation of the complexes at 420 (Figure 5a) and 624 nm (Figure 5b). To prevent any potential shift in the spectrum due to high pH, all samples were acidified with 20 μL of 1 mM HCl, as outlined in Section 3.3. The full spectra of commercial PD samples after complexation with FG are shown in Figure S3 in the Supplementary Material. To verify the similarity of the measured spectra to that of FG, a Pearson's correlation of all spectra with FG in the 300–800 range nm was evaluated using the CORREL function in Excel©, yielding R > 0.993.

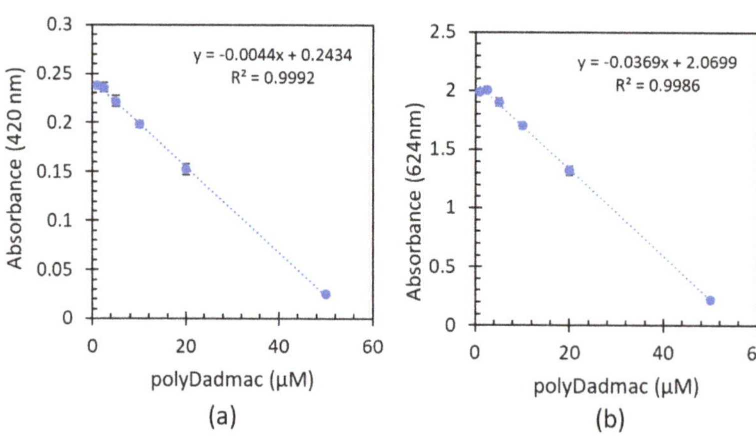

Figure 5. Absorbance values at (**a**) 420 nm and (**b**) 624 nm, as a function of PD concentrations (1–50 μM) after complexation with 20 μM FG.

Consistent linear relationships between absorbance and PD concentration were observed at both wavelengths (R2 = 0.9992 for 420 nm and R2 = 0.9986 for 624 nm), indicating a strong and reliable correlation, as illustrated in Figure 2 for analytical PD. It is noteworthy that after the examination of the linear regression equations in Figures 2b and 4b, a pronounced similarity emerges, both in terms of the intercept and slope values. Furthermore, the slope in Figures 2a and 5a are identical, although the intercept differs due to levels of FG concentrations. These findings highlight the applicability of the complexation method for quantifying commercial PD concentrations.

3.5. Quantification of Poly-DADMAC in Coagulation Treatment Samples

The method's applicability in practical field conditions was assessed by determining residual PD concentrations in cyanobacteria-contaminated water and cowshed effluents during coagulation and clarification treatments.

The cyanobacteria water samples were treated with commercial PD and kaolinite-PD nanocomposites (DKG). As mentioned in Section 2.5, coagulants were added in amounts equivalent to the following PD concentrations: 14, 28, 42, 53, and 71 μM (either in free form or adsorbed onto kaolinite clay particles). The initial turbidity value for the cyanobacteria suspension was 98.2 ± 3.67 NTU. Following the coagulant addition as detailed in Section 2.5, 10 mL test tubes were used to sample the supernatants. These samples were then acidified, and a concentrated FG solution was added, resulting in an FG concentration of 20 μM. The absorbance was measured at 420 nm, and the residual PD concentration was determined using the calibration curve established in the previous section (Figure 5). Residual PD concentrations and turbidity values are shown in Figure 6. Based on the measured results, minimal turbidity is observed across the entire range of added PD-kaolinite, and up to 53 μM of commercial PD. Additionally, there is a substantial clear phase, accounting for approximately 90% of the volume, at the top of the vessels. However, increasing residual PD is clearly observed as the added amounts of coagulant increase. Furthermore, significant differences between the remaining PD concentrations were observed between the two coagulants, with the lowest amounts of 1.9 and 4.1 PD μM for the nanocomposites and commercial PD, respectively. The lower residual PD values observed in the DKG samples are likely attributed to the PD binding to the kaolinite clay particles and the higher density of these clays. This, in turn, leads to the enhanced precipitation of excess coagulant materials. As the coagulant concentrations increased, no significant changes were observed in the turbidity values. However, higher turbidity was observed in the addition of 71 μM commercial PD indicating coagulant over-dose that disrupts the neutralization of the colloidal charge of the suspension [39].

As an additional test for the method, we conducted clarification and coagulation experiments on cowshed effluents. Such effluents are heavily polluted with a total suspended solid (TSS) value that may reach 10 g L^{-1}, and turbidity >4000 NTU [42]. TSS removal is required as pretreatment since wastewater treatment plants limit the maximum inflow value to <0.4 g L^{-1} [43]. In the samples collected at the Kfar Blum cowshed during July 2023, the initial turbidity of the effluents was 3355 ± 226 NTU. Optimal coagulant doses were determined by measuring the colloidal particle charge as described in Section 2.5. Following the coagulation process, turbidity decreased to 1.2–4.5% of the initial values (77, 152, and 43 NTU in the kaolinite-PD, sepiolite-PD, and commercial PD coagulant treatments, respectively). The figure below presents the residual PD concentrations in the three treatments (Figure 7). As in the cyanobacteria-polluted effluents, residual PD was considerably lower in the clay-PD coagulants (~25 μM = ~4.04 mg L^{-1}) than in the commercial PD (37 μM = 5.98 mg L^{-1}), probably due to the high density of the clays and precipitation of excess coagulants.

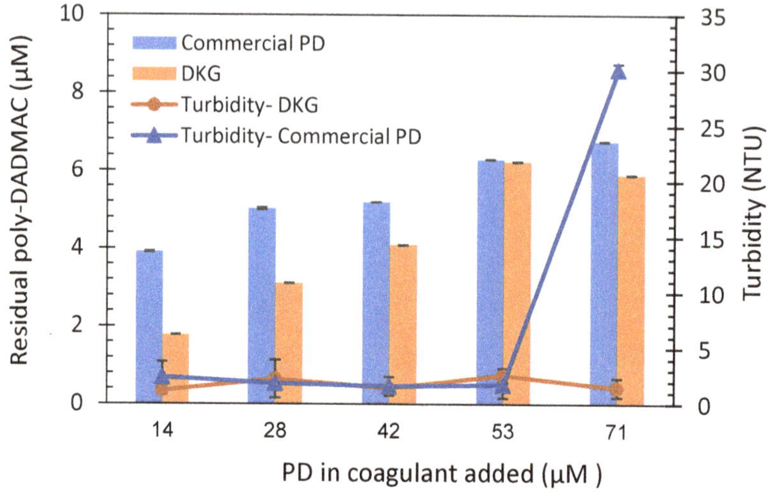

Figure 6. Evaluated PD residual concentrations (bars) and measured turbidity (lines and dots) in cyanobacteria-polluted water at different added PD doses, using commercial PD (blue, triangles) or kaolinite-PD nanocomposite (orange, circles).

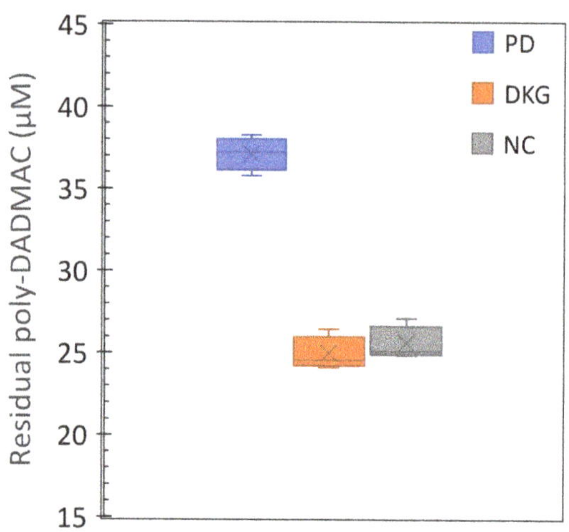

Figure 7. Residual PD in cowshed effluent coagulation experiments with commercial PD (blue), kaolinite-PD nanocomposites-DKG (orange), and sepiolite-PD nanocomposites-NC (grey).

Normalized spectra of cyanobacteria and cowshed water samples after treatment compared to the spectrum of FG with commercial PD in clear water are presented in Figure S4 of the Supplementary Material. FG spectrum in cyanobacteria sample is almost identical to clear water (Pearson's correlation in the range 300–800 nm = 0.999). On the other hand, FG in the cowshed effluents differs from that in clear water (Pearson's correlation = 0.938) due to the presence of remaining color and turbidity. However, limiting the range to 600–800 nm, where colloids light dispersion is lower, increases Pearson's correlation to 0.977, indicating that differences do not interfere with the measurement of absorbance at 624 nm.

The results of both demonstration experiments highlight the method's applicability in real conditions, offering vital insights into optimal coagulation treatments in a convenient, cost-effective manner. It should be mentioned that the removal of residual PD concentration from post-coagulation wastewater is essential for the prevention of extensive membrane fouling and negative environmental impact. Several studies examined different approaches in order to reduce the negative impact of PD by inhibiting the formation of carcinogenic by-products [19,44], but additional studies are required to examine possible complete removal techniques, such as adsorption and ion exchange. Nevertheless, the first step to approach the problem is to improve the monitoring ability of PD in wastewater plant facilities.

4. Conclusions

- In this study, a colorimetric quantification method for poly-DADMAC through complexation with fast green dye was developed and evaluated.
- The method exhibited high sensitivity with a detection limit of 0.02 μM (0.0032 mg L^{-1}), meeting regulatory demands according to the world standard limitations.
- Quantification experiments of analytical PD concentrations, including low, medium, and high ranges demonstrated a linear correlation between the absorbance and PD concentrations.
- The influence of pH on the quantification process was tested, revealing that the absorbance spectrum of FG remains unchanged within the pH range of 2 to 7, exhibiting a hypsochromic and hyperchromic shift at higher pH. Therefore, an acidification step (or accurate pH measurements) is essential before quantification to ensure the FG spectrum falls within the required range.
- The quantification method's feasibility was demonstrated in coagulation experiments, estimating remaining PD concentrations in treated water samples.
- The method's feasibility in estimating PD concentrations under actual coagulation conditions highlights its potential for efficient monitoring and control of coagulation processes. It offers a rapid, cost-effective, and sensitive tool for accurate PD concentration measurement. The colorimetric PD quantification holds promising potential for advancing water treatment systems and quality control.
- Further research may concentrate on upscaling the application of the quantification method for large-scale water treatment systems, as well as refining its adaptation and addressing limitations. This may include examining factors such as temperature, reaction time, and ionic strength.

Supplementary Materials: The following supporting information can be downloaded at: https://www.mdpi.com/article/10.3390/w15193352/s1, Figure S1: Scheme of the experimental setup. Figure S2: ATR-FTIR spectra of PD-FG complexes. Figure S3: Spectra of an initial 20 μM FG upon complexation with the increasing concentrations of commercial PD; Figure S34: Normalized spectra of an initial 20 μM FG upon complexation with commercial PD in distilled water, cyanobacteria polluted water and cowshed effluents.

Author Contributions: Conceptualization, G.R.; methodology, G.R., I.L., C.B. and I.M.; software, I.L. and Y.K.; validation, G.R., I.L. and Y.K.; formal analysis, I.L.; investigation, G.R., I.L., C.B. and I.M.; resources, G.R.; data curation, I.L. and G.R.; writing—original draft preparation, G.R., I.L. and C.B.; writing—review and editing, I.L. and G.R.; visualization, I.L. and Y.K.; supervision, G.R.; project administration, G.R.; funding acquisition, G.R. All authors have read and agreed to the published version of the manuscript.

Funding: This research was partially funded by CSO-MOH (Israeli Ministry of Health), in the frame of the collaborative international consortium (REWA) financed under the 2020 AquaticPollutants Joint call of the AquaticPollutants ERA-NET Cofund (GA No. 869178).

Data Availability Statement: All raw data are available from the authors.

Acknowledgments: We would like to dedicate this manuscript to the memory of the late Ido Maor. The authors would like to thank the European Commission and AKA (Finland), CSO-MOH (Israel), IFD (Denmark), and WRC (South Africa) for funding in the frame REWA international consortium (additional details in "funding" paragraph). REWA is an integral part of the activities developed by the Water, Oceans and AMR JPIs. The authors are also thankful to the Yehezkel Zwecher for growing the cyanobacteria suspension, Simmah Hendrickson for the English editing, and the whole team of the Hydrogeology and Examination of Soil Fertility Lab at MIGAL Research Institute.

Conflicts of Interest: The authors declare no conflict of interest.

Abbreviations

PD	poly-dadmac (Poly-diallyl-dimethylammonium chloride)
FG	Fast-green
LOD	limit of detection
DKG	kaolinite-PD nanocomposites
NC	sepiolite-PD nanocomposites
PCD	particle charge detector
NTU	nephelometric turbidity unit
TSS	total suspended solids

References

1. Amuda, O.S.; Amoo, I.A. Coagulation/flocculation process and sludge conditioning in beverage industrial wastewater treatment. *J. Hazard. Mater.* **2007**, *141*, 778–783. [CrossRef]
2. Dautzenberg, H.; Görnitz, E.; Jaeger, W. Synthesis and characterization of poly(diallyldimethylammonium chloride) in a broad range of molecular weight. *Macromol. Chem. Phys.* **1998**, *199*, 1561–1571. [CrossRef]
3. Becker, N.S.; Bennett, D.; Bolto, B.; Dixon, D.; Eldridge, R.; Le, N.; Rye, C. Detection of polyelectrolytes at trace levels in water by fluorescent tagging. *React. Funct. Polym.* **2004**, *60*, 183–193. [CrossRef]
4. Razali, M.A.A.; Ahmad, Z.; Ahmad, M.S.B.; Ariffin, A. Treatment of pulp and paper mill wastewater with various molecular weight of polyDADMAC induced flocculation. *Chem. Eng. J.* **2011**, *166*, 529–535. [CrossRef]
5. Ariffin, A.; Razali, M.A.A.; Ahmad, Z. PolyDADMAC and polyacrylamide as a hybrid flocculation system in the treatment of pulp and paper mills waste water. *Chem. Eng. J.* **2012**, *179*, 107–111. [CrossRef]
6. Hamad, M.J.A.; Chirwa, E.M.N. Forward osmosis for water recovery using polyelectrolyte PolyDADMAC and DADMAC draw solutions as a low pressure energy saving process. *Desalination* **2019**, *453*, 89–101. [CrossRef]
7. Sampl, C.; Schaubeder, J.; Hirn, U.; Spirk, S. Interplay of electrolyte concentration and molecular weight of polyDADMAC on cellulose surface adsorption. *Int. J. Biol. Macromol.* **2023**, *239*, 124286. [CrossRef] [PubMed]
8. Ramos, P.; Singh Kalra, S.; Johnson, N.W.; Khor, C.M.; Borthakur, A.; Cranmer, B.; Dooley, G.; Mohanty, S.K.; Jassby, D.; Blotevogel, J.; et al. Enhanced removal of per- and polyfluoroalkyl substances in complex matrices by polyDADMAC-coated regenerable granular activated carbon. *Environ. Pollut.* **2022**, *294*, 118603. [CrossRef]
9. Bhuyan, C.; Konwar, A.; Bora, P.; Rajguru, P.; Hazarika, S. Cellulose nanofiber-poly(ethylene terephthalate) nanocomposite membrane from waste materials for treatment of petroleum industry wastewater. *J. Hazard. Mater.* **2023**, *442*, 129955. [CrossRef] [PubMed]
10. Gogoi, M.; Goswami, R.; Borah, A.; Sarmah, H.; Rajguru, P.; Hazarika, S. Amide functionalized DWCNT nanocomposite membranes for chiral separation of the racemic DOPA. *Sep. Purif. Technol.* **2021**, *279*, 119704. [CrossRef]
11. Gabelich, C.J.; Ishida, K.P.; Bold, R.M. Testing of water treatment copolymers for compatibility with polyamide reverse osmosis membranes. *Environ. Prog.* **2005**, *24*, 410–416. [CrossRef]
12. Wang, S.; Liu, C.; Li, Q. Fouling of microfiltration membranes by organic polymer coagulants and flocculants: Controlling factors and mechanisms. *Water Res.* **2011**, *45*, 357–365. [CrossRef] [PubMed]
13. Zhang, B.; Tang, H.; Shen, Y.; Liu, G.; Shi, W. Comparative analysis of membrane fouling mechanisms induced by colloidal polymer: Effects of sodium and calcium ions. *J. Colloid Interface Sci.* **2022**, *608*, 780–791. [CrossRef] [PubMed]
14. Guo, W.; Ngo, H.H.; Li, J. A mini-review on membrane fouling. *Bioresour. Technol.* **2012**, *122*, 27–34. [CrossRef]
15. Uzun, H.; Kim, D.; Karanfil, T. Removal of wastewater and polymer derived N-nitrosodimethylamine precursors with integrated use of chlorine and chlorine dioxide. *Chemosphere* **2019**, *216*, 224–233. [CrossRef]
16. Park, S.H.; Padhye, L.P.; Wang, P.; Cho, M.; Kim, J.-H.; Huang, C.-H. N-nitrosodimethylamine (NDMA) formation potential of amine-based water treatment polymers: Effects of in situ chloramination, breakpoint chlorination, and pre-oxidation. *J. Hazard. Mater.* **2015**, *282*, 133–140. [CrossRef]
17. Park, S.H.; Wei, S.; Mizaikoff, B.; Taylor, A.E.; Favero, C.; Huang, C.H. Degradation of amine-based water treatment polymers during chloramination as N-nitrosodimethylamine (NDMA) precursors. *Environ. Sci. Technol.* **2009**, *43*, 1360–1366. [CrossRef]
18. Huang, C.H.; Padhye, L.P.; Park, S.H. PolyDADMAC and dimethylamine as precursors of *N*-nitrosodimethylamine during ozonation. *Water Qual. Technol. Conf. Expo.* **2011**, *2011*, 2304–2315.
19. Atkinson, A.J.; Fischer, N.; Donovan, S.; Bartlett, J.; Alrehaili, O.; Sinha, S.; Kommineni, S.; Herckes, P.; Westerhoff, P. Purification and removal of the low molecular weight fraction of polyDADMAC reduces *N*-nitrosodimethylamine formation during water treatment. *Environ. Sci. Water Res. Technol.* **2020**, *6*, 2492–2498. [CrossRef]

20. Manickum, T. Occurrence, Fate and Preliminary Environmental Risk Assessment of Residual Poly-Diallyldimethyl Ammonium Chloride, and Some Disinfection By-Products, in Treated (Potable), and Environmental, Waters in the Umgeni Water Catchment in Kwazulu-Natal. *SM J. Public Health Epidemiol.* **2017**, *3*, 1043.
21. Al Momani, F.A.; Örmeci, B. Measurement of polyacrylamide polymers in water and wastewater using an in-line UV-vis spectrophotometer. *J. Environ. Chem. Eng.* **2014**, *2*, 765–772. [CrossRef]
22. Di Gaudio, F.; Barreca, S.; Orecchio, S. Diallyldimethylammonium Chloride (DADMAC) in Water Treated with Poly-Diallyldimethylammonium Chloride (PDADMAC) by Reversed-Phase Ion-Pair Chromatography—Electrospray Ionization Mass Spectrometry. *Separations* **2023**, *10*, 311. [CrossRef]
23. John, W. *Synthesis, Properties and Analysis of Polydadmac for Water Purification*; Stellenbosch University: Stellenbosch, South Africa, 2008.
24. Mwangi, I.W.; Ngila, J.C.; Ndungu, P.; Msagati, T.A.M. Method Development for the Determination of Diallyldimethylammonium Chloride at Trace Levels by Epoxidation Process. *Water Air Soil Pollut.* **2013**, *224*, 1638. [CrossRef] [PubMed]
25. Mwangi, I.W.; Catherine Ngila, J.; Ndungu, P.; Msagati, T. Preconcentration and spectrophotometric determination of poly-DADMAC in treated water by in situ co-precipitation with naphthalene. *Phys. Chem. Earth Parts A/B/C* **2014**, *72–75*, 54–60. [CrossRef]
26. Mwangi, I.; Ngila, J.; Ndungu, P. A new spectrophotometric method for determination of residual polydiallyldimethylammonium chloride flocculant in treated water based on a diazotization-coupled ion pair. *Water SA* **2012**, *38*, 707–714. [CrossRef]
27. Gumbi, B.; Ngila, J.C.; Ndungu, P.G. Gold nanoparticles for the quantification of very low levels of poly-diallyldimethylammonium chloride in river water. *Anal. Methods* **2014**, *6*, 6963. [CrossRef]
28. Manickum, T.; John, W.; Toolsee, N.; Rajagopaul, R. Preliminary Performance Evaluation of the Gold Nanoparticle Method for Quantification of Residual Poly-(Diallyldimethyl Ammonium Chloride) in Treated Waters in the Umgeni Water Catchment, Kwazulu-Natal (South Africa). *Hydrol. Curr. Res.* **2015**, *6*, 1000206.
29. Magubane, S.E.; Ntlhoro, S.; Sabela, M.; Kanchi, S.; Mlambo, M.; Onwubu, S.C.; Mdluli, P.S.; Inamuddin; Asiri, A.M. Novel on-site residual screening of poly-diallyldimethylammonium chloride in treated potable water using gold nanoparticle based lovibond color filters. *J. Taiwan Inst. Chem. Eng.* **2019**, *101*, 159–166. [CrossRef]
30. Majam, S.; Thompson, P.A. Polyelectrolyte determination in drinking water. *Water SA* **2006**, *32*, 705–707. [CrossRef]
31. Thilak Kumar, R.; Umamaheswari, S. FTIR, FTR and UV-Vis analysis of carbamazepine. *Res. J. Pharm. Biol. Chem. Sci.* **2011**, *2*, 685–693.
32. Mendelovits, A.; Prat, T.; Gonen, Y.; Rytwo, G. Improved colorimetric determination of chitosan concentrations by dye binding. *Appl. Spectrosc.* **2012**, *66*, 979–982. [CrossRef]
33. Venkatesh, N.; Murugadoss, G.; Mohamed, A.A.A.; Kumar, M.R.; Peera, S.G.; Sakthivel, P. A Novel Nanocomposite Based on Triazine Based Covalent Organic Polymer Blended with Porous g-C3N4 for Photo Catalytic Dye Degradation of Rose Bengal and Fast Green. *Molecules* **2022**, *27*, 7168. [CrossRef]
34. Rytwo, G.; Lavi, R.; Rytwo, Y.; Monchase, H.; Dultz, S.; König, T.N. Clarification of olive mill and winery wastewater by means of clay-polymer nanocomposites. *Sci. Total Environ.* **2013**, *442*, 134–142. [CrossRef]
35. Rytwo, G. The Use of Clay-Polymer Nanocomposites in Wastewater Pretreatment. *Sci. World J.* **2012**, *2012*, 498503. [CrossRef]
36. Rytwo, G. Hybrid Clay-Polymer Nanocomposites for the Clarification of Water and Effluents. *Recent Pat. Nanotechnol.* **2017**, *11*, 181–193. [CrossRef] [PubMed]
37. Rytwo, G. Method for Production of Potable Water. U.S. Patent US20190152813, 23 May 2019.
38. Sukenik, A.; Beardall, J.; Hadas, O. Photosynthetic characterization of developing and mature akinetes of Aphanizomenon ovalisporum (Cyanoprokaryota). *J. Phycol.* **2007**, *43*, 780–788. [CrossRef]
39. Rytwo, G.; Lavi, R.; König, T.N.; Avidan, L. Direct Relationship Between Electrokinetic Surface-charge Measurement of Effluents and Coagulant Type and Dose. *Colloids Interface Sci. Commun.* **2014**, *1*, 27–30. [CrossRef]
40. Kozak, M.; Domka, L. Adsorption of the quaternary ammonium salts on montmorillonite. *J. Phys. Chem. Solids* **2004**, *65*, 441–445. [CrossRef]
41. Coates, J. Interpretation of infrared spectra, a practical approach. In *Encyclopaedia of Analytical Chemistry*; Meyers, R.A., Ed.; John Wiley & Sons Ltd.: Chichester, UK, 2000; pp. 10815–10837.
42. Rytwo, G.; Malka, H. A pilot plant for the treatment of cowshed effluents. *Water Irrig.* **2013**, *530*, 6–9.
43. Inbar, Y. New Standards for Treated Wastewater Reuse in Israel. In *Proceedings of the Wastewater Reuse—Risk Assessment, Decision-Making and Environmental Security*; Zaidi, M.K., Ed.; Springer: Dordrecht, The Netherlands, 2007; pp. 291–296.
44. Zeng, T.; Li, R.J.; Mitch, W.A. Structural Modifications to Quaternary Ammonium Polymer Coagulants to Inhibit N-Nitrosamine Formation. *Environ. Sci. Technol.* **2016**, *50*, 4778–4787. [CrossRef]

water

MDPI

Article

Reuse of Treated Wastewater for Crop Irrigation: Water Suitability, Fertilization Potential, and Impact on Selected Soil Physicochemical Properties

Solomon Ofori [1,*], David Kwesi Abebrese [2], Iveta Růžičková [1] and Jiří Wanner [1]

[1] Department of Water Technology and Environmental Engineering, Faculty of Environmental Technology, University of Chemistry and Technology, Technická 5, Prague 6–Dejvice, 166 28 Prague, Czech Republic; ruzickoi@vscht.cz (I.R.); wannerj@vscht.cz (J.W.)

[2] Department of Water Resources, Faculty of Agrobiology, Food and Natural Resources, Czech University of Life Sciences, Kamýcká 129, Prague 6–Suchdol, 165 00 Prague, Czech Republic; abebrese@af.czu.cz

* Correspondence: oforis@vscht.cz

Abstract: This study evaluates the suitability of treated wastewater (TWW: secondary effluent and membrane effluent) for crop irrigation and the resultant impact on crop growth and soil physicochemical characteristics. Carrot seeds (*Daucus carota* subsp. *sativus*) were grown on loam soil and irrigated with tap water (Tap), secondary effluent (SE), and membrane effluent (ME) until maturity. Bacteriological analyses showed four log counts of *E. coli* and thermotolerant coliforms for secondary effluent, making it unsafe for the irrigation of carrots. Tap water and membrane effluent fulfilled the microbial limit for water reuse and were suitable for irrigation. The sodium absorption ratio, Kelly index, and magnesium hazard assessments indicated that all three irrigation water streams were suitable for irrigation. The average mass of carrot fruits for Tap, SE, and ME was 2.14 g, 3.96 g, and 3.03 g, respectively. A similar trend was observed for the dry matter composition: Tap had 15.9%, SE had 18.3%, and ME had 16.6%. The soil pH increased from 7.08 to 7.26, 7.39, and 7.33 for tap water-, secondary effluent-, and membrane effluent-irrigated soils, respectively. Nitrate-nitrogen and potassium levels increased in the TWW-irrigated soil, while that of the tap water-irrigated soil decreased. Sodium levels in the TWW-irrigated soil increased significantly but did not induce soil sodicity. The application of the TWW enhanced the growth of the carrot plants and increased the soil nutrient levels. Hence, using TWW in agricultural irrigation could promote food production and also limit the overdependency on freshwater resources. However, TWW should be disinfected by using UV disinfection and ozonation to reduce the risk of microbial contamination. Such disinfection methods may not lead to the formation of toxic byproducts, and therefore secondary pollution to crops is not anticipated.

Keywords: wastewater treatment; water reuse; irrigation; sodium adsorption ratio; crop growth; biomass production

1. Introduction

Water is considered an essential natural commodity that is required for the development of life on Earth [1]. Its (freshwater) availability has been acknowledged as a critical environmental control factor for driving anthropogenic activities to maintain socioeconomic strategies [2]. However, freshwater availability is becoming scarce, especially in arid regions of the globe [3]. Global warming, climate change, uneven water resource distribution, and high population growth have led to this water scarcity [2,4,5]. According to estimates, 2.4 billion people currently reside in water-scarce basins because of climate change, and predictions show that this deficit will grow due to the high sensitivity of water scarcity to climate change patterns [6–8]. The agriculture sector is most likely to be the sector that will be affected the most due to its high dependency on water resources. Water scarcity has been identified as the principal constraint to the sustainability of agriculture production [5].

The world's population is projected to reach approximately 10 billion in 2050, which will require an increase of more than 60% in agricultural production (from the 2005 production baseline) [9,10]. The increased demand for food will translate into higher agricultural water requirements, especially for the irrigation of food crops [4]. The continuous reliance on rainfed agriculture or/and the irrigation of crops by conventional water sources, such as rivers, streams, lakes, groundwater, springs, etc., may not be sustainable. An alternative nonconventional water source for crop irrigation is essential for the sustainability of agricultural production. Treated wastewater (TWW) or recycled water has emerged as a viable alternative nonconventional water source for crop irrigation, particularly in countries experiencing water stress [3]. TWW could supply important nutrients to crops, improve soil fertility, and improve soil physical and chemical properties to promote plant growth and crop production, as well as reduce farmers' dependency on commercial fertilizers [1,11]. Additionally, TWW has the potential to remediate and restore degraded and contaminated sandy soils [12–14]. The use of TWW for irrigation could come with some demerits, such as soil salinization, the risk of heavy metals, and the microbial contamination of food crops.

The irrigation of crops with TWW has been studied by several authors under different climatic conditions. Studies have involved crops like alfalfa, radish, wheat, maize, citrus, acacias, eucalyptus, cotton, and vegetable crops [1,15]. Al-Lahham et al. [16] studied the impact of TWW irrigation on the quality attributes of tomato fruits. The tomato fruits increased in diameter by 2 cm and in weight by up to 78.8 g. The authors concluded that TWW could be an alternative water source for irrigating tomatoes and care should be taken to avoid bacteriological contamination. In another study, the authors found that the eggplant yield under TWW irrigation was twice that of the yield under freshwater irrigation [17]. Other works involving barley and maize plants all recorded higher dry matter content and yields (respectively) under wastewater irrigation [18,19]. The increase in the fruit diameter, weight, length, and yield is due to the fertilization effect of TWW, since it is a reservoir of nutrients, such as nitrogen, phosphorus, and potassium [11].

One vegetable crop that has gained attention under TWW irrigation is carrot (*Daucus carota* subsp. *sativus*). Carrot is a root vegetable belonging to the *apiaceae* family and is the most important crop of the family [20]. Several studies involving the irrigation of carrot plants with TWW have been conducted. However, most of the studies focused on health and environmental risks, such as bacteriological contamination, pharmaceutical compound uptake and accumulation, and heavy metal bioaccumulation and bioaccessibility [15,21–25]. Few works have focused on the impact of the TWW on the physiological and morphological traits of the carrot, and the suitability assessment of TWW using water quality criteria [15,24,25]. The suitability of irrigation water is key to the effective growth of crops and the sustainability of arable lands. Higher levels of exchangeable sodium in irrigation water could increase the soil salinity and reduce the soil permeability, thereby affecting crops' nutrient uptake and osmotic activities. Elevated levels of magnesium, on the other hand, could cause the deterioration of the soil structure and quality, thereby affecting crop production [26]. Also, the suitability of irrigation water is key to protecting public health. Bacteriological characteristics or quality is one of the most important parameters in the evaluation of the suitability of TWW for crop irrigation. The presence of pathogenic microorganisms could pose serious health risks to farm employees and consumers through the inhalation and ingestion of these pathogenic organisms [27]. Several guidelines and legislations have been enacted globally to ensure public safety [28–30]. Notable among them are the World Health Organization Guidelines for the Safe Use of Wastewater, Excreta and Grey Water [29], the California Water Recycling Criteria [30], the United States Environmental Protection Agency Guidelines on Water Reuse [31], and the European Union Regulation 2020/741 on the minimum requirements for water reuse [32]. The latter mandates member countries across Europe to evaluate the microbial quality of TWW before application for reuse in agriculture.

In this study, we hypothesize that TWW is suitable for the irrigation of crops (carrots) and will not adversely impact the physicochemical properties of the soil. The objectives are, therefore, (i) to assess the suitability of TWW (secondary effluent and membrane effluent) for crop irrigation using irrigation water quality assessment criteria, such as the sodium adsorption ratio (SAR), Kelly index (KI), and magnesium hazard (MH); (ii) to evaluate the bacteriological quality of TWW; (iii) to investigate the variation in the fertilization potential of TWW and tap water; (iv) to evaluate the impact of TWW on some selected physicochemical properties of the irrigated soil. To the best of our knowledge, studies incorporating SAR, KI, and MH in the evaluation of treated wastewater suitability for irrigation is scarce. These techniques are usually used in groundwater and surface water quality assessment. This study contributes to closing the knowledge gap on wastewater reuse for irrigation. The outcome of this study will be very useful for the promotion and implementation of treated wastewater reuse for irrigation.

2. Materials and Methods

2.1. Experimental Design

This study was conducted in a greenhouse facility at the Czech University of Life Sciences, Czech Republic. It involved irrigating carrot plants (*Daucus carota* subsp. *sativus*) with two different streams of TWW and tap water (control). The TWW irrigation streams consisted of secondary effluent and membrane effluent. Crop treatments were made up of secondary effluent-irrigated carrots (SE), membrane effluent-irrigated carrots (ME), and tap water-irrigated carrots (Tap). Tap water-irrigated carrots (Tap) represented the control group. Carrot seeds were purchased from a commercial supermarket and sown directly into soil contained in pots. Each pot had a volume of 0.0034 m^3 (3.4 L) and was filled with soil to about 85% of the volume. The soil was classified as loam, composed of clay (14.0%), silt (37.2%), and sand (48.7%) fractions. The pots were placed on a greenhouse growth bench and under a Growth Spectrum Advanced 600 W lamp (GIB Lighting, Berlin, Germany) with a 12 h daytime setting. After 66 days, the time setting was changed to a 9 h daytime setting. This was to reduce the high evaporation rate of water from the soil due to the heat from the lamp. The lamp has a nominal power of 600 W, a light intensity of 48,000 lm, a color temperature of 8000 K, and a photon flux (100 h) of 740 μmols/s. Each pot was adequately spaced from the other to ensure the effective circulation of air to the plants.

On average, a once-a-day irrigation regime was employed to ensure an adequate supply of water due to the high evaporation rate caused by the lamp. The mean temperature and relative humidity in the greenhouse were 26 °C and 35%, respectively. Sprinkling of water on the greenhouse floor was regularly performed to reduce the temperature and increase the humidity. Throughout the experiment, no fertilizer, manure, compost, or any other form of soil amendment was applied. The sowing of the seeds occurred in September 2021, and the matured carrots were harvested in February of the following year.

2.2. Irrigation Water Streams

Secondary effluent was obtained from a municipal wastewater treatment plant (WWTP) in Prague, which operates a conventional activated sludge system. The treatment tray consists of screening, grit removal, primary sedimentation, biological removal of nutrients and organics, and secondary sedimentation. Influent to the WWTP was initially treated using physical and mechanical processes. The partially treated wastewater was treated further using an activated sludge process to remove nutrients and organics. After this stage, the effluent was sent to the secondary clarifier, and afterwards discharged as secondary effluent. An amount of 0.02 m^3 of the secondary effluent was collected and stored in a reservoir for irrigation. Another 0.02 m^3 of secondary effluent was collected and post-treated with a laboratory-scale submersible hollow fiber ultrafiltration membrane (Figure 1). The membrane treatments were performed in batches using ZeeWeed-1, a polyvinylidene difluoride membrane from GE Water (Oakville, ON, Canada). The module had a nominal membrane surface area of 0.093 m^2 and a maximum transmembrane pressure of 62 kPa (manufacturer's manual). Effluent from the post-treatment was referred to as membrane effluent. Tap water was collected from the Biological Wastewater Treatment Laboratory (University of Chemistry and Technology-Prague) into a 0.01 m^3 reservoir and used for irrigation. All the irrigation water streams were frozen to minimize changes in the water quality and applied in batches to irrigate the carrot plants.

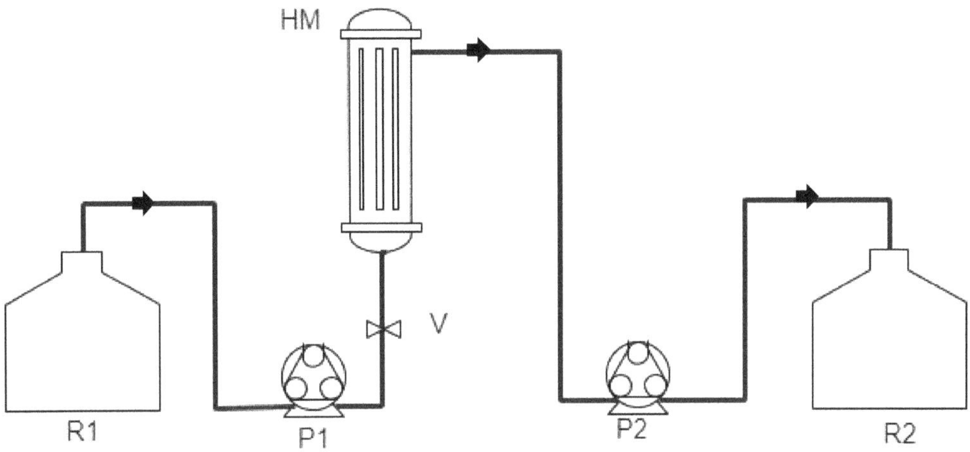

Figure 1. Schematic diagram of the ultrafiltration membrane laboratory set-up for the further treatment of the secondary effluent into the membrane effluent. R1 is the feed reservoir (secondary effluent); R2 is the permeate reservoir (membrane effluent); HM is the hollow tube ultrafiltration membrane; P1 and P2 are peristaltic pumps; V is a valve.

2.3. Determination of the Physicochemical Characteristics of the Different Irrigation Water Streams

The physicochemical quality of the irrigation streams was analyzed using Thermo Fischer's Gallery Analyzer (Thermo Fischer Scientific Inc., Vantaa, Finland), except for the total suspended solids (TSS), chemical oxygen demand (COD), and total nitrogen. Water samples were collected into new and clean cuvettes, inserted into the sample rack, and placed inside the Gallery analyzer for the analyses. The limit of detection for the methods of phosphate (PO_4), magnesium (Mg), chloride (Cl), and calcium (Ca) was 3.6 µg/L, 1.5 mg/L, 0.35 mg/L, and 1.16 mg/L, respectively. The TSS was determined by the gravimetric method at 105 °C and the COD was determined by the colorimetry method [33]. The COD was measured by using a photoLab 7100 Vis series spectrophotometer (WTW GmbH, Wellheim, Germany) at a limit detection of 10 mg/L. Total nitrogen was analyzed photometrically using a spectroquant nitrogen cell test kit (Merck, Darmstadt, Germany). The determination followed the manufacturer's protocol, which was analogous to ISO and DIN standards. The test kit had a measurement range of 0.5–15 mg/L.

Analyses of potentially toxic elements (lead (Pb), zinc (Zn), cadmium (Cd), essential metals (sodium (Na), potassium (K)), and boron (B)) were performed using atomic absorption spectroscopy (AAS). The limit of detection for all the analytes was 0.01 mg/L, except Pb, which was 0.05 mg/L. Water samples for such analyses were collected in clean plastic containers, preserved (pH < 2) with nitric acid, and refrigerated until the measurement. For each 0.02 m^3 of irrigation water (irrigation cycle), the water quality test was performed twice, before the start of the irrigation and in the latter part of the cycle, except for the boron and the metals. The physical and chemical quality characteristics of the different irrigation water streams are presented in Table 1.

2.4. Determination of the Irrigation Water Quality Characteristics according to Salt and Ion Toxicity

The evaluation of the suitability of the different irrigation water streams for irrigation was performed using the sodium absorption ratio (SAR), Kelly index (KI), and magnesium hazard (MH). These indicators provide vital information on the potential risk of sodium and magnesium hazards posed by irrigation water to soil and crops. Below are the equations of the indicators, which were adopted from [34]:

$$SAR = \frac{Na^{+}}{\sqrt{\frac{Ca^{2+}+Mg^{2+}}{2}}} \tag{1}$$

$$KI = \frac{Na^{+}}{Ca^{2+} + Mg^{2+}} \tag{2}$$

$$MH = \frac{Mg^{2+} \times 100}{Ca^{2+} + Mg^{2+}} \tag{3}$$

The ions Na, Ca, and Mg are very essential for the evaluation of the suitability of water for irrigation. Sodium ions could induce plant toxicity and reduce soil permeability, while Ca and Mg could reduce the crop yield by making the soil alkaline [35]. The FAO guideline stipulates a limit of >3 meq/L for Na. For Ca and Mg, the guideline provides the usual range of these ions in irrigation water as $\leq$20 meq/L and $\leq$5 meq/L, respectively [28].

2.5. Determination of the Microbiological Quality of the Irrigation Water Streams

Water samples were collected into sterilized bottles for the determination of the presence of total coliforms, thermotolerant coliform, *Escherichia coliform* (*E. coli*), and *Clostridium perfringens*. The determination of total coliform bacteria was performed according to the Czech Standard. An amount of 100 cm^3 of the diluted water sample was filtered through a 0.45 μm membrane filter. The membrane filter was transferred to a selective culture agar medium containing lactose (Endo agar) in a Petri dish. The Petri dish containing the membrane filter was placed upside down in a thermostat and cultured at a temperature of 36 °C for about 18–24 h. The filter was then transferred from the Petri dish to a piece of filter paper saturated with cytochrome oxidase solution. After 2 min, the colonies of coliform bacteria formed were counted [36,37].

The detection of *Clostridium perfringens* in the irrigation water streams followed the Czech Decree 252/2004. An amount of 100 cm^3 of the diluted water sample was filtered through a membrane filter and the filter was transferred to a solid m-CP agar medium. Cultivation was then carried out in an anaerobic environment at 44 °C for about 21 h. After cultivation, the grown colonies, yellow in color, were counted. The colonies were exposed to ammonia vapor for 20–30 s. A change in the color from yellow to pink to red was observed, confirming the colonies as *Clostridium perfringens* [38,39].

In the evaluation of the presence of thermotolerant coliform and *E. coli* in the irrigation water, 100 cm^3 of the diluted water sample was filtered with a 0.45 μm membrane filter. The filter was then transferred to m-FC agar in a Petri dish. Cultivation of the bacteria occurred at a temperature of 44 °C for 18–24 h, after which the number of thermotolerant coliform bacteria colonies formed was counted. The membrane filter was later transferred to a culture mat saturated with liquid culture medium for 2–4 h in the dark at a temperature of 36 °C. After this step, the filter was visualized under a UV lamp at 360 nm, and the fluorescent colonies were counted. The method of detection was based on the Czech Standard ČSN 75 7835 [40,41].

Table 1. Physicochemical analyses of the three streams of irrigation water (tap water, secondary effluent, and membrane effluent) in comparison with the Food and Agriculture Organization (FAO) guidelines on irrigation water quality.

Irrigation Water Quality Parameters	Tap Water	Secondary Effluent	Membrane Effluent	FAO Irrigation Water Quality Guidelines [28]
pH	7.68 ± 0.69	7.90 ± 0.53	8.01± 0.43	6.5–8.4
TSS (mg/L)	0.14 ± 0.06	4.33 ± 0.67	0.29 ± 0.14	≤10.0 **
Conductivity (dS/m)	0.37 ± 0.14	0.74 ± 0.10	0.72 ± 0.13	<0.7
Alkalinity (mg/L $CaCO_3$)	77.84 ± 8.62	120.76 ± 24.44	118.12 ± 20.20	n.p
Total Nitrogen (mg/L)	4.12 ± 1.00	12.27 ± 0.50	12.57 ± 1.17	n.p
CODcr (mg/L)	<20	<20	<20	n.p
NO_3-N (mg/L)	3.36 ± 0.57	9.00 ± 2.01	8.96 ± 2.33	<5.0
PO_4 (mg/L)	0.08 ± 0.02	0.72 ± 0.12	0.77 ± 0.19	n.p
B (mg/L)	0.02 ± 0.003	0.07 ± 0.002	0.07 ± 0.001	<0.7
Ca (mg/L)	58.92 ± 4.64	79.89 ± 12.25	79.41±10.37	n.p
Mg (mg/L)	7.56 ± 0.40	14.77 ± 2.85	14.99 ± 2.59	n.p
Cl (meq/L)	0.74 ± 0.05	3.23 ± 0.72	3.24 ± 0.74	<4.0
Pb (mg/L)	<0.05	<0.05	<0.05	5.0
Zn (mg/L)	0.04 ± 0.01	0.13 ± 0.04	0.12 ± 0.05	2.0
Cd (mg/L)	<0.01	<0.01	<0.01	0.01
K (mg/L)	4.97 ± 0.34	26.76 ± 2.72	27.86 ± 6.01	n.p
Na (meq/L)	0.75 ± 0.01	3.77 ± 0.32	3.80 ± 0.17	<3.0

Notes: Values are mean plus standard deviation (±); "meq/L" is milliequivalent per liter; "n.p" means the limit is not provided by the FAO guidelines; ** limit from EU Regulation 2020/741 on the minimum requirement for water reuse. Part of the data in the table are presented in another manuscript which is under review [42].

2.6. Determination of the Physical and Chemical Properties of Soil

Soil samples were air-dried and sieved with a 2 mm sieve. The pH was measured with a WTW Multi 9420 pH meter (WTW GmbH, Wellheim, Germany) in a 0.01 M calcium chloride solution ($CaCl_2$). A 2:1 (v/w) ratio of $CaCl_2$ solution to soil was used, following the procedure outlined by Motsara and Roy [43]. Soil extract for nitrate-nitrogen (NO_3-N), total nitrogen (Total-N), phosphate (PO_4), sulfate (SO_4), lead (Pb), zinc (Zn), cadmium (Cd), potassium (K), sodium (Na), and magnesium (Mg) analyses were obtained by using the same 0.01 M $CaCl_2$ extractant in a 10:1 (v/w) ratio. The soil samples were weighed, and a measured volume of the $CaCl_2$ solution was added. The mixture was shaken mechanically for about 2 h and then filtered with filter paper to obtain the extract. The extraction procedure was based on the protocol by Houba et al. [44] and Motsara and Roy [43]. The NO_3, PO_4, SO_4, and Mg were analyzed in the soil extract using Thermo Fischer's Gallery Analyzer (Thermo Fischer Scientific Inc., Vantaa, Finland), and the total nitrogen was measured by the spectroquant nitrogen cell test kit (Merck, Darmstadt, Germany). Soil extract analyses for Pb, Zn, Cd, K, and Na were preserved (pH < 2) with nitric acid and refrigerated below 4 °C. The measurements were performed using atomic absorption spectroscopy (Agilent 280FS AA, Agilent Technologies, Mulgrave, Australia) with the flame atomization technique. The flame type used was acetylene–air and the wavelengths were 217 nm, 213.9 nm, 228.8 nm, 766.5 nm, and 589 nm for Pb, Zn, Cd, K, and Na, respectively. The choice of extractant ($CaCl_2$) was to enable the extraction of the bioavailable fraction of the elements of interest.

The soil particle density was calculated from the ratio of the mass to the volume of the soil particles. The mass was determined by weighing the dry soil sample, whereas the volume was determined by calculating the mass and density of the water displaced by the soil sample in a 100 cm^3 pycnometer [45]. Four replicate tests were carried out and averaged as 2.44 g/cm^3 (standard deviation < 0.03).

The hydrometer method (ČSN EN ISO 17892-4) was used in determining the fraction of particles for fitting the particle size distribution curve. The density of the soil sample suspended in distilled water was measured with a calibrated hydrometer at predefined time intervals (30, 60, 120, 300, 900, 2700, 7200, 18,000, 97,200, and 172,800 s after the start of sedimentation). Temperature readings of the samples were simultaneously recorded with the hydrometer readings [46].

2.7. Measurement of Plant Height, Fresh Mass, and Dry Matter Content of Biomass

Plant height was measured against a white background using a measuring ruler. The measurement involved only the shoot (upper vegetative part), excluding the root or fruit. The determination of the fresh mass/weight and percentage dry matter content was based on the Organization for Economic Cooperation and Development guidelines (OECD) [47]. The edible part (fruit) was washed with distilled water several times to remove soil particles. It was then dried quickly with tissue paper and weighed to obtain the mass. Samples for the dry matter content determination were oven-dried at 70 °C for about 5 h after the initial mass was recorded. The difference in the mass of the dried sample and the fresh sample was used to compute the dry matter content using Equation (4).

$$\text{Percentage Dry Matter} = \frac{(C - A)}{(B - A)} \times 100\% \quad (4)$$

where A is the mass of the Petri dish; B is the mass of the fresh sample and Petri dish; C is the mass of the dried sample and Petri dish [47].

2.8. Statistical Analyses of Data

Statistical analyses and graphical representations of the obtained data were performed using Microsoft Excel 2019 and Statistica 13.5 by TIBCO Software Inc., Palo Alto, CA, USA. The test of significance of the means was performed by the Student t-test (paired) at a confidence interval of 95% ($p < 0.05$) for the investigated parameters, except for the fresh mass and dry matter composition. An ANOVA was used to analyze difference in the means of fresh mass at a 95% confidence level. Pearson correlation was used to evaluate the relationship between biomass mass and plant height for SE and Tap.

3. Results and Discussion

3.1. Evaluation of the Physicochemical Suitability of the Different Water Streams for Irrigation

The SAR results showed that all three irrigation water streams had values below the 15-value threshold limit (Table 2), classifying the TWW and tap water as suitable for irrigation. This implies that the use of these streams of water for irrigation may not pose a high risk of soil salinization. This was also confirmed by the KI, which classified all the irrigation water as acceptable. Water with a very high SAR (>26) and KI ($\geq$1) may cause the deposition and accumulation of sodium ions in the soil, and therefore is considered unsuitable. The build-up of sodium in the soil could lead to the clogging of the soil pores and permeability problems. Such a phenomenon could adversely alter the uptake of nutrients and water by the crops [26]. The soil could become sodic due to the excess sodium ions, and negatively impact the physical, structural, and nutritional characteristics of the soil [48]. The SAR of the secondary effluent (2.34) and membrane effluent (2.36) were higher than the tap water (0.57) due to their relatively high sodium content (Table 1). The high sodium content is attributed to their source, which was municipal wastewater. Municipal wastewater is usually high in sodium, and the conventional biological treatment process may not be effective in removing sodium, thereby leading to the relatively high sodium content in the effluent. The works of Kalavrouziotis et al. [49] and Bedbabis et al. [50] reported sodium levels above 150 mg/L in the treated wastewater effluent. Notwithstanding, the use of the secondary effluent and the membrane effluent for crop irrigation may not have a significantly adverse effect on crop growth due to the low SAR value. This was evident during the experiment, where no negative impact was observed on the carrot plants, nor was the biomass production adversely affected.

Tap water had a very low KI value (0.26), indicating a lower sodium content and making it more suitable for irrigation. Irrigation water with a KI value of $\geq$1 is considered unsuitable because such waters contain excess sodium ions (salt) than magnesium and calcium, and therefore have the potential to transform alkaline soil into saline soil [26,34]. The KI values of the secondary effluent and membrane (0.92 and 0.93, respectively) were below the threshold limit of 1. This indicates that the TWW did not contain excess sodium ions and is suitable for irrigation. The level of sodium in the TWW (Table 1) was well within the usual range of 0–40 meq/L typical of irrigation water [28]. Both the SAR and KI analyses suggested that the TWW used in this study may not pose a high salinity or sodicity risk, and therefore could be used for agricultural irrigation. The pH of the tap water- and secondary and membrane effluent-irrigated soils were below 8.5, confirming no occurrence of soil sodicity [51].

Table 2. Irrigation water quality classification according to sodium absorption ratio (SAR), Kelly index (KI), and magnesium hazard (MH). Description of the quality classification was adopted from Gorfie et al. [52] and Mukherjee et al. [26].

Quality Criteria	Tap Water	Secondary Effluent	Membrane Effluent
SAR	0.57	2.34	2.36
Quality Classification	Suitable	Suitable	Suitable
KI	0.26	0.92	0.93
Quality Classification	Acceptable	Acceptable	Acceptable
MH (%)	11.37	15.6	15.88
Quality Classification	Suitable	Suitable	Suitable

Notes: The computation was performed using the mean concentrations of sodium, calcium, and magnesium from Table 1. The concentrations of sodium, magnesium, and calcium were converted to milliequivalent per liter (meq/L) for the SAR computation. For a full description of the water quality classification of the SAR, KI, and MH, refer to Gorfie et al. [52] and Mukherjee et al. [26].

The magnesium hazard (MH) assessment, which is a measure of excess calcium and magnesium in water, showed that the tap water and the TWW were all suitable for agricultural application. None of the irrigation water exhibited higher MH values. The MH of the tap water, secondary effluent, and membrane effluent were 11.37%, 15.60%, and 15.66% respectively, far below the threshold limit of 50% [52]. This implies that none of the irrigation water poses a risk of clay particle dispersion or decreasing the hydraulic conductivity of the soil [26]. The order of quality or suitability according to the SAR, KI, and MH assessments was tap water > secondary effluent > membrane effluent in all the cases.

However, the order is different when the quality of the water is considered from the point of view of the classical nutrient supply. In Table 1, both the secondary effluent and membrane effluent exhibited significantly ($p < 0.05$) higher nitrate-nitrogen, phosphate, and potassium content than the tap water. Therefore, the TWW is of higher quality and has a higher potential to supply nutrients for crop growth than tap water. This is evident in the higher biomass production of carrots that were irrigated with the TWW. Also, the relative increase in the nitrate-nitrogen, potassium, and magnesium content of the secondary effluent-irrigated soil and membrane effluent-irrigated soil attests to the fertilization potential of the TWW. No significant difference was observed between the fertilization potential of the secondary effluent and the membrane effluent.

The pH of the irrigation water ranged between 7.68 and 8.01 for the tap water and TWW (Table 1). Irrigation water with extreme acidity or alkalinity affects the bioavailability of nutrients, such as phosphates. At a high pH, the precipitation of phosphates occurs, while fixation occurs at a low pH [52,53]. The tap water, secondary effluent, and membrane effluent were slightly alkaline and did not contain excessive bicarbonates. All three irrigation water streams do not pose a high risk of turning nonalkaline soil into alkaline soil. With the appropriate irrigation management strategy, the reuse of the TWW over a longer period may not significantly alter the acidic or alkaline status of the soil. The pH of the secondary effluent (7.90) and membrane effluent (8.01) were well within the acceptable range of 6.5–8.4 [28,54]. Potentially toxic elements (PTEs), such as lead, zinc, and cadmium, were all within the acceptable limit stipulated by the Food and Agricultural Organization

(FAO) guidelines on irrigation water quality [28]. Except for nitrate-nitrogen and sodium, all the physicochemical quality limits provided in the FAO guidelines were fulfilled by the two streams of the TWW used in the present study. Generally, the TWW used in this study could therefore be considered suitable for agricultural irrigation from the point of view of the potential risk of soil salinity or sodicity and the promotion of crop growth. However, irrigation management practices, such as the periodic dilution of the TWW by freshwater, are recommended to prevent groundwater salinization by nitrates and soil salinization by salts.

3.2. *Evaluation of the Bacteriological Suitability of the Different Irrigation Water Streams*

None of the selected indicator microorganisms were detected in the tap water, as expected (Table 3). The tap water had a high purity and its quality was consistent with typical drinking water quality. It fulfilled the criteria set out in the EU guideline, Regulation (EU) 2020/741, on water reuse for irrigation. On the other hand, the two streams of the TWW showed varied bacteriological quality characteristics. Secondary effluent had a relatively high microbial load in the order of four logs for the total coliform, thermotolerant coliform, and *E. coli*, and three logs for *Clostridium perfringens*. The *E. coli* count per 100 mL of water sample was higher than the $\leq$10 CFU/100 mL limit stipulated in the EU Regulation (EU) 2020/741 [32]. Thermotolerant coliform counts were also several thousand higher than the recommended 0–200 CFU/100 mL count [29]. Secondary effluent is therefore not considered suitable for irrigating carrots and other edible crops, and poses considerable risk to human health. The membrane effluent had different bacteriological quality characteristics when compared to the secondary effluent. Just like tap water, *Clostridium perfringens* was not detected in the membrane effluent. Thermotolerant coliform and *E. coli* counts were below 10 CFU/100 mL, fulfilling the limit stated in the EU Regulation (EU) 2020/741. The membrane effluent is therefore suitable for the irrigation of carrot plants and other edible crops, and poses little or no risk to human health.

The significant difference observed between the two streams of the TWW was due to the extra treatment or polishing by the ultrafiltration membrane module. Even though both streams came from the same source, further treatment by the membrane led to the removal or filtering of the microorganisms from the water. The outcome of the bacteriological analyses affirms the assertion that, for the safe use of TWW for crop irrigation, the disinfection or polishing of the water is required. The new EU Regulation 2020/741, which came into force in 2023, stipulates that TWW should be disinfected before being used for crop irrigation [32]. This is to ensure the safe use of TWW and to protect public health. It is evident from the microbiological results that the secondary effluent was not safe for crop irrigation and will require further treatment. Chlorination could be effective in significantly reducing the bacteriological or pathogenic loads of the secondary effluent to safe limits. Also, a combination of membrane technology coupled with UV disinfection could produce an effluent of high quality that meets the requirements for reuse [55]. Ozonation, an advanced treatment process, could also sanitize the TWW, making it safe for crop irrigation. The use of ozonation or UV disinfection may be preferred over chlorination due to the formation of toxic byproducts, such as trihalomethanes and haloacetic acids, by chlorination [56]. The effluent disinfected by these preferred processes may not pose a threat to plants, as well as soil, since no toxic byproducts will be formed.

Table 3. Microbiological water quality characteristics of the tap water, secondary effluent, and membrane effluent compared with the European Union (EU) Regulation 2020/741 on water reuse.

Indicator Microorganism	Tap Water	Secondary Effluent	Membrane Effluent	EU Water Reuse Guidelines [32]
Coliform [CFU/100 mL]	nd	$9.85 \pm 6.52 \times 10^4$	12.75 ± 7.93	≤1000 **
Thermotolerant Coliform [CFU/100 mL]	nd	$5.80 \pm 4.15 \times 10^4$	<10	≤10 *
Escherichia coli [CFU/100 mL]	nd	$2.76 \pm 2.33 \times 10^4$	<10	≤10
Clostridium perfringens [CFU/100 mL]	nd	$1.22 \pm 0.87 \times 10^3$	nd	>4log

Notes: "CFU" means colony-forming unit; "nd" implies not detected; * ISO guidelines cited from Drechsel et al. [57]; ** FAO guidelines [58].

3.3. Impact on Plant Growth and Biomass Production

The TWW-irrigated plants (SE and ME) showed better growth and had a higher biomass production than carrot plants irrigated with tap water (Tap). The fresh mass (mean) of the carrot fruits produced under tap water, secondary effluent, and membrane effluent irrigation were 2.14 g, 3.96 g, and 3.03 g, respectively (Figure 2). A similar pattern was observed for the percentage dry mass composition. The SE had the highest dry matter composition (18.3%), followed by the ME (16.6%), and then the Tap (15.9%), as can be seen in Figure 2. In Figure 3, the growth trend was not different from the aforementioned; the SE and ME exhibited better growth characteristics in terms of the height than the Tap. However, the difference in the growth indicators among the different irrigation treatments was not significant. A very strong correlation existed between the plant height and the corresponding fresh mass of the fruit produced (Table 4); the carrot plants with longer shoots produced heavier fruits. The observed trend in the growth characteristics and biomass production could be attributed to the higher fertilization potential of the secondary effluent and membrane effluent. The high levels of NPK in these two streams of the TWW promoted the supply of nutrients to the carrot plants and boosted their growth and biomass production.

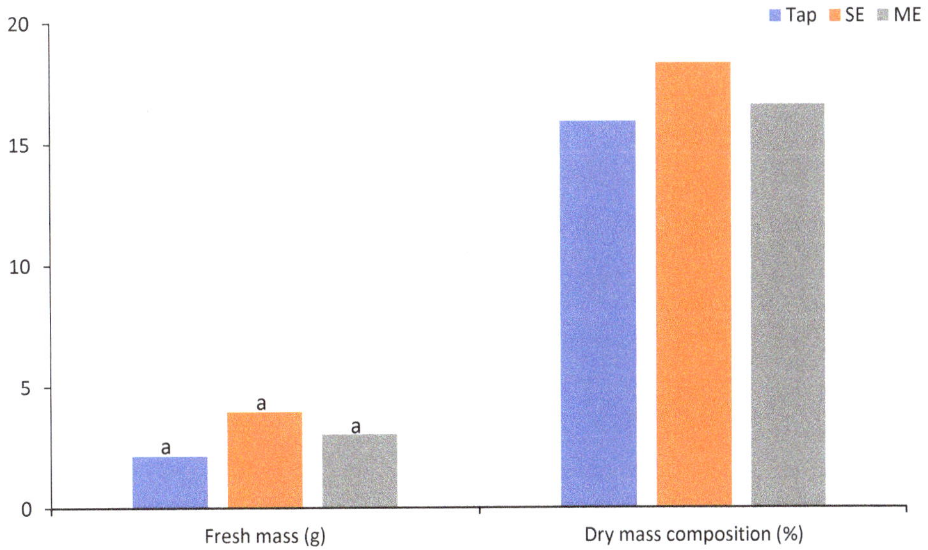

Figure 2. Effect of the TWW and tap water irrigation on the fresh mass (mean) and percentage dry matter composition/content of the carrot plants. The same letter indicates no significant difference ($p > 0.05$) was observed among the different treatments. Fresh mass is expressed in grams and dry matter content is in percentages.

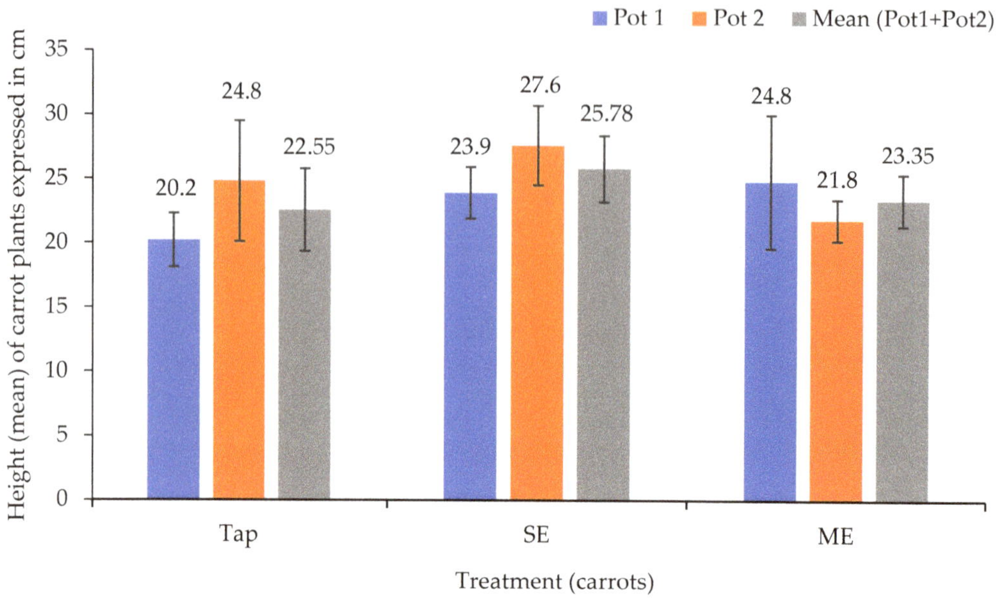

Figure 3. Effect of the TWW and tap water on the height (mean) of the carrot plant. Only the vegetative part of the carrot plant was used in the measurement. The blue and brown bar graphs represent the mean height of the individual pot treatments, and the gray bar graph represents the mean for six plants per treatment. Error bars are the standard deviation.

Table 4. Pearson correlation shows a strong relationship between the plant height and the fresh mass of the carrot plants under tap water and secondary effluent irrigation. Tap 1 and SE 2 are the fresh mass and Tap 3 and SE 4 are the height, respectively.

	Tap 1	SE 2	Tap 3	SE 4
Tap 1	1			
SE 2	0.478449	1		
Tap 3	0.814232	0.818247	1	
SE 4	0.522865	0.972042	0.798666	1

The findings of the study are not in isolation from similar studies that have been conducted. Alkhamisi et al. [3] reported a significant increase in the height of wheat and cowpeas irrigated with the TWW compared to groundwater-irrigated plants. The yield of cowpeas was significantly higher for the TWW-irrigated cowpeas. Another study also found that TWW-irrigated vegetables exhibited better growth and had higher biomass production [22].

3.4. *Impact of Irrigation Water on Selected Physical and Chemical Properties of Soil*

3.4.1. pH

Tap water- and TWW-irrigated soils changed from neutral (7.08) to moderately alkaline after irrigation. Tap water-irrigated soil increased to 7.26, while secondary and membrane effluent-irrigated soils increased to 7.39 and 7.33, respectively (Figure 4). This corresponds to an increase of 0.18, 0.31, and 0.25 units, accordingly. The irrigation water of the present study may have altered the pH status of the soil, changing it to moderately alkaline. The magnitude of the change in the soil pH corroborates with the alkaline status of the different irrigation water streams (Table 1). The pH of the TWW-irrigated soil was higher than tap water-irrigated soil. The increase may have been caused by the release of exchangeable cations during organic matter mineralization and the relatively high cation content of the TWW [52,59,60]. Also, the relatively high level of bicarbonates in the TWW might be responsible for the higher pH elevation of the TWW-irrigated soil, since bicarbonates have a pH-increasing effect on soil. This may have precipitated cations, such as calcium and magnesium, out of the soil solution and formed calcium bicarbonates and magnesium bicarbonates [61]. These formed products tend to increase the pH of soils.

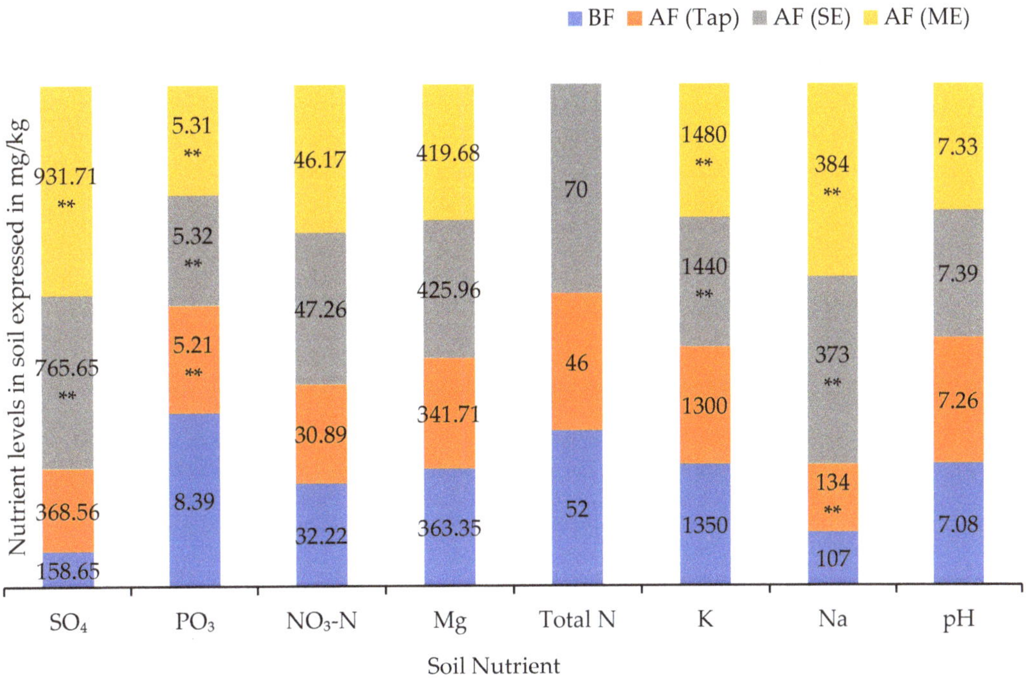

Figure 4. Soil pH and nutrient levels of the soil before and after the irrigation. Values indicated are nutrient contents in mg/kg, except pH. ** indicate a significant difference ($p < 0.05$) in nutrient levels before and after irrigation. BF refers to the level of nutrients in the soil before irrigation; AF refers to the level of nutrients after irrigation.

Research has shown that wastewater high in bicarbonates has the potential to increase the soil pH [62,63]. In a study, Tarchouna et al. [60] observed an increase in the soil pH after irrigation with slightly alkaline TWW. The soil changed from slightly acidic (6.85) to moderately alkaline (8.35). The pH increases in the present study did not have a significantly adverse impact on the availability of nutrients.

3.4.2. Nutrients

The nitrate-nitrogen increased from 32.22 mg/kg to 47.26 mg/kg and 46.17 mg/kg for the secondary effluent and membrane effluent-irrigated soil, respectively, but decreased to 30.89 mg/kg for the tap water-irrigated soil. A similar trend was observed for potassium and magnesium. This affirms the assertion that the TWW has a higher fertilization potential than tap water. Nitrate-nitrogen and potassium levels in the tap water were below the crop nutrient requirement, leading to a depletion of the already-existing levels in the soil. Phosphate decreased from 8.39 mg/kg to 5.21 mg/kg, 5.32 mg/kg, and 5.31 mg/kg for the tap water-, secondary effluent-, and membrane effluent-irrigated soils, respectively. Phosphate levels in all the irrigation water might have been below the crop requirement and were insufficient. Since no additional supply of phosphate in the form of fertilizer or manure was made, the carrot crops made use of the phosphate present in the soil, leading to the reduction in the initial amount. Crop nutrient requirement assessment is therefore necessary before the application of TWW irrigation to ensure an adequate supply of nutrients. Sulfate, a readily available form of sulfur for plant uptake, was significantly higher in the soil irrigated with the TWW than the tap water. Tabatabai [64] suggested that irrigation water influences the concentration of water-soluble sulfate in soil.

Tap water-irrigated soil increased from 158 mg/kg to 368 mg/kg, while secondary and membrane effluent-irrigated soil increased to 765 mg/kg and 931 mg/kg, respectively (Figure 4). The significant increase ($p < 0.05$) in sulfate in the TWW soil is attributed mainly to the fertilization potential of the TWW. Several studies have reported similar results of improvement in soil nutrients after TWW irrigation [51,63,65].

3.4.3. Sodium Accumulation in Irrigated Soil

Sodium levels in the soil increased from 107 mg/kg to 134 mg/kg, 373 mg/kg, and 383 mg/kg for tap water-, secondary effluent-, and membrane effluent-irrigated soil, respectively (Figure 4). The accumulation of sodium in the TWW-irrigated soil was significantly higher than in the tap water-irrigated soil. This increase corroborated with the quantity of sodium present in the respective irrigation water (Table 1) and is in line with the existing literature [63,65]. The increase was caused by the deposition of sodium by the irrigation water. The relatively high calcium content of the TWW may have enhanced the selectivity of the potassium uptake and transport over sodium, leading to the deposition. Also, the antagonistic activity of potassium may have caused a reduction in the sodium adsorption capacity [63].

Even though the accumulation of sodium in the TWW-irrigated soil was significantly higher, the carrot plants did not show signs of toxicity, nor was the soil sodic (pH < 8.5) [51]. The use of the TWW for long-term irrigation would require salinity control measures, such as the dilution of TWW with less-saline water and the leaching of salts beyond the root zone, to maintain the integrity of the soil.

3.4.4. Distribution of Potentially Toxic Elements (PTEs) in Irrigated Soil

Potentially toxic elements (Cd and Pb) were below the limit of detection for both tap water- and the TWW-irrigated soils (Table 5). The concentration of Cd was below 0.01 mg/kg, and that of Pb was below 0.04 mg/kg. This result is consistent with what has been reported by Bedbabis et al. [63]. After 5 years of irrigation with TWW, the authors found that the Cd and Pb contents in the soil were below the detection limit. In this present study, there was no increase in the Zn, Cd, and Pb contents of the soil after irrigation with secondary effluent and membrane effluent. The concentration of the PTEs remained the same for all the treatments, except for Zn, which increased to 0.02 mg/kg from the initial level of 0.01 mg/kg for the tap water-irrigated soil. The application of the TWW to the soil did not lead to the deposition and accumulation of Zn, Cd, and Pb in the soil. Hence, the TWW may not pose a risk of PTE dissemination in soil, and therefore its use for crop irrigation will not lead to the uptake of the PTEs by the crops. A comparison of the results to the WHO guidelines on the maximum tolerable soil concentrations of toxic chemicals based on human health protection revealed that Cd and Pb concentrations were far below the guideline limits of 4 mg/kg for Cd and 7 mg/kg for Pb. Therefore, both the secondary effluent- and membrane effluent-irrigated soils pose little or no risk to humans and are safe for food cultivation.

Table 5. Distribution of selected potentially toxic elements (PTEs) in the soil before and after irrigation compared to the maximum tolerable soil concentrations of toxic chemicals based on human health protection.

Potentially Toxic Elements	PTE Levels in Soil before Irrigation (mg/kg)	PTE Levels in Soil after Irrigation (mg/kg)			Maximum Tolerable Soil Concentration (mg/kg) [29]
		Tap	SE	ME	
Pb	<0.04	<0.04	<0.04	<0.04	7.00
Zn	0.1	0.02	0.01	0.01	n.p
Cd	<0.01	<0.01	<0.01	<0.01	4.00

Notes: The bioavailable fraction of the PTEs was considered for the study and not the total fraction. This was based on the consideration that, under normal environmental conditions, the bioavailable fraction is the fraction that could be available for crop uptake (expert opinion).

4. Conclusions

This study shed light on the suitability of TWW for crop irrigation and the resultant impact on crop growth and soil characteristics. Of the two streams of TWW that were used for the irrigation, secondary effluent did not meet the bacteriological quality standard for irrigating carrot plants. The use of the TWW for irrigation led to better plant growth in terms of height and biomass production in carrots, and significantly improved the nutrient levels in the soil. These improvements were due to the high fertilization potential of the TWW. The accumulation of potentially toxic elements in the soil and the deterioration of the soil quality, such as salinization or sodicity, were not observed after the TWW irrigation. This study concludes that TWW is suitable for crop irrigation, but must be disinfected to reduce the bacteriological load to an acceptable limit. The use of ozonation or UV disinfection methods may be preferred over chlorination due to the formation of toxic byproducts by chlorination. Also, salinization or sodicity minimization measures, such as leaching and the periodic dilution of the irrigation water, should be incorporated into the irrigation plan when TWW is used for long-term irrigation.

Author Contributions: Conceptualization, S.O.; Methodology, S.O. and D.K.A.; Formal Analysis, S.O. and D.K.A.; Resources, I.R. and J.W.; Writing—Original Draft Preparation, S.O.; Writing—Review and Editing, S.O., D.K.A., I.R., and J.W.; Supervision, I.R. and J.W.; Project Administration, S.O., I.R., and J.W.; Funding Acquisition, I.R. and J.W. All authors have read and agreed to the published version of the manuscript.

Funding: This research was funded by the European Union Horizon 2020 project: Achieving Wider Uptake of Water-smart Solutions (H2020-SC5-2019-2) grant number: 869283, and the University of Chemistry and Technology Internal Grant, grant number: 217 88 2205.

Data Availability Statement: The data presented in this study are available in the manuscript. Any additional data will be submitted upon request.

Acknowledgments: The support of the UCT-Prague Internal Grant Agency and the UCT-Prague Biological Wastewater Treatment Group for the study is duly acknowledged.

Conflicts of Interest: The authors declare no conflicts of interest.

References

1. Martınez, S.; Suay, R.; Moreno, J.; Segura, M.L. Reuse of tertiary municipal wastewater effluent for irrigation of *Cucumis melo* L. *Irrig. Sci.* **2012**, *31*, 661–672. [CrossRef]
2. Theerawitaya, C.; Pipatsitee, P.; Taota, K.; Praseartkul, P.; Tisarum, R.; Samphumphuang, T.; Pal Singh, H.; Cha-um, S. Impact of irrigation regime on morpho-physiological and biochemical attributes and centelloside content in Indian pennywort (*Centella asiatica*). *Irrig. Sci.* **2023**, *41*, 23–34. [CrossRef]
3. Alkhamisi, S.A.; Ahmed, M.; Al-Wardy, M.; Prathapar, S.A.; Choudri, B.S. Effect of reclaimed water irrigation on yield attributes and chemical composition of wheat (*Triticum aestivum*), cowpea (*Vigna sinensis*), and maize (*Zea mays*) in rotation. *Irrig. Sci.* **2016**, *35*, 87–98. [CrossRef]
4. Ahmadi, L.; Merkley, G.P. Wastewater reuse potential for irrigated agriculture. *Irrig. Sci.* **2017**, *35*, 275–285. [CrossRef]

5. Gong, J.; He, L.; Liu, X.; Wang, S. Optimizing the allocation of irrigation water for multiple crops based on the crop water allocation priority. *Irrig. Sci.* **2022**, *41*, 49–68. [CrossRef]
6. de Nicola, E.; Aburizaiza, O.; Siddique, A.; Khwaja, H.; Carpenter, D.O. Climate change and water scarcity: The case of Saudi Arabia. *Ann. Glob. Health* **2015**, *81*, 342–353. [CrossRef] [PubMed]
7. Gosling, S.N.; Arnell, N.W. A global assessment of the impact of climate change on water scarcity. *Clim. Chang.* **2016**, *134*, 371–385. [CrossRef]
8. Berríos, P.; Temnani, A.; Zapata, S.; Forcen-Munoz, M.; Franco, J.A.; Pastor, A.P. Sensitivity to water deficit of the second stage of fruit growth in late mandarin trees. *Irrig. Sci.* **2023**, *41*, 35–47. [CrossRef]
9. United Nations Environment. *Global Environment Outlook—GEO-6, Healthy Planet, Healthy People*; Cambridge University Press: Cambridge, UK, 2019; pp. 34–37. [CrossRef]
10. Mays, C.; Garza, G.L.; Waite-Cusic, J.; Radniecki, T.S.; Navab-Daneshmand, T. Impact of biosolids amendment and wastewater effluent irrigation on enteric antibiotic-resistant bacteria—A greenhouse study. *Water Res. X* **2021**, *13*, 100119. [CrossRef] [PubMed]
11. Makhadmeh, I.M.; Gharaiebeh, S.F.; Albalasmeh, A.A. Impact of Irrigation with treated domestic wastewater on squash (*Cucurbita pepo* L.) fruit and seed under semi-arid conditions. *Horticulturae* **2021**, *7*, 226. [CrossRef]
12. Reis, M.M.; da Silva, A.J.; Santos, L.D.T.; Lopes, E.M.G.; Barros, R.E.; Donato, L.M.S. Millet irrigation with treated wastewater: Gas exchange response and nutrient accumulation. *Agric. Sci.* **2019**, *43*. [CrossRef]
13. Hashem, M.S.; Qi, X. Treated wastewater irrigation-a review. *Water* **2021**, *13*, 1527. [CrossRef]
14. Singh, G.; Nagora, P.R.; Haksar, P.; Rani, A. Species influenced growth, biomass allocation and productivity in wastewater irrigated plants in sandy soils of Indian desert. *Irrig. Sci.* **2022**, *40*, 829–843. [CrossRef]
15. Zavadil, J. The effect of municipal wastewater irrigation on the yield and quality of vegetables and crops. *Soil Water Res.* **2009**, *4*, 91–103. [CrossRef]
16. Al-Lahham, O.; El Assib, N.M.; Fayyad, M. Impact of treated wastewater irrigation on quality attributes and contamination of tomato fruit. *Agric. Water Manag.* **2003**, *61*, 51–62. [CrossRef]
17. Al-Nakshabandi, G.A.; Saqqar, M.M.; Shatanawi, M.R.; Fayyad, M.; Al-Horani, H. Some environmental problems associated with the use of treated wastewater for irrigation in Jordan. *Agric. Water Manag.* **1997**, *34*, 81–94. [CrossRef]
18. Alkhamisi, S.A.; Abdelrahman, H.A.; Ahmed, M.; Goosen, M.F.A. Assessment of reclaimed water irrigation on growth, yield, and water-use efficiency of forage crops. *Appl. Water Sci. J.* **2011**, *1*, 57–65. [CrossRef]
19. Alkhamisi, S.A. Maximizing the Use of Reclaimed Water for Crop Production in Arid Regions. Doctoral Thesis, Sultan Qaboos University, Muscat, Oman, 2013.
20. El-Batal, A.I.; Ismail, M.A.; Amin, M.A.; El-Sayyad, G.S.; Osman, M.S. Selenium nanoparticles induce growth and physiological tolerance of wastewater-stressed carrot plants. *Biologia* **2023**, *78*, 2339–2355. [CrossRef]
21. Malchi, T.; Maor, Y.; Tadmor, G.; Shenker, M.; Chefetz, B. Irrigation of root vegetables with treated wastewater: Evaluating uptake of pharmaceuticals and the associated human health risks. *Environ. Sci. Technol.* **2014**, *48*, 9325–9333. [CrossRef] [PubMed]
22. Hussain, A.; Priyadarshi, M.; Dubey, S. Experimental study on accumulation of heavy metals in vegetables irrigated with treated wastewater. *Appl. Water Sci.* **2019**, *9*, 122. [CrossRef]
23. Letshwenyo, M.W. Accumulation of heavy metals and bacteriological contaminations in carrots and okra irrigated with tertiary wastewater. *Environ. Qual. Manag.* **2021**, *30*, 47–59. [CrossRef]
24. Naz, S.; Anjum, M.A.; Sadiq, B.; Ahmad, R.; Altaf, M.A.; El-Sheikh, M.A.; Shakoor, A. Purification of sewage wastewater through sand column filter for lessening of heavy metals accumulation in lettuce, carrot, and cauliflower. *Water* **2022**, *14*, 3770. [CrossRef]
25. Jorge, M.F.; da Silva, L.D.B.; Hüther, C.M.; Cecchin, D.; de Melo, A.C.F.; Francisco, J.P.; Nascentes, A.L.; Alves, D.G.; Guerra, J.G.M. Potential use of treated wastewater from a cattle operation in the fertigation of organic carrots. *Braz. J. Environ. Sci.* **2022**, *57*, 542–554. [CrossRef]
26. Mukherjee, I.; Singh, U.K.; Chakma, S. Evaluation of groundwater quality for irrigation water supply using multi-criteria decision-making techniques and GIS in an agroeconomic tract of Lower Ganga basin, India. *J. Environ. Manag.* **2022**, *309*, 114691. [CrossRef] [PubMed]
27. Khalid, S.; Shahid, M.; Natasha Bibi, I.; Sarwar, T.; Shah, A.H.; Niazi, N.K. A review of environmental contamination and health risk assessment of wastewater use for crop irrigation with a focus on low and high–income countries. *Int. J. Environ. Res. Public Health* **2018**, *15*, 895. [CrossRef]
28. Food and Agriculture Organisation (FAO). Irrigation and Drainage Paper 29. In *Water Quality for Agriculture*; Food and Agriculture Organization of the United Nations: Rome, Italy, 1985.
29. World Health Organization (WHO). Volume 2: Wastewater use in agriculture. In *Guidelines for the Safe Use of Wastewater, Excreta and Greywater*; World Health Organization (WHO): Geneva, Switzerland, 2006; Volume 8.
30. California Department of Public Health. *Water Recycling Criteria*; California Code of Regulations: Sacramento, CA, USA, 2009.
31. United States Environmental Protection Agency. *Guidelines for Water Reuse*; United States Environmental Protection Agency: Washington, DC, USA, 2012.
32. European Union (EU). *Regulation (EU) 2020/741 of the European Parliament and the Council on Minimum Requirements for Water Reuse*; Official Journal of the European Union: Brussels, Belgium, 2020.

33. APHA/AWWA/WEF. *Standard Methods for the Examination of Water and Wastewater*, 22nd ed.; American Public Health Association: Washington, DC, USA; American Water Works Association: Washington, DC, USA; Water Environment Federation: Washington, DC, USA, 2012.
34. Azhari, H.E.; Cherif, E.K.; Sarti, O.; Azzirgue, E.M.; Dakak, H.; Yachou, H.; Esteves da Silva, J.C.G.; Salmoun, F. Assessment of surface water quality using the water quality index (IWQ), multivariate statistical analysis (MSA) and geographic information system (GIS) in Oued Laou Mediterranean Watershed, Morocco. *Water* **2022**, *15*, 130. [CrossRef]
35. Hwang, J.Y.; Park, S.; Kim, H.-K.; Kim, M.-S.; Jo, H.-J.; Kim, J.-I.; Lee, G.-M.; Shin, I.-K.; Kim, T.-S. Hydrochemistry for the Assessment of Groundwater Quality in Korea. *J. Agric. Chem. Environ.* **2017**, *6*, 73576. [CrossRef]
36. *ČSN 75 7837*; Jakost Vod–Stanovení Koliformních Bakterií v Nedesinfikovaných Vodác (Water Quality–Detection and Enumeration of Coliform Bacteria in Non-Disinfected Waters). Office for Technical Standardization, Metrology and State Testing: Prague, Czech Republic. Available online: https://shop.normy.biz/detail/85345#zakladni-informace (accessed on 19 December 2023).
37. Ambrožová, J.R.; Vejmelková, D.; Škopová, V.; N217019—Laboratoř Hydrobiologie a Mikrobiologie (Laboratory of Hydrobiology and Microbiology). *Kultivační Stanovení: Stanovení Koliformních Bakterií (Cultivation Determination: Determination of Coliform Bacteria)*; Ústav Technologie Vody a Prostředí (Institute of Water and Environmental Technology, University of Chemistry and Technology): Prague, Czech Republic, 2015.
38. Vyhláška č. 252/2004 Sb., Příloha č. 6. (Decree 252/2004, Annex Number 6). Available online: https://www.zakonyprolidi.cz/cs/2004-252 (accessed on 19 December 2023).
39. Ambrožová, J.R.; Vejmelková, D.; Škopová, V.; N217019—Laboratoř Hydrobiologie a Mikrobiologie (Laboratory of Hydrobiology and Microbiology). *Kultivační Stanovení: Stanovení Clostridium perfringens (Cultivation Determination: Determination of Clostridium perfringens)*; Ústav Technologie Vody a Prostředí (Institute of Water and Environmental Technology, University of Chemistry and Technology): Prague, Czech Republic, 2015.
40. *ČSN 75 7835*; Jakost Vod–Stanovení Termotolerantních Koliformních Bakterií a *Escherichia coli* (Water Quality—Detection and Enumeration of Thermotolerant Coliform Bacteria and *Escherichia coli*). Office for Technical Standardization, Metrology and State Testing: Prague, Czech Republic. Available online: https://www.tzb-info.cz/normy/csn-75-7835-2009-03 (accessed on 19 December 2023).
41. Ambrožová, J.R.; Vejmelková, D.; Škopová, V. *N217019—Laboratoř Hydrobiologie a Mikrobiologie (Laboratory of Hydrobiology and Microbiology). Kultivační Stanovení: Stanovení Termotolerantních Koliformních Bakterií a Escherichia coli (Cultivation Determination: Determination of Thermotolerant Coliform Bacteria and Escherichia coli)*; Ústav Technologie Vody a Prostředí (Institute of Water and Environmental Technology, University of Chemistry and Technology): Prague, Czech Republic, 2015.
42. Ofori, S.; Abebrese, D.K.; Klement, A.; Provazník, D.; Tomášková, I.; Růžičková, I.; Wanner, J. Impact of treated wastewater on plant growth: Leaf fluorescence, reflectance and biomass-based assessment (unpublished work). *Water Sci. Technol.* 2023, *revised version to be submitted.*
43. Motsara, M.R.; Roy, R.N. Guide to laboratory establishment for plant nutrient analysis. In *FAO Fertilizer and Plant Nutrition Bulletin 19*; Food and Agriculture Organization of the United Nations: Rome, Italy, 2008.
44. Houba, V.J.G.; Temminghoff, E.J.M.; Gaikhorst, G.A.; van Vark, W. Soil analysis procedures using 0.01 M calcium chloride as extraction reagent. *Commun. Soil Sci. Plant Anal.* **2000**, *31*, 1299–1396. [CrossRef]
45. Blake, G.R.; Hartge, K.H. Particle Density. In *Methods of Soil Analysis*; Soil Science Society of America, Inc.: Wisconsin, DC, USA, 1986; pp. 377–382. [CrossRef]
46. *ČSN EN ISO 17892-4*; Geotechnický Průzkum a Zkoušení–Laboratorní Zkoušky Zemin–Část 4: Stanovení Zrnitosti; Třídící Znak, 72 1007; Office for Technical Standardization, Metrology and State Testing: Prague, Czech Republic, 2017.
47. Organization for Economic Co-Operation and Development (OECD). Fruit and Vegetables Scheme: Guidelines on Objective Tests to Determine Quality of Fruit and Vegetables, Dry and Dried Produce. 2018. Available online: http://www.oecd.org/agriculture/fruit-vegetables/ (accessed on 10 May 2022).
48. Queensland Government. Soil Sodicity. 2014. Available online: https://www.qld.gov.au/environment/land/management/soil/soil-properties/sodicity (accessed on 26 August 2023).
49. Kalavrouziotis, I.K.; Koukoulakis, P.; Kostakioti, E. Assessment of metal transfer factor under irrigation with treated municipal wastewater. *Agric. Water Manag.* **2012**, *103*, 114–119. [CrossRef]
50. Bedbabis, S.; Rouina, B.B.; Boukhris, M.; Ferrara, G. Effect of irrigation with treated wastewater on soil chemical properties and infiltration rate. *J. Environ. Manag.* **2014**, *133*, 45–50. [CrossRef] [PubMed]
51. Jahan, K.M.; Khatun, R.; Islam, M.Z. Effects of wastewater irrigation on soil physico-chemical properties, growth and yield of tomato. *Prog. Agric.* **2020**, *30*, 352–359. [CrossRef]
52. Gorfie, B.N.; Tuhar, A.W.; Keraga, A.S.; Woldeyohannes, A.B. Effect of brewery wastewater irrigation on soil characteristics and lettuce (*Lactuca sativa*) crop in Ethiopia. *Agric. Water Manag.* **2022**, *269*, 107633. [CrossRef]
53. Jiménez, B. Irrigation in developing countries using wastewater. *Int. Rev. Environ. Strateg.* **2006**, *6*, 229–250.
54. Food and Agriculture Organisation. *User Manual for Irrigation with Treated Wastewater*; RNE 2003. TC/D/Y5009F/1/10.03/100; Food and Agriculture Organisation of the United Nations: Cairo, Egypt, 2003.
55. Ofori, S.; Puškáčová, A.; Růžičková, I.; Wanner, J. Treated wastewater reuse for irrigation: Pros and cons. *Sci. Total Environ.* **2021**, *760*, 144026. [CrossRef] [PubMed]

56. Foglia, A.; Gonzalez-Camejo j Radini, S.; Sgroi, M.; Li, K.; Eusebi, A.L.; Fatone, F. Transforming wastewater treatment plants into reclaimed water facilities in water-unbalanced regions. An overview of possibilities and recommendations focusing on the Italian case. *J. Clean. Prod.* **2023**, *410*, 137264. [CrossRef]
57. Drechsel, P.; Marjani Zadeh, S.; Pedrero, F. (Eds.) *Water Quality in Agriculture: Risks and Risk Mitigation*; FAO & IWMI: Rome, Italy, 2023. [CrossRef]
58. Pescod, M. *Wastewater Treatment and Use in Agriculture*; Irrigation and Drainage Paper 47; Food and Agriculture Organization of the United Nations (FAO): Rome, Italy, 1992.
59. Woomer, P.L.; Martin, A.; Albrecht, A.; Reseck, D.V.S.; Scharpenseel, H.W. The importance and management of soil organic matter in the tropics. In *The Biological Management of Tropical Soil Fertility*; Woomer, P.L., Swift, M.J., Eds.; John Wiley and Sons: Chichester, UK, 1994; pp. 47–80.
60. Tarchouna, L.G.H.; Merdy, P.; Raynaud, M.; Pfeifer, H.R.; Lucas, Y. Effects of long-term irrigation with treated wastewater. Part I: Evolution of soil-chemical properties. *Appl. Geochem.* **2010**, *25*, 1703–1710. [CrossRef]
61. Mason, B. Soil Doctor. Bicarbonates: Is it a Problem in Your Soil? 2020. Available online: https://www.soildoctorconsulting.com/pro-tips/bicarbonate-soil (accessed on 15 August 2023).
62. Kumar, A.; Christen, E. *Developing a Systematic Approach to Winery Wastewater Management*; Report CSL05/02; Final Report to Grape and Wine Research & Development Corporation (CSIRO Land and Water Science Report Adelaide); Grape and Wine Research & Development Corporation: Adelaide, Australia, 2009.
63. Bedbabis, S.; Trigui, D.; Ahmeda, C.B.; Clodoveo, M.L.; Camposeo, S.; Vivaldi, G.A.; Rouina, B.B. Long-terms effects of irrigation with treated municipal wastewater on soil, yield and olive oil quality. *Agric. Water Manag.* **2015**, *160*, 14–21. [CrossRef]
64. Tabatabai, M.A. Physicochemical fate of sulfate in soils. *J. Air Pollut. Control Assoc.* **1987**, *37*, 34–38. [CrossRef]
65. Ayoub, S.; Al-Shdiefat, S.; Rawashdeh, H.; Bashabsheh, I. Utilization of reclaimed wastewater for olive irrigation: Effect on soil properties, tree growth, yield and oil content. *Agric. Water Manag.* **2016**, *176*, 163–169. [CrossRef]

water

MDPI

Article

From District to City Scale: The Potential of Water-Sensitive Urban Design (WSUD)

Joachim Schulze *, Simon Gehrmann, Avikal Somvanshi and Annette Rudolph-Cleff

Technical University of Darmstadt, Faculty of Architecture, Urban Design and Development, 64287 Darmstadt, Germany; gehrmann@stadt.tu-darmstadt.de (S.G.); avikals@gmail.com (A.S.); rudolph@stadt.tu-darmstadt.de (A.R.-C.)

* Correspondence: schulze@stadt.tu-darmstadt.de; Tel.: +49-6151-16-22162

Abstract: The summer of 2022 was one of the hottest and driest summers that Germany experienced in the 21st century. Water levels in rivers sank dramatically with many dams and reservoirs running dry; as a result, fields could not be irrigated sufficiently, and even power generation and supply were affected. The impact of abnormally high temperatures for extended periods (heatwaves) is not restricted to nature and the economy but is also a considerable public health burden. Experts worldwide agree that these extreme weather events are being driven by climate change and will increase in intensity and frequency in the future. The adverse impact of these extreme weather events multiplies among dense urban environments, e.g., through heat islands. This calls for cities to take action to heat-proof and water-secure their urban developments. Water-Sensitive Urban Design (WSUD) is one such approach to mitigate the aforementioned challenges by leveraging the urban water ecosystem with special attention to the subject of water reclamation, retention, treatment and distribution. This paper introduces and builds upon a prototype of WSUD that centers around an artificial lake as an integrated water resource management system (IWRMS) fed by treated grey water and storm water obtained from two housing blocks flanking the water reservoir. Based on the specifications of this prototype, indicators of site suitability are derived and applied to identify potential locations for replicable projects in the city of Darmstadt. The results confirm the impact WSUD can have: a total of 22 sites with 2527 apartments are found suitable for prototype implementation in Darmstadt. Savings in town water consumption from these 22 sites would add up to 147 million liters. Further benefits include the provision of 24 million liters of irrigation water, storm water retention, adiabatic cooling during heatwave, increased biodiversity and the improvement in livability of the sites and the city.

Keywords: water-sensitive urban design; WSUD; integrated water resource management system; IWRMS; climate adaptive cities; climate resilience

1. Introduction

Climate change is considered as one of the greatest challenges of the 21st century [1]. Weather conditions in Europe, the United States of America and Asia have increasingly become unusual with both frequency and intensity of heatwaves, droughts and flash floods on the rise [2–4]. These extreme weather conditions have impacted agricultural production, irrigation, energy generation and led to a water supply crisis for cities. This is adversely affecting and even endangering human life. Based on data from the German federal office for statistics during summer months of June to August in 2022, which were characterized by record heat, the number of deaths was significantly higher than the median values from 2018 to 2021 by 9% to 13%. The number of deaths were particularly high in calendar week 29 (18–24 July 2022) at 25%, which was an exceptionally hot week [5]. It is not only extreme heat; even torrential rain and severe flooding are exacting a toll on human life. For instance, the severe flooding catastrophe in the Ahrtal region of Germany in 2021 resulted in devastating damage to houses and infrastructure with a loss of 133 lives [6]. The latest report from The Intergovernmental Panel on Climate Change (IPCC) states that these adverse conditions are not a temporary phenomenon, but the intensity and frequency will increase with "high confidence" [7] (p. 43) in the coming centuries. The 2015 Paris Agreement signed by 196 parties at the 21st UN Climate Change Conference acknowledges that "climate change is a common concern of humankind" [8] (p. 2) and the need for action to limit the increase in global temperature to 1.5 °C. Measures to achieve this objective are directed toward the reduction in greenhouse gas emissions, which are the driving force behind climate change [8] (p. 4). But as recent events show and the IPCC report predicts, climate change is already taking its toll on human life. This calls for tailored ad hoc to medium-term measures to reduce the immediate impact of climate-change-related shocks and stress. The urban settings of cities are more exposed [9] to extremes that culminate in "heat, flooding, water scarcity and droughts" [10] (p. 18). This is due to the dense built environment that characterizes cities and is known to exacerbate the prevailing adverse conditions induced by climate change. The city microclimate, for example, is more prone to heat waves because of the Urban Heat Island (UHI) effect [11]. Studies have also shown that the UHI effect can induce heavy rainfall events [12]. These can subsequently lead to flooding because the sealed (concretized) surfaces lead to heavy runoff, creating a storm water overload that regularly overwhelms the city drainage systems. Droughts also exacerbate the flooding as they damage the urban greenery and dry out the ground, reducing its natural ability to soak in storm water, which adds to heavy runoff when rainfall eventually does set in. In light of these challenges, state agencies have begun to promote climate change mitigation and adaption, e.g., through heat action plans or directives. These include measures referring to the transformation of the built environment such as "conserve and create shaded green spaces and parks, preferably with cooling evaporation areas such as bodies of water [...], reduce heat by creating or keeping clear air channels [...] or reduce degree of soil sealing in open and public squares [...]" [13] (p. 23). Others aim for the protection of vulnerable population groups [13] or water saving incentives [14]. Cities in particular need to take action, a fact which is stressed by studies on urban water scarcity and security [15,16]. After all, an estimated 60% of the world's population will be living in urban environments by 2030 [17] and 80% of Europeans by 2050 [10].

One of the possible adaptation measures that urban settlements can apply is Water-Sensitive Urban Design (WSUD) which incorporates a number of the aspects outlined in the German state heat action plan referenced above [13]. The concept of WSUD originated in the 1990s in Australia [18] and Australian government agencies defined it as "an approach to urban planning and design that aims to integrate the management of the urban water cycle into the urban development process" [19] (p. 1). This paper refers to WSUD as a design principle that, in its essence, reconsiders the management of urban water streams such as storm water (precipitation induced runoff), town water (from tap), greywater (from sinks, showers, bathtubs and laundry) and black water (from toilettes or urinals) and intends to integrate these in landscape, building and infrastructure design by means like "reducing potable water demand, minimizing wastewater generation, optimizing the use of water sources" or "promoting a significant degree of water related self-sufficiency" [20] (p. 1). WSUD represents a mitigative measure in the adaptation process of urban environments toward climate change. This is underlined by a report of the European Environment Agency entitled *Urban adaptation in Europe: how cities and towns respond to climate change* [21].

In this paper, a prototype of WSUD is introduced that includes a grey-to-service water treatment, storm water catchment and an artificial lake as a reservoir and central element of landscape design of a residential housing development. Building upon the specifications of this WSUD prototype and investigating the case of the city of Darmstadt, the research in this paper is directed toward the following key questions:

1. What indicators describe the suitability of a city site for application of a replicable WSUD?
2. How many potential sites for the application of WSUD exist in the city of Darmstadt?
3. What impact on the city water cycle of Darmstadt would the application of all WSUD have?
4. What relevance would this application have in regard to climate change adaptation?

By addressing these questions, this study conducts a multi-scale assessment method based on a district-scale prototype that is multiplied on a city scale. In consideration of the fact that the IWRMS design of the WSUD prototype is one-of-a-kind in Germany and that this study uses real-world as opposed to simulated data, it contributes to filling a gap in research: the quantification of town water savings through district-based WSUD on a city scale.

In the following section on Materials and Methods, the principles of WSUD are outlined, followed by a detailed description of how they are integrated in the WSUD prototype. Next, the city of Darmstadt is introduced. Federal, state and city assessments on climate change are presented in order to evaluate to what extent Darmstadt is already affected by climate change and what climate projections indicate for the future. The last section names the indicators that are derived from the WSUD prototype in order to assess potential WSUD sites. Based on the number of possible WSUD sites in Darmstadt, a grand total of reclaimed and retained water is calculated in order to estimate the overall impact that citywide application could have on the city water cycle.

2. Materials and Methods

2.1. Water-Sensitive Urban Design

There are multiple definitions for WSUD, but for this paper, it is defined as a design principle that in its essence reconsiders the management of urban water streams such as storm water, town water, grey water and black water and intends to integrate these in landscape, building and infrastructure design. In Germany, WSUD projects referring to the aforementioned definition are rarely found. However, it is worth noting that other design approaches, labeled as "Decentralization" or "Sponge City", can be found in Germany, which are similar to WSUD in objective but differ in scale, design and realization. "Decentralization" or "Sponge City" usually focus on flood protection in the scope of localized storm water management that is mostly designed and realized by engineers and hydrologists with no or little contribution from architects and landscape designers. In the authors' understanding, this is a significant difference and qualifies WSUD as a relatively young new discipline that bridges the gap between urban design and urban water management for building urban climate resilience.

2.2. The WSUD Prototype

A WSUD project, which was designed based on an interdisciplinary understanding and the principles of WSUD [22–24], is under construction in the city of Mannheim at the time of the writing of this paper and expected to be finished by 2024. The project contains 74 residential units distributed in two multi-story family houses and offers a variety of different apartment types from 1.5 to 4.5 rooms with a total living space of 5888 m^2. The two buildings are arranged in such a way that they form a courtyard with a naturally designed but technically controlled waterbody referred to as the WSUD lake from here onwards (Figure 1). The WSUD lake is a technical water reservoir fed by local storm water and surplus service water reclaimed from greywater from the apartments. It is designed to hold 120 m^3 of water on average under normal operating conditions. It also has a provision for extension which enables the holding of an additional 50 m^3 of water, which corresponds to a 100-year storm event. In order to allow unlimited access to the lakefront and to prevent any form of fencing, the level of the lake is limited to a depth of 0.4 m. The WSUD lake maintains a steady water level as it receives 100% of the storm water which is collected from the roofs and the façade of the development and the surplus of an integrated grey-to-service water system of the apartments. Though it is strongly dependent on the season, the storm water inflow accounts for between 60% and 90% of the overall reservoir capacity while service water accounts for the remaining 10% to 40% under normal operation conditions. The grey-to-service water system in the households is designed to collect grey water from showers, sinks and laundry. The grey water is then processed with the help of contemporary water reclamation technology such as ultrafiltration membrane to remove contaminants. This is followed by ultraviolet light exposure (UV) for disinfection, before the majority of it is pumped back to the apartments as service water, while the surplus of service water is spilled in the WSUD lake (Figure 2).

Figure 1. Site plan WSUD prototype.

In order to maintain a steady current and to avoid stagnated areas, the WSUD lake is designed in a length to width ratio of 1:7 as a result of specific flow simulations. This also refers to assessments in the field of environmental engineering, in particular, lake management [25]. Furthermore, the lake water is constantly pumped through a sand filter for treatment purposes. The WSUD lake is used for the irrigation of the surrounding green areas between March and October—the extended summer season. The regular capacity is sufficient to meet the irrigation requirements of the development during the normal dry season, i.e., without any storm water inflow into the WSUD lake. In case of extreme weather events (heat wave or drought) the regular irrigation can be continued until the water level reaches a critical minimum of 25%, after which irrigation needs to be stopped to protect the ecosystem of the WSUD lake. In this situation, the critical minimum of 25% water level of the WSUD lake is maintained by the daily inflow of the surplus of service water which is determined to be sufficient to overcome evaporative losses. In the case of an extreme rain event, the lake extension is activated to hold an additional 50 m^3. Water held in the lake extension is designed to be slowly released through infiltration trenches without triggering flooding. This additional water is not retained for the rest of the year for irrigation or reuse purposes.

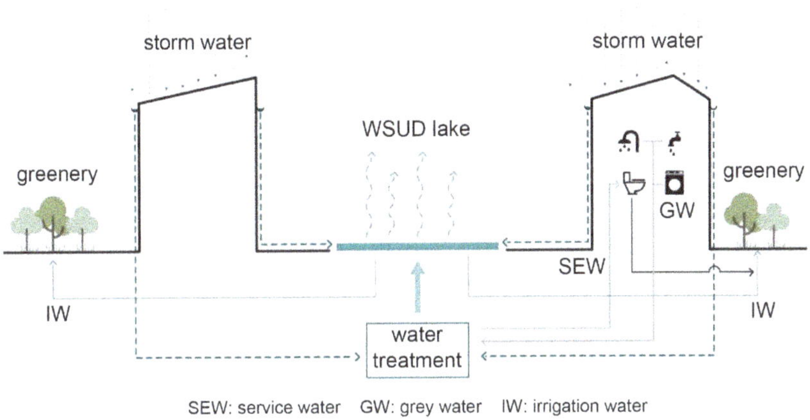

Figure 2. Scheme WSUD prototype.

WSUD by closing the water cycle at the decentralized scale can lead to a significant reduction in town water supply. An estimated reduction of up to 40% can be achieved per household. WSUD also eliminates the burden of irrigating common greens within the development with town water, which is estimated to further lower town water demand from the site by 20%. The estimated payback period for the initial investment in WSUD is about 25 years based on the savings in town water consumption and wastewater fee payments. While the decentralized water management technology itself is not new [26–30], the WSUD prototype stands out by combining treated grey water, i.e., service water and storm water, in the WSUD lake. This approach distinguishes it from other notable WSUD projects in Germany such as the "Hamburg WATER CYCLE" [31] or the "BUGA Heilbronn" [32]. In addition, the way the technology is applied under WSUD and connects to people via design and architecture is unique in this project. WSUD also provides an improvement in quality of life and a reduction in vulnerability to extreme weather, which is not easily captured in the aforementioned technical and financial details. It is critical to acknowledge the importance of design and architecture in unlocking the full potential of technical intervention like decentralized water management. Design can help leverage it to increase biodiversity by creating a constantly green oasis while providing the community with a highly resilient structure which can mitigate the effects of climate change.

2.3. *The City of Darmstadt*

The city of Darmstadt is a mid-sized university town located in the federal state of Hesse 30 km south of Frankfurt am Main. Darmstadt has a population of 163,435 and an area of 122 km^2 [33], making it the fourth biggest city within the federal state of Hesse. The city is an important center of education, research and science. Darmstadt has major employers in the information and communication, healthcare and manufacturing sectors [34]. Darmstadt is part of the Frankfurt Rhine-Main metropolitan region which includes the cities of Aschaffenburg, Offenbach and Frankfurt am Main. Frankfurt Rhine-Main is one of the eleven metropolitan regions in Germany and is home to a population of 5.8 million people and the location of numerous influential political institutions and corporate headquarters [35]. Climate change adaptation has gained traction within the municipality of Darmstadt with multiple strategies being developed and implemented in the last decade. In 2013, an integrated concept for climate protection was set up [36] which was followed by a 25-point immediate action plan for climate protection in 2020. In 2021, the city of Darmstadt established the office for climate protection and climate adaptation, which meanwhile has grown to a staff of over ten employees.

Exposure to Climate Change

In 2021, the German Environment Agency (UBA) published its report, *The Risks of Climate in Germany,* where it clearly confirms that Germany will be affected by climate change nationwide [37,38]. It also noted that adverse effects such as heatwaves or heavy storm water will not be distributed evenly but impact differs regionally [38]. The overview map in Figure 3 illustrates how different parts of Germany, e.g., the coasts, the northwest and the mountainous regions, are exposed to either a minor, strong or very strong increase in four categories of risk, i.e., average temperature, heat, heavy storm water and drought.

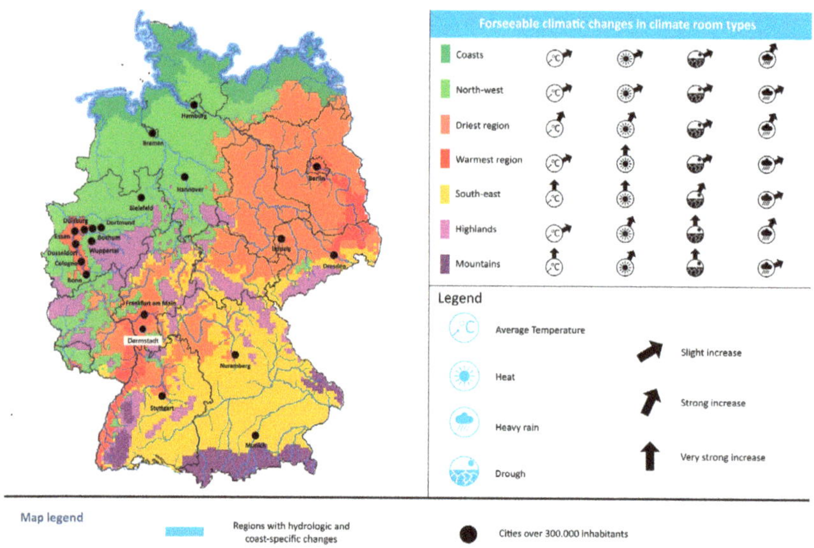

Figure 3. Regional impact of climate change in Germany.

Climate projections for the city of Darmstadt predict a very strong increase in heat risk. While a minor increase in the remaining three risk categories is also predicted. Figure 3 also shows that Darmstadt is located in the warmest region of Germany. In addition, the report states that densely populated urban areas such as the "Ruhrgebiet" or the "Rhein-Main-Neckar Region" will be affected significantly more by weather extremes [38].

The findings of UBA have been seconded by climate change studies performed at the state and city levels. The 2017 integrated climate protection plan of the state of Hesse that incorporates climate projections carried out by the Potsdam Institute for Climate Impact Research confirms that the degree and characteristics of climate-change-related effects vary regionally, and the south of the state of Hesse including the region around the city of Darmstadt will be especially affected by high temperatures and long-lasting heat [39]. Studies on climate change carried out or commissioned by the municipality of Darmstadt complement the federal and the state reports by providing data of higher granularity on a district scale. The semi-annual report on statistics in 2020, for example, includes a special issue on climate change in Darmstadt [40]. This report makes use of data provided by a number of temperature metering stations positioned throughout the city. From 1961 to 2019, the annual mean temperature has risen from 10 °C to 11 °C with summer and winter seasons experiencing the greatest increase in monthly average temperatures. The number of summer days with temperatures exceeding 25 °C has increased from an average of 43 days between 1961 and 1990 to 76 days in the time span from 2015 to 2019. The number of hot days characterized by a temperature that exceeds 35 °C has tripled while average frequency of heat waves has increased eight-fold for the same reference time frames. A heat waves is defined as "a multi-day period with extraordinary thermal stress" [40] (p. 11). The climate reports paint a slightly different concern regarding storm water risk to the city of Darmstadt. The number of rainy days per year has not changed significantly, but the mean annual storm water has decreased. From 2015 to 2019 average storm water per year amounted to 639 mm, which is 13% less than from 1961 to 1990. Furthermore, the distribution of storm water has shifted. Spring and summer are getting dryer, whereas the winter season is receiving more storm water. From 2015 to 2019, storm water volume dropped by more than 20 mm in the months of July and August, which is significant. Nevertheless, the Hessian Agency for Nature Conservation, Environment and Geology labels the city of Darmstadt as "increasingly at risk" [40] (p. 18) to heavy storm water events. A 2016 study commissioned by the city of Darmstadt also provides further details on the city microclimate and its exposure to climate-change-related effects [41]. All climate risk and vulnerability studies are unanimous that the city of Darmstadt is already suffering ill-effects of climate change, which are predicted to worsen in the coming future. The risk to the city is predominantly posed by increase in temperature and decrease in storm water during the summer. The city already has a tough baseline, given its location in the hottest region of Germany, and shows pronounced UHI effects in its inner core. The predicted further increases in heat and decreases in storm water will lead to numerous problems in the city. One of the most obvious will be drought, as high heat leads to water evaporation that in turn aggravates drought [40]. Thus, dryness will also pose a major problem in the future.

2.4. Further Data and Sources

To map and assess potential WSUD sites in Darmstadt following data were acquired:

1. For the assessment of potential WSUD sites, a leading municipal housing association provides data on their real estate located in Darmstadt. Altogether, its building stock includes 17,000 units, which is comparable to the size of the housing association realizing the WSUD prototype in Mannheim that maintains 15,000 units. This is an important indicator for transferability from the setting of Mannheim to Darmstadt. And both operate in mid-size cities. The data provided include address, number of buildings per site, number of building levels, number of apartments, total living space, date of erection and status of refurbishment.
2. For the assessment and the mapping of potential WSUD sites, the spatial data of the city of Darmstadt including infrastructure, plots, buildings, topographic and aerial photos are provided by the land surveyors office of the city of Darmstadt.
3. Data on monthly, daily and hourly storm water for Darmstadt are retrieved online from the German Weather Service (DWD) [42,43] and assessed for a ten-year period.

3. Results

3.1. Indicators for WSUD Application

The WSUD prototype presented in the previous section serves as a template for the assessment of potential sites in Darmstadt that are suited for the citywide application of a replicable WSUD. To ensure citywide coverage, the paper is focused on identifying existing building sites with suitability for WSUD application rather than green field developments. The main intention for focusing on the existing building stock is simply that the ratio of existing buildings is much higher than those which are being newly built in Darmstadt. Thus, emphasizing existing sites will, in all probability, lead to the identification of more WSUD applications than focusing on sites that are currently in development. This approach to maximize impact is supported by studies carried out in the context of the German energy transition, which have concluded that the biggest reduction in energy consumption can be achieved by the energy-focused refurbishment of the existing building stock [44]. The energy-focused refurbishment of the building stock in Germany is an ongoing process that is driven by state and federal subsidization as well as legal requirements with the long-term goal of achieving climate neutrality by 2050 [45]. From an economic, strategic and logistic point of view, it will be reasonable to combine WSUD application with energy-focused refurbishment measures, because there is an overlap in the overall climate objectives of both programs. Moreover, "resource efficiency" in a holistic sense [46] (p. 15) as defined by the Sustainable Development Goals (SDG) to "make cities and human settlements inclusive, safe, resilient and sustainable" [46] (p. 15) includes both water and energy. Hence, non-refurbished sites awaiting energy-focused renovation are particularly suited for WSUD application.

Indicators (financial and O and M viability) for site section derived from the prototype specifications are (Table 1):

Table 1. Indicators for WSUD site assessment.

Reference to Prototype	
Indicator	**Specification**
Building use	Residential
Ownership	Housing association
Operation	Housing association
Number of units and/or grand total of living space	≥70 and or ≥5900 m^2
Reference to existing site application	
State of refurbishment (one of the three specifications must apply)	Building stock that was erected at least 40 years ago and that has never been renovated
	Building stock that was renovated at least 40 years ago
	Building stock that by the year 2030 will have been unrenovated for 40 years
Available open space	Must be sufficient to position the WSUD lake

1. Building use, ownership and operation

These indicators refer to the usage of the buildings in the prototype which are 100% residential and the building cluster being owned and operated by a city housing association. The building stock assessed in this study comply with both indicators.

2. Number of units and/or grand total of living space

The WSUD prototype is connected to a total of 74 apartment units located in the two buildings flanking the courtyard. Altogether, these provide 5888 m^2 of living space.

Both figures are indicators of financial feasibility. This is a necessity for the housing association executing the prototype, otherwise the project would not have been approved by the executive board. With this number of units, or rather living space, involved, the payback for additional costs of the WSUD prototype is estimated to be within 25 years from the savings in town water use for household and irrigation purposes and by a reduction in wastewater fee payments.

Indicators (logistical viability) for site selection derived from existing site surveys are outline below (Table 1):

1. State of refurbishment

WSUD application in existing building sites would require extensive civil work which includes landscaping, concrete work and the installation of the water treatment unit that additionally requires separate piping in the buildings for grey water and service water [47]. In order to mount this system, the installation shafts in the buildings need to be opened. This is an expensive undertaking, but costs can be rationalized if they are aligned with other refurbishment measures or incorporated within the scope of a full renovation. This can be achieved by clubbing WSUD applications with federal- and state-mandated and subsidized energy-focused refurbishments. Therefore, to ensure the city-scale implementation of WSUD, it is deemed important to link site selection with the refurbishment status of the potential sites. But the timespan when a building is due for renovation can only be defined as an approximation because the lifespan of building materials depends on numerous factors [48] that can lead to disparities between the expected and the actual lifespan, and even though a building might be in a state of neglect, it may not mean that the building operator will take immediate action. In this paper, a 40-year timespan is defined as the age for refurbishment based on the average lifespan of different piping materials [49] when they have reached a state of "major damage that cannot be repaired due to technical or financial reasons" [49] (p. 12). Based on the aforementioned rationale, indicators of state of refurbishment are assessed in three ways: (1) building stock that was erected at least 40 years ago and that has never been renovated, (2) building stock that was renovated at least 40 years ago and (3) building stock that by the year 2030 will have been unrenovated for 40 years.

2. Available open space

The site must have sufficient open space for the WSUD lake to be positioned.

3.2. Shortlisted Sites for WSUD Application

A total of 496 sites were examined in Darmstadt (Figure 4) against the indicators as described in Table 1. In the initial assessment, 19 sites met all the indicators. In the follow-up assessment, three additional sites were shortlisted despite not meeting all the stated indicators. These three sites, namely #3, #7 and #9, fell short of the minimum threshold indicator "number of units and/or grand total of living space" by only less than 9%, which the authors feel is not a significant enough shortfall to disregard their potential for inclusion in the study. Finally, a grand total of 22 sites were shortlisted for assessing the city-scale WSUD application. These sites account for 4.4% of all sites examined in the city of Darmstadt. Since most of the shortlisted sites are significantly large housing developments, they together represent 19% of all apartment units and 20% of total living space examined in the city with a population of about 6300 tenants.

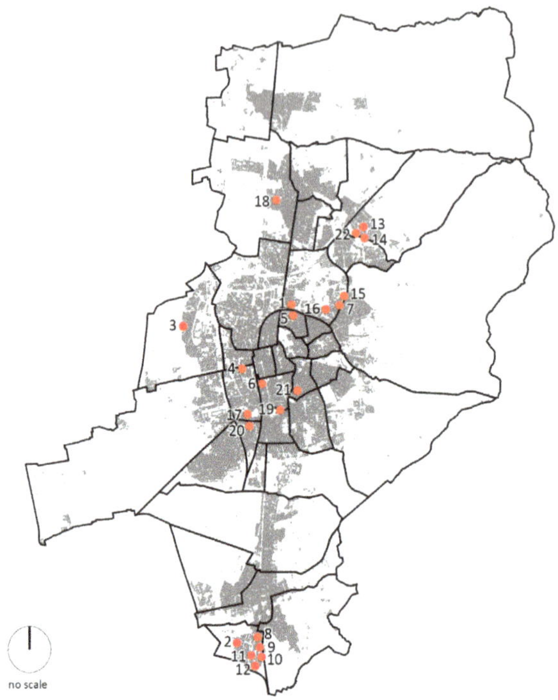

Figure 4. Distribution of 22 sites in Darmstadt.

3.3. WSUD Lake Dimensioning and Water Balance

In order to assess the overall impact of citywide WSUD prototype implementation, a WSUD lake is dimensioned for each of the 22 shortlisted sites. Based on the dimensions and the water balance of each WSUD lake, a grand total of reused service water, gained irrigation water and retained storm water has been calculated. The lake design and dimensioning calculations refer to specifications and average values applied in the prototype, such as averages of town water consumption per capita, averages of the amount of grey water and black water per capita or averages of the number of tenants for each unit. The lake balancing method refers to standards used in hydrology studies [50]. Climatology data on storm water in Darmstadt from 2011 to 2021 has been sourced from DWD. Data on urban morphologies (Table A1) are used to calculate unsealed space, i.e., the greenery, based on average ratios. Further values include standard irrigation rates (Table A2). Water inflow includes storm water and the surplus of service water from the households; water outtake refers to evaporation and the water provided for irrigation purposes. It must be noted that water inflow by runoff from greenery and water outtake in the form of infiltration from the lake bank and bottom are not taken into consideration because the prototype design specifications do not do so. Furthermore, calculations in this study are performed based on monthly water inflow and outtake in the WSUD lake in order to achieve a monthly water balance. This is the appropriate granularity to meet the objectives of this paper, but the application of approximations, standard values and monthly as opposed to daily or hourly balancing lead to limitations. The precise calculation of roofs and greenery and the application of exact figures on town water consumption or irrigation demands would refine the results. The conduction of hourly simulations would lead to an even better understanding of the inflow and outflow of water and could consider the holiday season with reduced water consumption per capita, which results in less service water inflow. This could further improve the lake design. The calculations consist of two preliminary steps leading to the main calculation for the monthly lake water balance. For conciseness, this section focuses on the description of the main calculation. The preliminary steps are only presented in brief but are outlined in Appendix A of this paper. Table 2 lists all variables and parameters included in the dimensioning and balancing of the WSUD lakes.

Table 2. WSUD lake variables and parameters.

Description		Additional Notes
Variables		
Total number of units [-]	**U**	-
Σ base area of buildings [m^2]	$\mathbf{A}_{base}$	Proxy for calculation of SW inflow
Σ open space of site [m^2]	$\mathbf{A}_{open\ space}$	Includes sealed (concretized) and unsealed area
Σ unsealed open space of site (greenery) [m^2]	$\mathbf{A}_{greenery}$	Is calculated based on average ratios
Σ irrigated open space of site [m^2]	$\mathbf{A}_{irrigation}$	Is calculated by subtraction of lake area
∑ litres of irrigation water (IW) [l]	($\sum_{\text{liters IW pre}}$)	Total liters IW from pre-dimensioning
∑ litres of irrigation water (IW) [l]	($\sum_{\text{liters IW lake}}$)	Total liters IW referring to lake
∑ litres of irrigation water (IW) [l]	($\sum_{\text{liters IW}}$)	Total liters IW
Σ greywater [L/per day]	**GW**	Total GW based on number of units
Σ reused greywater as service water [L/per day]	$\mathbf{SEW}_{reuse}$	Total GW that is reused
Σ surplus service water [L/per day]	$\mathbf{SEW}_{surplus}$	Is GW minus SEW_{reuse}
Area lake pre-dimension	$\mathbf{L}_{areapre}$	Pre-dimension of main water body
Area lake retention	$\mathbf{L}_{arearetention}$	Area of lake extension for SW retention
Area lake	$\mathbf{L}_{area}$	Area of main water body
Width of lake [m]	$\mathbf{L}_{width}$	-
Length of lake [m]	$\mathbf{L}_{length}$	-
Parameters		
Depth of main water body [m]	$\mathbf{L}_{depth\ main\ water\ body}$	Is 0.,4 m according to the prototype
Depth of lake extension for SW retention [m]	$\mathbf{L}_{depth\ retention}$	Is 0.2 m
Ratio lake width to lake length [-]	$\mathbf{L}_{ratio}$	Is 1:7 according to the prototype
Ø Σ greywater [L/per resident and day]	**ØGW**	Average value
Ø Σ reused GW as SEW [L/per resident and day]	$\mathbf{ØSEW}_{reuse}$	Average value
Ø Σ surplus SEW [L/per resident and day]	$\mathbf{ØSEW}_{surplus}$	Is ØGW minus ØSEW_{reuse}
Ø Σ surplus SEW [L/per resident and mth]	$\mathbf{ØSEW}_{surplusmth}$	Monthly SEW surplus
Ratios unsealed to sealed open space [%]	$\mathbf{R}_{open\ space}$	Are taken from a study on urban morphologies
Duration of irrigation [-]	$\mathbf{IR}_{duration}$	Is defined as 15 days
Ø daily irrigation demand [mm]	$\mathbf{ØIR}_{day}$	Average values
Ø monthly irrigation demand [mm]	$\mathbf{ØIR}_{mth}$	Average values
Daily storm water maximum [mm]	$\mathbf{SW}_{max}$	Based on SW data from 2011–2021
Ø monthly storm water [mm]	$\mathbf{ØSW}_{mth}$	Based on SW data from 2011–2021
Ø monthly evaporation [mm]	$\mathbf{ØEV}_{mth}$	Average values

3.3.1. Preliminary Steps

Preliminary steps define the size of the main water body of the WSUD lake that is to be used for irrigation purposes and an extension area intended to temporarily retain storm water from extreme rainfall events. The irrigation capacity refers to the amount of water that is necessary to fully irrigate the standard greenery of each site for a period of 15 days. Retention, on the other hand, is based on the amount of storm water that would add to the lake level in the case of a 100-year heavy storm water event. The calculations incorporate average values on urban typologies [51] (Table A1) and irrigation rates [52] (Table A2). Retention capacity refers to the maximum daily storm water event for Darmstadt (Table A3). The dimensions of the lakes are restricted to a depth of 0.4 m and a width to length ratio of 1:7, as outlined in the description of the prototype.

3.3.2. Monthly Water Balance

The preliminary steps are the basis for the monthly lake water balance (Figure 5). Water inflow through storm water refers to monthly averages (Table 3). Because site data do not include information on the nature of the roofs, SW_{mth} is multiplied with the base area (A_{base}) of the buildings to approximate the total of storm water inflow. Additionally, inflow includes the surplus of service water per month ($SEW_{surplus}$) based on average consumption values applied in the design of the prototype and multiplied with the number of units (U). Lake outtake includes irrigation water and evaporation. The German weather service (DWD) provides data on evaporation for a lake depth of two meters. The deviation in the depth of the prototype is negligible because lake depth is not taken into account for the calculation of evaporation [53]. The total amount of irrigation water withdrawn from the lake is calculated by multiplying monthly irrigation averages ($ØIR_{mth}$) with the irrigation area ($A_{irrigation}$) of the site.

$$\mathbf{A}_{\text{base}} \times \mathbf{ØSW}_{\text{mth}} = \sum \text{litres from SW inflow} \tag{1}$$

$$\mathbf{U} \times \mathbf{ØSEW}_{\text{surplusmth}} = \sum \text{litres from SEW surplus} \tag{2}$$

$$\mathbf{A}_{\text{irrigation}} \times \mathbf{ØIR}_{\text{mth}} = \sum \text{litres of irrigation} \tag{3}$$

$$\mathbf{L}_{\text{area}} \times \mathbf{ØEV}_{\text{mth}} = \sum \text{litres of evaporation} \tag{4}$$

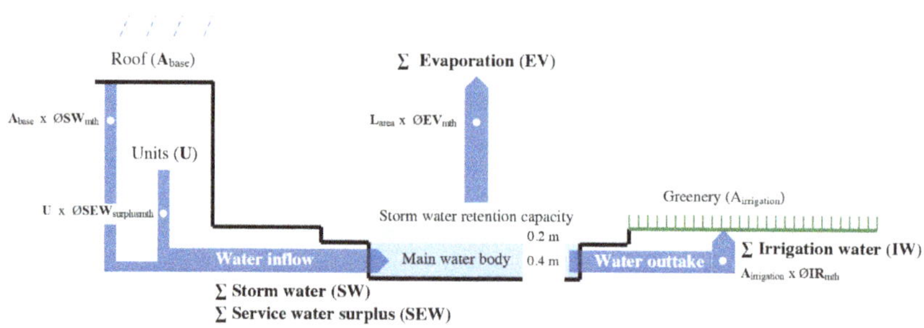

Figure 5. Scheme monthly lake water balance.

Table 3. Monthly storm water average Darmstadt.

Month	Ø Σ SW [MM] per Month 2011–2021	Month	Ø Σ SW [MM] per Month 2011–2021
January	62	July	59
February	41	August	66
March	39	September	49
April	38	October	52
May	68	November	50
June	62	December	68

As an example, Figure 6 shows the lake water balance sheets for sites 03 and 04 alongside the corresponding site plan. Irrigation rates are reduced in gradations of $\frac{3}{4}$, $\frac{1}{2}$, $\frac{1}{3}$, $\frac{1}{4}$ and $\frac{1}{8}$ referring to the greenery that is watered. As stated in the description of the prototype, lake levels must not fall below a capacity of 25%. Apart from this requirement lake levels must reach 100% when the irrigation phase begins in April.

Number of site	03	
Base area of buildings [m²]	888	Color matches site plan
Grand total of open space [m²]	4242	Color matches site plan
Grand total of greenery [m²]	3478	
Area of main water body [m²]	256	
Capacity of main water body [m³]	102	
Capacity of extension [m³]	52	

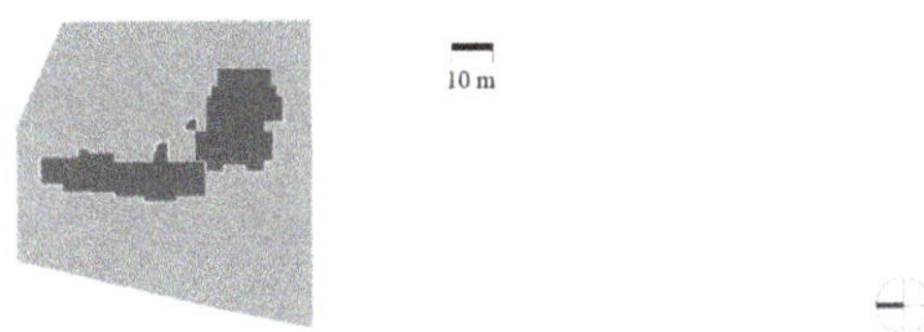

Site plan

Irrigation rate	0%	0%	0%	100%	100%	33%	25%	25%	50%	0%	0%	0%
	January	February	March	April	May	June	July	August	September	October	November	December
Storm water inflow [L/mth]	54,744	36,138	34,211	33,713	60,446	55,304	52,817	58,430	43,829	46,094	44,353	60,615
Irrigation outtake [L/mth]	0	0	0	64,442	161,105	74,431	80,553	80,553	80,553	0	0	0
Service water inflow [L/mth]	58,522	58,522	60,472	58,522	60,472	58,522	60,472	60,472	58,522	60,472	58,522	60,472
Total litre [L]	102,000	102,000	102,000	102,000	32,604	40,484	36,582	42,648	45,742	102,000	102,000	102,000
Evaporation [L/mth]	1025	3075	7174	18,960	29,209	31,515	36,639	32,283	18,704	8968	2818	1537
Lake level	100%	100%	100%	100%	[illegible]	[illegible]	[illegible]	[illegible]	[illegible]	100%	100%	100%

Number of site	04	
Base area of buildings [m²]	1490	Color matches site plan
Grand total of open space [m²]	7078	Color matches site plan
Grand total of greenery [m²]	6299	
Area of main water body [m²]	459	
Capacity of main water body [m³]	184	
Capacity of extension [m³]	87	

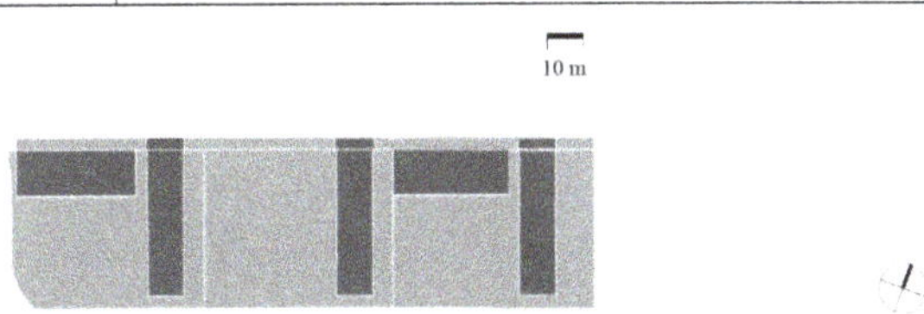

Site plan

Irrigation rate	0%	0%	0%	100%	75%	33%	25%	25%	50%	0%	0%	0%
	January	February	March	April	May	June	July	August	September	October	November	December
Storm water inflow [L/mth]	91,858	60,637	57,404	56,569	101,425	92,797	88,624	98,042	73,543	77,343	74,422	101,708
Irrigation outtake [L/mth]	0	0	0	116,800	219,000	134,904	146,000	146,000	146,000	0	0	0
Service water inflow [L/mth]	93,269	93,269	96,378	93,269	96,378	93,269	96,378	96,378	93,269	96,378	93,269	96,378
Total litre [L]	184,000	184,000	184,000	183,052	109,498	104,169	77,495	68,046	55,331	184,000	184,000	184,000
Evaporation [L/mth]	1837	5511	12,860	33,986	52,357	56,490	65,676	57,868	33,527	16,074	5052	2756
Lake level	100%	100%	100%	99%	60%	57%	[illegible]	[illegible]	[illegible]	100%	100%	100%

Figure 6. Example site sheet 03 and 04.

The remaining 21 site sheets (Figures A1–A10) can be found in the Appendix B of this paper alongside data on monthly storm water in Darmstadt (Table A4) and site assessment (Table A5).

3.4. Impact on the City Water Cycle

Based on the final calculations, altogether (a) 123 million liters of town water could be saved by WSUD prototype application per year from the 22 sites. This refers to the reuse of service water in toilets and washing machines in the households. Irrigation water provided by the lake would account for (b) 24 million liters per year. Two million liters is the total storm water retention capacity. Thus, the total savings in town water add up to 147 million liters (a + b) per year, which corresponds to the consumption of 3200 inhabitants. Under normal circumstances, this water would need to be retrieved from groundwater. However, it must be noted that despite all the designing and engineering permutations and combinations, only four sites can maintain full irrigation throughout the entire year with WSUD application. These four sites have a relatively higher ratio of built-up area to open space ranging between 0.35 and 0.51. This means they have much higher storm water runoff and less greenery to irrigate compared to other sites. In these cases, the supply of storm water and surplus of service water exceed the irrigation demands from April to September. The remaining 18 sites must lower irrigation levels between May and September ranging from 25% to 75%, otherwise the lake level would fall below the critical 25%. This is due to their ratio of built-up area to open space being low, and they range between 0.09 and 0.21. The reduced amount of irrigation water can either be distributed evenly or can be allocated to priority irrigation spots.

3.5. Further Limitations and Obstacles

Apart from the indicators listed in Table 1 applied for WSUD site assessment, there are a few caveats that need to be named and considered. These do not affect the application but may have implications on the WSUD lake design. These caveats would need further investigation and consideration in case a WSUD project is ultimately executed on site. The first caveat refers to the area that needs to be provided for the fire brigade so it can access the site in case of a fire. Length, width and position of this area are based on set requirements but to a limited extent are negotiated with the local fire brigade and tailored according to the site [54]. Nevertheless, this area cannot overlap with the WSUD lake; therefore, lake design has to be adjusted to accommodate this. The second caveat relates to the existing trees on a site. It can be a challenge to preserve old and full-grown trees while positioning the WSUD lake on the site. It is important to point out that trees are beneficial for climate resilience in an urban context [55] and should be preserved; therefore, WSUD lake design has to be adjusted to accommodate such trees. As noted earlier, splitting the lake into more than one basin is possible and could help address both aforementioned caveats.

Other noteworthy obstacles were witnessed in the planning process of the WSUD prototype. Generally speaking, storm water management and in particular the infiltration into the soil has to be approved by the environmental agency. The management of grey and service water, on the other hand, has to be approved by the health agency. Since the IWRMS combines storm, grey and service water, there was no clear responsibility. Only a special permit signed by both units enabled the continuation of planning and realization. In addition, the housing association in Mannheim insisted on a proof-of-concept ensuring that lake levels do not fall below the critical minimum in summertime. These examples demonstrate two challenges: it was new for the city administration to handle a project that did not comply with water management standards, and on the client side, there was recurring insecurity in the prototype's flawless operation.

4. Conclusions and Discussion

Climate change is already impacting weather patterns across the world, and the scientific consensus is that it will only get worse. Urban environments, due to their inherent form, are aggravating the impact of climate change on their residents, increasing the urgency to implement adaptation measures. Assessments carried out for Germany and the city of Darmstadt in particular underline this necessity. The WSUD prototype presented in this paper could be one of numerous interconnected measures taken in order to mitigate the impact of climate change. The assessment carried out in this paper provides evidence that WSUD can help the city during drought by reducing dependency on town water for local irrigation and household use. It also helps provide thermal relief during heatwave due to the adiabatic cooling effect of the WSUD lake on the development sites. During extreme storm water events, the WSUD can soften the pressure on the city storm water drainage system by locally managing the extra storm water and reducing the possibility of flooding. Apart from these climate adaptation benefits, the WSUD also improves quality of life. It enhances landscape design, it helps to preserve biodiversity and it provides adiabatic cooling.

The other question that this paper attempted to explore was that of scalability and application to an existing city. The paper finds through its investigation of the city of Darmstadt that a considerable number of housing developments already have the logistical suitability—physical space and appropriate vintage warranting renovation—for implementing WSUD. The paper found that 22 WSUD projects can be implemented in the city serving an impressive 2527 apartment units covering 155,153 m^2 of living space in the city of Darmstadt. Annual savings of about 147 million liters of town water and local storm water retainment capacity of 2 million liters are achieved in this assessment. These results are despite the scope of this assessment being limited to sites owned and serviced by one party—municipal housing associations—which limited the evaluation to 496 sites. Additionally, the paper adopted very strict indicators that further limited selection to existing sites that have not been renovated for at least 40 years.

Another critical finding of the paper is that even with WSUD application it would not be possible to maintain 100% irrigation through WSUD of standard greeneries during the driest portion of the summer. This is because standard irrigation rates applied in this paper [51] lead to rather high irrigation outtakes, as they are based on water-intensive greenery. This means that to obtain 100% irrigation for the whole summer, there needs to be additional tweaking in landscape design by moving away from the standard water-intensive greenery toward native drought-tolerant plants and trees. Further, the viability of the greenery can be supported by increasing the saturation capacity of the soil. These two options were not investigated in this paper as they were beyond the study scope but could be explored as part of future research. Other research opportunities include the quantification of the added benefits of adiabatic cooling, the improvement of local biodiversity and overall quality of life from this exercise, the investigation of other potential sites for WSUD prototype application incorporating mixed forms of ownership and the refinement of WSUD prototype lake design through simulations. Additionally, investigating other cities could provide further insight on the question of scalability. In this case, transferability relies first and foremost on matching the indicators described in Chapter 3.1. City size, on the other hand, can vary.

The benefits and the water saving potential of WSUD outlined in this study suggest a change in city water politics in Germany. As mentioned earlier, administrations need to develop procedures that can handle new innovative approaches like the WSUD prototype that do not correspond to the well-established understanding of water management. In addition, WSUD should be adopted as a climate adaptation policy mandate and receive support with subsidies as performed with energy conservation renovations. But this calls for a more frequent consideration of WSUD by planners which, in contrast to known design principles, also emphasizes improving the quality of life. Both could pave the way for further WSUD projects in German cities. In conclusion, this paper demonstrates that WSUD has clear potential to help build a city's climate resilience and reduce the pressure on town water systems. But it must also be noted that this paper is based on a specific WSUD prototype application, and it is only one element in the transformation process toward climate resilience. Other measures need to be investigated.

Author Contributions: Conceptualization, J.S.; methodology, J.S. and S.G.; investigation, J.S. and S.G.; data curation, J.S.; writing—original draft preparation, J.S. and S.G.; writing—review and editing, A.S. and A.R.-C.; visualization, J.S. and S.G. All authors have read and agreed to the published version of the manuscript.

Funding: This research received no external funding.

Data Availability Statement: Data are found in the manuscript and Appendices A and B.

Acknowledgments: This work has been funded by the LOEWE initiative (Hesse, Germany) within the emergenCITY center.

Conflicts of Interest: The authors declare no conflicts of interest.

Appendix A

$$\mathbf{A}_{\text{open space}} \times \mathbf{R}_{\text{open space}} = \mathbf{A}_{\text{greenery}} \quad \text{(A1)}$$

$$\mathbf{A}_{\text{greenery}} \times \mathbf{ØIR}_{\text{day}} \times 15 = \sum \text{liters IW pre} \quad \text{(A2)}$$

$$\mathbf{L}_{\text{width}} \times \mathbf{L}_{\text{length}} \times 0.4 = \mathbf{L}_{\text{areapre}} \quad \text{(A3)}$$

(with a ratio of 1:7 and in accordance with (A2))

$$\mathbf{A}_{\text{base}} \times \mathbf{SW}_{\text{max}} = \sum \text{litres from SW inflow} \quad \text{(A4)}$$

$$\mathbf{L}_{\text{width}} \times \mathbf{L}_{\text{length}} \times 0.2 = \mathbf{L}_{\text{areaextension}} \quad \text{(A5)}$$

(with a ratio of 1:7 and in accordance with (A4))

Table A1. Ratios of greenery to open space of common urban morphologies.

Type	Detached Houses	Row Houses	Row Structures with Low Density	Row Structures with High Density	Block Structure	Rural Houses	Historical Center	City Center
$R_{\text{open space}}$	0.9	0.97	0.89	0.82	0.54	0.62	0.57	100

Table A2. Average irrigation demands.

	J	F	M	A	M	J	J	A	S	O	N	D
mm per mth	0	0	0	20	50	70	100	100	50	0	0	0
mm per day	0	0	0	0.67	1.61	2.33	3.23	3.23	1.67	0	0	0
Ø mm per day	2.12 (from April to September)											

$$\mathbf{L}_{\text{areapre}} \times \mathbf{ØIR}_{\text{day}} \times 15 = \sum \text{liters IW lake} \quad \text{(A6)}$$

$$\sum \text{liters IW pre} - \sum \text{litres IW lake} = \sum \text{liters IW} \quad \text{(A7)}$$

$$\mathbf{L}_{\text{width}} \times \mathbf{L}_{\text{length}} \times 0.4 = \mathbf{L}_{\text{area}} \quad \text{(A8)}$$

(with a ratio of 1:7 and in accordance with (A7))

$$\mathbf{A}_{\text{greenery}} - \mathbf{L}_{\text{area}} = \mathbf{A}_{\text{irrigation}} \quad \text{(A9)}$$

Table A3. Maximum daily storm water event Darmstadt.

Day	Hour	SW [mm]	Day	Hour	SW [mm]
26 August 2011	00:00	0	26 August 2011	12:00	0
26 August 2011	01:00	0	26 August 2011	13:00	0
26 August 2011	02:00	21.4	26 August 2011	14:00	0
26 August 2011	03:00	22.8	26 August 2011	15:00	0
26 August 2011	04:00	0	26 August 2011	16:00	0
26 August 2011	05:00	0	26 August 2011	17:00	0
26 August 2011	06:00	0	26 August 2011	18:00	0
26 August 2011	07:00	0	26 August 2011	19:00	1.1
26 August 2011	08:00	0	26 August 2011	20:00	10.4
26 August 2011	09:00	0	26 August 2011	21:00	1.6
26 August 2011	10:00	0	26 August 2011	22:00	0.6
26 August 2011	11:00	0	26 August 2011	23:00	0.6
Sum total: 58.5					

Appendix B

Table A4. Monthly storm water Darmstadt.

Year	Month	Σ SW [mm]	Year	Month	Σ SW [mm]
2011	January	59.2	2017	January	25.5
2011	February	18.4	2017	February	27.2
2011	March	16.1	2017	March	48.6
2011	April	12.3	2017	April	14.9
2011	May	15.3	2017	May	80.8
2011	June	66.9	2017	June	52.6
2011	July	75.8	2017	July	130.3
2011	August	151	2017	August	93.6
2011	September	42.1	2017	September	61.9
2011	October	28.2	2017	October	50.1
2011	November	1.2	2017	November	108.6
2011	December	104.6	2017	December	74.9
2012	January	70.6	2018	January	79.3
2012	February	10.5	2018	February	15.8
2012	March	19.3	2018	March	55.8
2012	April	42.5	2018	April	54.4
2012	May	48.5	2018	May	43.5
2012	June	89.6	2018	June	22.6
2012	July	123.9	2018	July	8
2012	August	62.7	2018	August	15.9

Table A4. *Cont.*

Year	Month	Σ SW [mm]	Year	Month	Σ SW [mm]
2012	September	42.3	2018	September	34.6
2012	October	60.5	2018	October	7.9
2012	November	45.2	2018	November	24.7
2012	December	94.1	2018	December	113
2013	January	47.3	2019	January	58.7
2013	February	47	2019	February	10.2
2013	March	29	2019	March	50.8
2013	April	78.1	2019	April	43.1
2013	May	136.3	2019	May	109.2
2013	June	69.4	2019	June	49
2013	July	26.8	2019	July	41.2
2013	August	72.1	2019	August	49.6
2013	September	83.2	2019	September	75.6
2013	October	90.3	2019	October	93.4
2013	November	68.6	2019	November	63.8
2013	December	42	2019	December	69
2015	January	88.6	2020	January	37.4
2015	February	25	2020	February	108.8
2015	March	31.2	2020	March	66
2015	April	25.2	2020	April	14.7
2015	May	10.1	2020	May	67.1
2015	June	24	2020	June	48.5
2015	July	28.2	2020	July	21.9
2015	August	40.4	2020	August	75.5
2015	September	62.1	2020	September	38.4
2015	October	16.8	2020	October	65.1
2015	November	74.7	2020	November	22.2
2015	December	30.5	2020	December	91.5
2016	January	72.4	2021	January	77.4
2016	February	99.8	2021	February	44.2
2016	March	45.6	2021	March	22.8
2016	April	72.2	2021	April	22.2
2016	May	81.4	2021	May	88.4
2016	June	103.1	2021	June	97
2016	July	43.5	2021	July	95.1
2016	August	35.9	2021	August	61.2
2016	September	27.4	2021	September	25.9
2016	October	53	2021	October	53.7
2016	November	48.4	2021	November	42
2016	December	9.8	2021	December	53.1

Table A5. Shortlisted sites.

Indicator			Value			Number
Number of units and or grand total of living space			$\geq$70 and or $\geq$5900 m^2			1
State of refurbishment			Building stock that was erected at least 40 years ago and that has never been renovated			2
			Building stock that was renovated at least 40 years ago			3
			Building stock that by the year 2030 will have been unrenovated for 40 years			4
Number of site [-]	Levels per building [-]	Erection year	State of refurbishment	Number of units [-]	Living space [m^2]	Indicator match
1	4	1929	Not refurbished	10	505	1 and 2
1	4	1929	Not refurbished	10	505	
1	4	1929	Not refurbished	10	517	
1	4	1929	Not refurbished	9	502	
1	4	1929	Not refurbished	10	502	
1	4	1929	Not refurbished	10	522	
1	4	1929	Not refurbished	10	520	
1	4	1929	Not refurbished	9	507	
1	4	1929	Not refurbished	9	505	
1	4	1929	Not refurbished	10	517	
1	4	1929	Not refurbished	10	511	
1	4	1929	Not refurbished	10	521	
Total				117	6135	
2	2	1950	Not refurbished	6	304	1 and 2
2	2	1950	Not refurbished	6	300	
2	2	1950	Not refurbished	6	301	
2	2	1950	Not refurbished	6	323	
2	2	1950	Not refurbished	6	323	
2	2	1950	Not refurbished	6	323	
2	2	1950	Not refurbished	6	301	
2	2	1950	Not refurbished	6	304	
2	2	1950	Not refurbished	6	300	
2	2	1950	Not refurbished	6	301	
2	2	1950	Not refurbished	6	301	
2	2	1950	Not refurbished	5	264	
2	2	1950	Not refurbished	6	309	
2	2	1950	Not refurbished	6	307	
2	2	1950	Not refurbished	6	323	
2	2	1950	Not refurbished	6	323	
2	2	1950	Not refurbished	6	316	
2	2	1950	Not refurbished	6	322	
2	2	1950	Not refurbished	6	382	

Table A5. *Cont.*

Indicator			Value			Number
2	2	1950	Not refurbished	6	322	
2	2	1950	Not refurbished	6	322	
2	2	1950	Not refurbished	6	323	
2	2	1950	Not refurbished	6	303	
2	2	1950	Not refurbished	6	303	
2	2	1950	Not refurbished	6	343	
2	2	1950	Not refurbished	5	291	
Total				154	8139	
3	9	1971	Not refurbished	38	3048	
3	9	1971	Not refurbished	16	1360	
3	9	1971	Not refurbished	10	608	1 and 2
Total				64	5015	
4	5	1951	Not refurbished	9	507	
4	5	1951	Not refurbished	10	619	
4	5	1951	Not refurbished	15	721	
4	5	1951	Not refurbished	9	507	
4	5	1951	Not refurbished	10	619	1 and 2
4	5	1951	Not refurbished	15	721	
4	5	1951	Not refurbished	9	507	
4	5	1951	Not refurbished	10	619	
4	5	1951	Not refurbished	15	721	
Total				102	5542	
5	4	1930	Not refurbished	10	394	
5	4	1930	Not refurbished	10	397	
5	4	1930	Not refurbished	9	374	
5	4	1930	Not refurbished	9	585	
5	4	1930	Not refurbished	10	533	
5	4	1930	Not refurbished	9	408	
5	4	1930	Not refurbished	9	460	
5	4	1930	Not refurbished	5	280	
5	4	1930	Not refurbished	4	271	1 and 2
5	4	1930	Not refurbished	10	384	
5	4	1930	Not refurbished	8	315	
5	4	1930	Not refurbished	10	396	
5	4	1930	Not refurbished	8	312	
5	4	1930	Not refurbished	10	391	
5	4	1930	Not refurbished	9	396	
5	4	1930	Not refurbished	4	281	
5	4	1930	Not refurbished	10	415	
Total				144	6591	

Table A5. *Cont.*

Indicator			Value			Number
6	5	1984	Not refurbished	11	837	1 and 4
6	5	1984	Not refurbished	10	692	
6	5	1984	Not refurbished	20	1198	
6	5	1984	Not refurbished	20	1255	
6	5	1984	Not refurbished	12	938	
Total				73	4920	
7	8	1969	Not refurbished	19	1965	1 and 2
7	8	1969	Not refurbished	25	1872	
7	8	1969	Not refurbished	18	1904	
Total				62	5741	
8	3	1966	Not refurbished	6	414	1 and 2
8	3	1966	Not refurbished	6	433	
8	3	1966	Not refurbished	6	433	
8	3	1966	Not refurbished	6	414	
8	3	1966	Not refurbished	6	433	
8	3	1966	Not refurbished	6	414	
8	12	1966	Not refurbished	60	3566	
8	4	1966	Not refurbished	8	573	
8	4	1966	Not refurbished	8	673	
8	4	1966	Not refurbished	8	618	
8	4	1966	Not refurbished	8	618	
Total				128	8586	
9	14	1967	Not refurbished	69	5161	1 and 2
Total				69	5161	
10	12	1968	Not refurbished	72	4293	1 and 2
Total				72	4293	
11	11	1974	Not refurbished	46	3277	1 and 2
11	11	1974	Not refurbished	61	4677	
Total				107	7954	
12	9	1975	Not refurbished	30	2296	1 and 2
12	5	1975	Not refurbished	13	764	
12	9	1975	Not refurbished	14	1049	
12	9	1975	Not refurbished	46	3401	
12	5	1975	Not refurbished	13	764	
12	9	1975	Not refurbished	46	3289	
12	5	1975	Not refurbished	13	764	
12	9	1975	Not refurbished	14	1049	
12	9	1975	Not refurbished	45	3214	
12	9	1975	Not refurbished	45	3238	
12	5	1975	Not refurbished	13	764	

Table A5. *Cont.*

Indicator			Value			Number
12	9	1975	Not refurbished	16	1198	
12	9	1975	Not refurbished	51	3819	
12	5	1975	Not refurbished	16	954	
12	9	1975	Not refurbished	27	2066	
12	9	1975	Not refurbished	45	3214	
12	5	1975	Not refurbished	13	764	
12	9	1975	Not refurbished	14	1049	
Total				474	33,655	
13	3	1982	Not refurbished	12	773	
13	3	1982	Not refurbished	10	817	
13	3	1982	Not refurbished	8	608	
13	3	1982	Not refurbished	10	817	
13	3	1982	Not refurbished	8	607	
13	3	1982	Not refurbished	12	772	
13	3	1982	Not refurbished	11	691	1 and 4
13	3	1982	Not refurbished	7	473	
13	3	1982	Not refurbished	11	693	
13	3	1982	Not refurbished	11	691	
13	3	1982	Not refurbished	12	714	
13	3	1982	Not refurbished	7	535	
Total				119	8190	
14	3	1983	Not refurbished	7	418	
14	3	1983	Not refurbished	8	463	
14	1	1983	Not refurbished	1	90	
14	1	1983	Not refurbished	1	90	
14	3	1983	Not refurbished	8	463	
14	1	1983	Not refurbished	1	90	
14	1	1983	Not refurbished	1	90	1 and 4
14	3	1983	Not refurbished	8	607	
14	3	1983	Not refurbished	8	696	
14	3	1983	Not refurbished	10	817	
14	3	1983	Not refurbished	11	693	
14	3	1983	Not refurbished	6	464	
14	3	1983	Not refurbished	8	584	
Total				78	5562	
15	3	1984	Not refurbished	60	2929	
15	1	1984	Not refurbished	35	1738	1 and 4
Total				95	4667	

Table A5. *Cont.*

Indicator			Value			Number
16	4	1990	Not refurbished	21	1209	1 and 4
16	4	1990	Not refurbished	29	1660	
16	4	1990	Not refurbished	29	1660	
Total				79	4530	
17	3	1986	Not refurbished	8	573	1 and 4
17	3	1986	Not refurbished	2	131	
17	3	1986	Not refurbished	2	131	
17	3	1986	Not refurbished	4	304	
17	3	1986	Not refurbished	8	601	
17	3	1986	Not refurbished	2	118	
17	3	1986	Not refurbished	2	144	
17	3	1986	Not refurbished	2	144	
17	3	1986	Not refurbished	2	118	
17	3	1986	Not refurbished	8	601	
17	3	1986	Not refurbished	4	304	
17	3	1986	Not refurbished	2	131	
17	3	1986	Not refurbished	2	131	
17	3	1986	Not refurbished	8	573	
17	3	1986	Not refurbished	2	144	
17	3	1986	Not refurbished	2	144	
17	3	1986	Not refurbished	2	118	
17	3	1986	Not refurbished	8	601	
17	3	1986	Not refurbished	8	601	
17	3	1986	Not refurbished	2	118	
Total				80	5731	
18	3	1989	Not refurbished	33	1576	1 and 4
18	3	1989	Not refurbished	27	1197	
18	3	1989	Not refurbished	35	1839	
Total				95	4612	
19	3	1960	Not refurbished	47	1395	1 and 2
19	4	1960	Not refurbished	23	848	
19	3	1960	Not refurbished	38	799	
19	3	1960	Not refurbished	24	1067	
Total				132	4109	
20	4	1926	Not refurbished	8	386	1 and 2
20	4	1926	Not refurbished	8	385	
20	4	1926	Not refurbished	8	386	
20	3	1926	Not refurbished	8	387	
20	4	1926	Not refurbished	8	387	
20	4	1926	Not refurbished	8	387	

Table A5. *Cont.*

Indicator			Value			Number
20	3	1926	Not refurbished	8	387	
20	3	1926	Not refurbished	8	387	
20	3	1926	Not refurbished	8	385	
20	3	1926	Not refurbished	8	372	
20	3	1926	Not refurbished	8	400	
20	3	1926	Not refurbished	8	383	
Total				96	4632	
21	1	1951	Not refurbished	6	366	1 and 2
21	1	1951	Not refurbished	6	345	
21	1	1951	Not refurbished	3	148	
21	1	1951	Not refurbished	6	368	
21	1	1951	Not refurbished	6	345	
21	1	1951	Not refurbished	8	452	
21	1	1951	Not refurbished	6	343	
21	1	1951	Not refurbished	8	451	
21	1	1951	Not refurbished	6	343	
21	1	1951	Not refurbished	8	450	
21	1	1951	Not refurbished	6	345	
21	1	1951	Not refurbished	8	451	
21	1	1951	Not refurbished	8	445	
21	1	1951	Not refurbished	10	322	
21	1	1951	Not refurbished	8	217	
Total				103	5390	
22	4	1983	Not refurbished	8	575	1 and 4
22	4	1983	Not refurbished	8	585	
22	4	1983	Not refurbished	8	585	
22	4	1983	Not refurbished	8	585	
22	4	1983	Not refurbished	8	585	
22	4	1983	Not refurbished	8	508	
22	3	1983	Not refurbished	6	438	
22	3	1983	Not refurbished	6	438	
22	3	1983	Not refurbished	6	438	
22	3	1983	Not refurbished	6	407	
22	3	1983	Not refurbished	6	415	
22	3	1983	Not refurbished	6	438	
Total				84	5996	

Number of site	**01**	
Base area of buildings [m^2]	2170	Color matches site plan
Grand total of open space [m^2]	4025	Color matches site plan
Grand total of greenery [m^2]	2174	
Area of main water body [m^2]	160	
Capacity of main water body [m^3]	64	
Capacity of extension [m^3]	127	

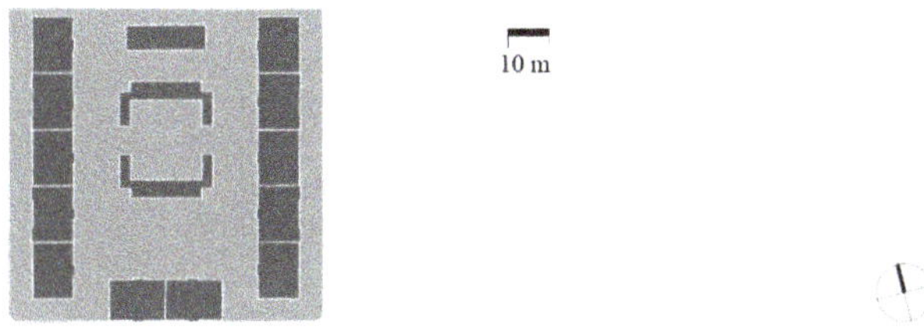

Site plan

Irrigation rate	0%	0%	0%	100%	100%	100%	100%	100%	100%	0%	0%	0%
	January	February	March	April	May	June	July	August	September	October	November	December
Storm water inflow[L/mth]	133,762	88,300	83,591	82,375	147,694	135,130	129,053	142,768	107,092	112,626	108,373	148,107
Irrigation outtake [L/mth]	0	0	0	40276	100690	140966	201379	201379	100690	0	0	0
Service water inflow[L/mth]	106,985	106,985	110,551	106,985	110,551	106,985	110,551	110,551	106,985	110,551	106,985	110,551
Total litre [L]	64,000	64,000	64,000	64,000	64,000	64,000	64,000	64,000	64,000	64,000	64,000	64,000
Evaporation [L/mth]	640	1919	4478	11,835	18,233	19,672	22,871	20,152	11,676	5598	1759	960
Lake level	100%	100%	100%	100%	100%	100%	100%	100%	100%	100%	100%	100%

Number of site	**02**	
Base area of buildings [m^2]	5608	Color matches site plan
Grand total of open space [m^2]	14935	Color matches site plan
Grand total of greenery [m^2]	13292	
Area of main water body [m^2]	973	
Capacity of main water body [m^3]	389	
Capacity of extension [m^3]	328	

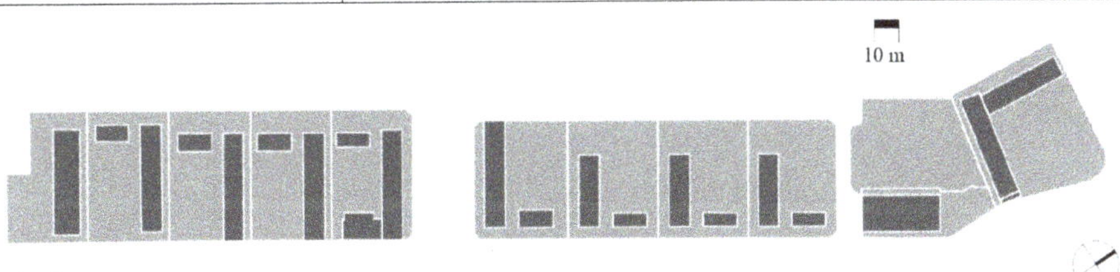

Site plan

Irrigation rate	0%	0%	0%	100%	100%	50%	25%	33%	50%	0%	0%	0%
	January	February	March	April	May	June	July	August	September	October	November	December
Storm water inflow[L/mth]	345,664	228,181	216,012	212,872	381,666	349,197	333,495	368,936	276,744	291,044	280,053	382,731
Irrigation outtake [L/mth]	0	0	0	246,382	615,956	431,169	307,978	406,531	307,978	0	0	0
Service water inflow[L/mth]	140,818	140,818	145,512	140,818	145,512	140,818	145,512	145,512	140,818	145,512	140,818	145,512
Total litre [L]	389,000	389,000	389,000	389,000	189,296	128,459	160,344	145,659	184,212	389,000	389,000	389,000
Evaporation [L/mth]	3892	11,676	27,245	72,004	110,925	119,683	139,143	122,602	71,031	34,056	10,703	5838
Lake level	100%	100%	100%	100%	[illegible]	[illegible]	[illegible]	[illegible]	[illegible]	100%	100%	100%

Figure A1. Site sheets 01 and 02.

Number of site	**05**	
Base area of buildings [m²]	5844	Color matches site plan
Grand total of open space [m²]	5464	Color matches site plan
Grand total of greenery [m²]	2950	
Area of main water body [m²]	218	
Capacity of main water body [m³]	87	
Capacity of extension [m³]	342	

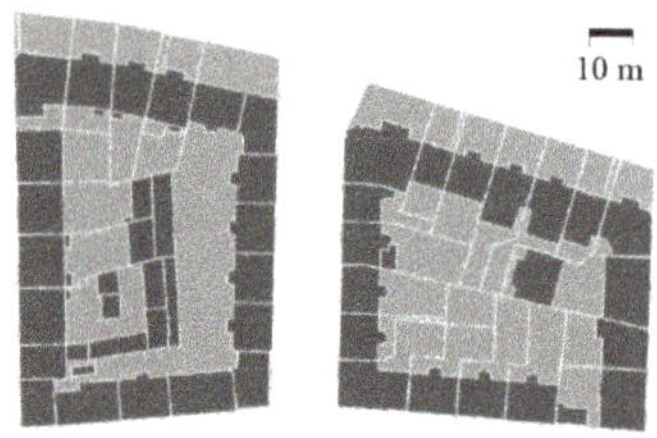

10 m

Site plan

Irrigation rate	0%	0%	0%	100%	100%	100%	100%	100%	100%	0%	0%	0%
	January	February	March	April	May	June	July	August	September	October	November	December
Storm water inflow[L/mth]	360,222	237,791	225,109	221,837	397,740	363,903	347,540	384,474	288,399	303,302	291,847	398,850
Irrigation outtake [L/mth]	0	0	0	54,650	136,624	191,273	273,248	273,248	136,624	0	0	0
Service water inflow[L/mth]	131,674	131,674	136,063	131,674	136,063	131,674	136,063	136,063	131,674	136,063	131,674	136,063
Total litre [L]	87,000	87,000	87,000	87,000	87,000	87,000	87,000	87,000	87,000	87,000	87,000	87,000
Evaporation [L/mth]	872	2615	6103	16,129	24,847	26,808	31,168	27,462	15,911	7628	2398	1308
Lake level	100%	100%	100%	100%	100%	100%	100%	100%	100%	100%	100%	100%

Number of site	**06**	
Base area of buildings [m²]	1270	Color matches site plan
Grand total of open space [m²]	3248	Color matches site plan
Grand total of greenery [m²]	1754	
Area of main water body [m²]	129	
Capacity of main water body [m³]	52	
Capacity of extension [m³]	74	

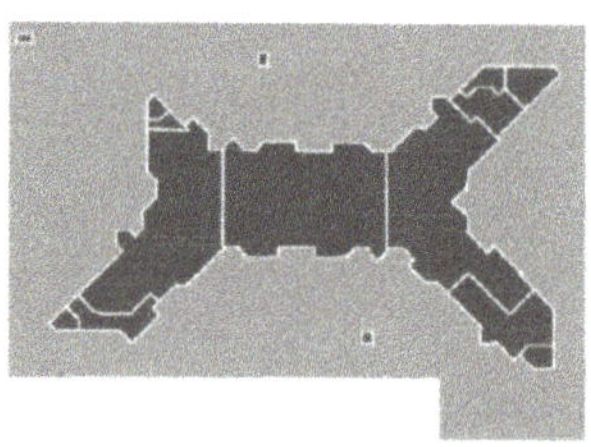

10 m

Site plan

Irrigation rate	0%	0%	0%	100%	100%	100%	100%	75%	100%	0%	0%	0%
	January	February	March	April	May	June	July	August	September	October	November	December
Storm water inflow[L/mth]	78,271	51,669	48,913	48,202	86,423	79,071	75,516	83,541	62,665	65,903	63,414	86,665
Irrigation outtake [L/mth]	0	0	0	32,486	81,215	113,701	162,431	121,823	81,215	0	0	0
Service water inflow[L/mth]	66,751	66,751	68,976	66,751	68,976	66,751	68,976	68,976	66,751	68,976	66,751	68,976
Total litre [L]	52,000	52,000	52,000	52,000	52,000	52,000	15,553	29,939	52,000	52,000	52,000	52,000
Evaporation [L/mth]	518	1553	3624	9578	14,755	15,920	18,508	16,308	9448	4530	1424	777
Lake level	100%	100%	100%	100%	100%	100%	[illegible]	58%	100%	100%	100%	100%

Figure A2. Sites 05 and 06.

Number of site	**07**	
Base area of buildings [m²]	1283	Color matches site plan
Grand total of open space [m²]	4216	Color matches site plan
Grand total of greenery [m²]	3753	
Area of main water body [m²]	276	
Capacity of main water body [m³]	110	
Capacity of extension [m³]	75	

Site plan

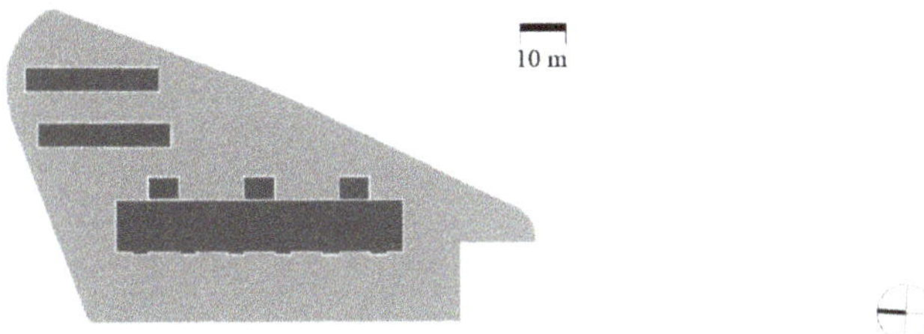

	January	February	March	April	May	June	July	August	September	October	November	December
Irrigation rate	0%	0%	0%	100%	100%	50%	25%	33%	50%	0%	0%	0%
Storm water inflow[L/mth]	79,084	52,205	49,421	48,702	87,320	79,892	76,299	84,408	63,316	66,587	64,073	87,564
Irrigation outtake [L/mth]	0	0	0	69,530	173,826	121,678	86.913	114,725	86,913	0	0	0
Service water inflow[L/mth]	56,693	56,693	58,583	56,693	58,583	56,693	58,583	58,583	56,693	58,583	56,693	58,583
Total litre [L]	110,000	110,000	110,000	110,000	50,605	31,555	40,046	33,527	46,469	110,000	110,000	110,000
Evaporation [L/mth]	1104	3313	7730	20,429	31,472	33,956	39,478	34,785	20,153	9662	3037	1656
Lake level	100%	100%	100%	100%	46%	[illegible]	[illegible]	[illegible]	[illegible]	100%	100%	100%

Number of site	**08**	
Base area of buildings [m²]	2716	Color matches site plan
Grand total of open space [m²]	10183	Color matches site plan
Grand total of greenery [m²]	9063	
Area of main water body [m²]	661	
Capacity of main water body [m³]	265	
Capacity of extension [m³]	159	

Site plan

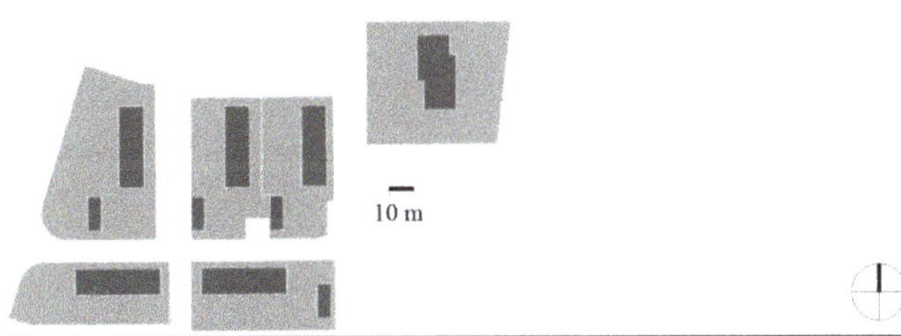

	January	February	March	April	May	June	July	August	September	October	November	December
Irrigation rate	0%	0%	0%	100%	100%	33%	12.5%	33%	50%	0%	0%	0%
Storm water inflow[L/mth]	167,438	110,530	104,635	103,114	184,877	169,149	161,543	178,711	134,053	140,980	135,656	185,393
Irrigation outtake [L/mth]	0	0	0	168,032	420,081	194,077	105,020	277,253	210,040	0	0	0
Service water inflow[L/mth]	117,043	117,043	120,945	117,043	120,945	117,043	120,945	120,945	117,043	120,945	117,043	120,945
Total litre [L]	265,000	265,000	265,000	265,000	75,347	86,116	169,011	108,083	100,861	265,000	265,000	265,000
Evaporation [L/mth]	2645	7936	18,518	48,940	75,394	81,346	94,573	83,330	48,278	23,147	7275	3968
Lake level	100%	100%	100%	100%	[illegible]	[illegible]	64%	[illegible]	[illegible]	100%	100%	100%

Figure A3. Sites 07 and 08.

Number of site	**09**	
Base area of buildings [m²]	464	Color matches site plan
Grand total of open space [m²]	4199	Color matches site plan
Grand total of greenery [m²]	3443	
Area of main water body [m²]	252	
Capacity of main water body [m³]	101	
Capacity of extension [m³]	27	

Site plan

Irrigation rate	0%	0%	0%	100%	75%	33%	12.5%	25%	33%	0%	0%	0%
	January	February	March	April	May	June	July	August	September	October	November	December
Storm water inflow[L/mth]	28,619	18,892	17,884	17,624	31,600	28,911	27,611	30,546	22,913	24,097	23,187	31,688
Irrigation outtake [L/mth]	0	0	0	63,827	119,676	73,721	39,892	79,784	52,658	0	0	0
Service water inflow[L/mth]	63,094	63,094	65,197	63,094	65,197	63,094	65,197	65,197	63,094	65,197	63,094	65,197
Total litre [L]	101,000	101,000	101,000	99,243	47,635	34,923	51,803	36,009	50,962	101,000	101,000	101,000
Evaporation [L/mth]	1008	3024	7056	18,648	28,728	30,996	36,036	31,752	18,396	8820	2772	1512
Lake level	100%	100%	100%	98%	[illegible]	[illegible]	51%	[illegible]	50%	100%	100%	100%

Number of site	**10**	
Base area of buildings [m²]	499	Color matches site plan
Grand total of open space [m²]	3797	Color matches site plan
Grand total of greenery [m²]	3114	
Area of main water body [m²]	227	
Capacity of main water body [m³]	91	
% of lake area to greenery	29	

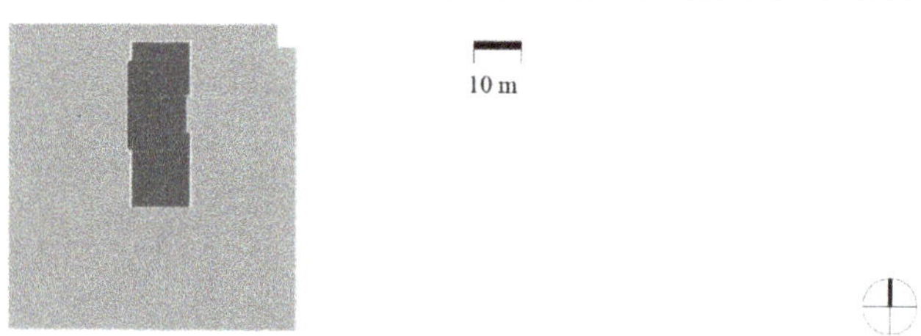

Site plan

Irrigation rate	0%	0%	0%	100%	100%	33%	12.5%	33%	50%	0%	0%	0%
	January	February	March	April	May	June	July	August	September	October	November	December
Storm water inflow[L/mth]	30,763	20,307	19,224	18,945	33,967	31,077	29,680	32,834	24,629	25,902	24,924	34,062
Irrigation outtake [L/mth]	0	0	0	57,722	144,305	66,669	36,076	95,241	72,152	0	0	0
Service water inflow[L/mth]	65,837	65,837	68,031	65,837	68,031	65,837	68,031	68,031	65,837	68,031	65,837	68,031
Total litre [L]	91,000	91,000	91,000	91,000	22,766	25,038	54,150	31,118	32,829	91,000	91,000	91,000
Evaporation [L/mth]	910	2729	6368	16,830	25,927	27,974	32,522	28,656	16,602	7960	2502	1365
Lake level	100%	100%	100%	100%	[illegible]	[illegible]	60%	[illegible]	[illegible]	100%	100%	100%

Figure A4. Sites 09 and 10.

Number of site	**11**	
Base area of buildings [m^2]	1292	Color matches site plan
Grand total of open space [m^2]	3250	Color matches site plan
Grand total of greenery [m^2]	2665	
Area of main water body [m^2]	194	
Capacity of main water body [m^3]	78	
Capacity of extension [m^3]	76	

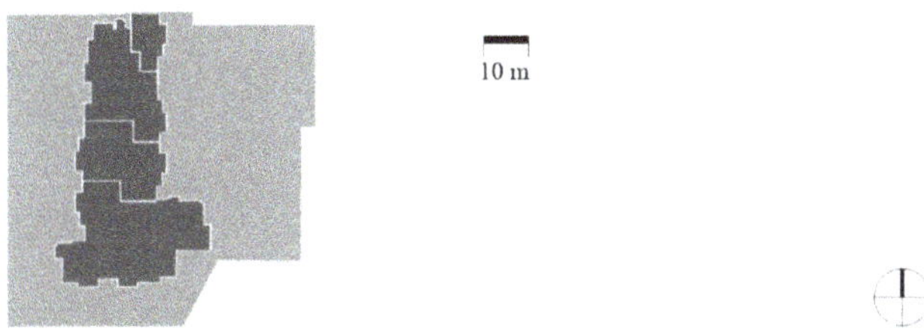

Site plan

Irrigation rate	0%	0%	0%	100%	100%	100%	75%	50%	100%	0%	0%	0%
	January	**February**	**March**	**April**	**May**	**June**	**July**	**August**	**September**	**October**	**November**	**December**
Storm water inflow[L/mth]	79,652	52,580	49,776	49,052	87,948	80,466	76,848	85,014	63,771	67,066	64,533	88,193
Irrigation outtake [L/mth]	0	0	0	49,415	123,538	172,953	185,307	123,538	123,538	0	0	0
Service water inflow[L/mth]	97,841	97,841	101,102	97,841	101,102	97,841	101,102	101,102	97,841	101,102	97,841	101,102
Total litre [L]	78,000	78,000	78,000	78,000	78,000	59,441	24,283	62,366	78,000	78,000	78,000	78,000
Evaporation [L/mth]	778	2333	5443	14,386	22,163	23,912	27,801	24,496	14,192	6804	2139	1166
Lake level	100%	100%	100%	100%	100%	76%		80%	100%	100%	100%	100%

Number of site	**12**	
Base area of buildings [m^2]	6027	Color matches site plan
Grand total of open space [m^2]	11952	Color matches site plan
Grand total of greenery [m^2]	9801	
Area of main water body [m^2]	718	
Capacity of main water body [m^3]	287	
Capacity of extension [m^3]	353	

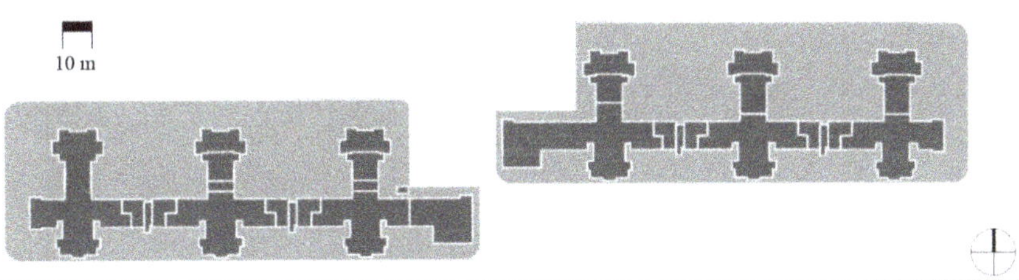

Site plan

Irrigation rate	0%	0%	0%	100%	100%	100%	100%	75%	100%	0%	0%	0%
	January	**February**	**March**	**April**	**May**	**June**	**July**	**August**	**September**	**October**	**November**	**December**
Storm water inflow[L/mth]	371,480	245,222	232,145	228,770	410,170	375,276	358,402	396,490	297,413	312,781	300,968	411,315
Irrigation outtake [L/mth]	0	0	0	181,646	454,116	635,763	908,232	681,174	454,116	0	0	0
Service water inflow[L/mth]	433,426	433,426	447,873	433,426	447,873	433,426	447,873	447,873	433,426	447,873	433,426	447,873
Total litre [L]	287,000	287,000	287,000	287,000	287,000	287,000	82,323	155,004	287,000	287,000	287,000	287,000
Evaporation [L/mth]	2873	8620	20,113	53,156	81,888	88,353	102,720	90,508	52,437	25,141	7902	4310
Lake level	100%	100%	100%	100%	100%	100%		54%	100%	100%	100%	100%

Figure A5. Sites 11 and 12.

Number of site	**13**	
Base area of buildings [m²]	2778	Color matches site plan
Grand total of open space [m²]	7667	Color matches site plan
Grand total of greenery [m²]	6824	
Area of main water body [m²]	500	
Capacity of main water body [m³]	200	
Capacity of extension [m³]	163	

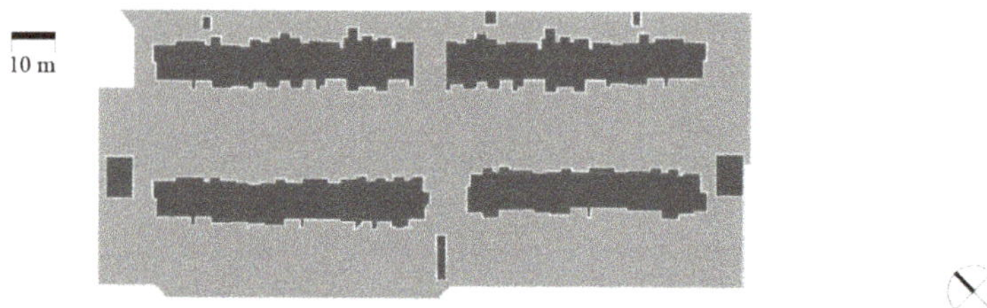

Site plan

Irrigation rate	0%	0%	0%	100%	100%	50%	33%	33%	75%	0%	0%	0%
	January	February	March	April	May	June	July	August	September	October	November	December
Storm water inflow[L/mth]	171,227	113,031	107,003	105,448	189,061	172,977	165,199	182,755	137,087	144,171	138,726	189,589
Irrigation outtake [L/mth]	0	0	0	126,477	316,192	221,334	208,687	208,687	237,144	0	0	0
Service water inflow[L/mth]	108,814	108,814	112,441	108,814	112,441	108,814	112,441	112,441	108,814	112,441	108,814	112,441
Total litre [L]	200,000	200,000	200,000	200,000	128,331	127,310	124,789	148,322	120,592	200,000	200,000	200,000
Evaporation [L/mth]	1999	5998	13,995	36,986	56,979	61,478	71,474	62,977	36,487	17,494	5498	2999
Lake level	100%	100%	100%	100%	[illegible]	[illegible]	[illegible]	74%	[illegible]	100%	100%	100%

Number of site	**14**	
Base area of buildings [m²]	2106	Color matches site plan
Grand total of open space [m²]	4292	Color matches site plan
Grand total of greenery [m²]	3820	
Area of main water body [m²]	281	
Capacity of main water body [m³]	112	
Capacity of extension [m³]	123	

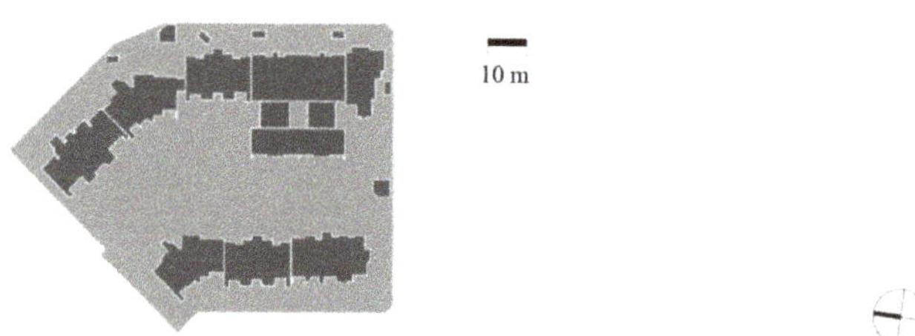

Site plan

Irrigation rate	0%	0%	0%	100%	100%	100%	33%	50%	100%	0%	0%	0%
	January	February	March	April	May	June	July	August	September	October	November	December
Storm water inflow[L/mth]	129,840	85,711	81,140	79,960	143,364	131,167	125,269	138,582	103,952	109,324	105,195	143,764
Irrigation outtake [L/mth]	0	0	0	70,786	176,964	247,750	116,796	176,964	176,964	0	0	0
Service water inflow[L/mth]	71,323	71,323	73,701	71,323	73,701	71,323	73,701	73,701	71,323	73,701	71,323	73,701
Total litre [L]	112,000	112,000	112,000	112,000	112,000	32,242	74,306	74,284	52,120	112,000	112,000	112,000
Evaporation [L/mth]	1122	3366	7854	20,756	31,975	34,499	40,109	35,341	20,475	9817	3085	1683
Lake level	100%	100%	100%	100%	100%	[illegible]	66%	66%	47%	100%	100%	100%

Figure A6. Sites 13 and 14.

Number of site	**15**	
Base area of buildings [m²]	2051	Color matches site plan
Grand total of open space [m²]	8613	Color matches site plan
Grand total of greenery [m²]	4651	
Area of main water body [m²]	340	
Capacity of main water body [m³]	136	
Capacity of extension [m³]	120	

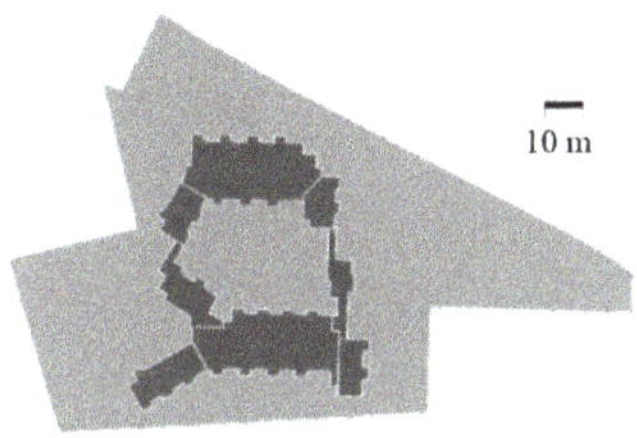

Site plan

Irrigation rate	0%	0%	0%	100%	100%	75%	33%	50%	75%	0%	0%	0%
	January	**February**	**March**	**April**	**May**	**June**	**July**	**August**	**September**	**October**	**November**	**December**
Storm water inflow[L/mth]	126,450	83,472	79,021	77,872	139,620	127,742	121,998	134,963	101,238	106,469	102,448	140,009
Irrigation outtake [L/mth]	0	0	0	86,216	215,540	226,317	142,256	215,540	161,655	0	0	0
Service water inflow[L/mth]	86,868	86,868	89,764	86,868	89,764	86,868	89,764	89,764	86,868	89,764	86,868	89,764
Total litre [L]	136,000	136,000	136,000	136,000	111,076	57,541	78,417	44,756	46,382	136,000	136,000	136,000
Evaporation [L/mth]	1360	4081	9522	25,165	38,768	41,828	48,629	42,848	24,825	11,902	3741	2040
Lake level	100%	100%	100%	100%	82%	[illegible]	58%			100%	100%	100%

Number of site	**16**	
Base area of buildings [m²]	1970	Color matches site plan
Grand total of open space [m²]	5909	Color matches site plan
Grand total of greenery [m²]	5259	
Area of main water body [m²]	385	
Capacity of main water body [m³]	154	
Capacity of extension [m³]	115	

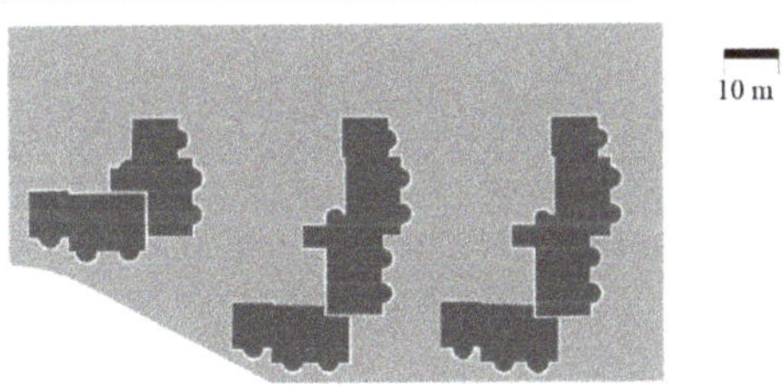

Site plan

Irrigation rate	0%	0%	0%	100%	100%	50%	25%	33%	50%	0%	0%	0%
	January	**February**	**March**	**April**	**May**	**June**	**July**	**August**	**September**	**October**	**November**	**December**
Storm water inflow[L/mth]	121,404	80,142	75,868	74,765	134,049	122,645	117,130	129,578	97,198	102,221	98,360	134,423
Irrigation outtake [L/mth]	0	0	0	97,464	243,659	170,562	121,830	160,815	121,830	0	0	0
Service water inflow[L/mth]	72,238	72,238	74,646	72,238	74,646	72,238	74,646	74,646	72,238	74,646	72,238	74,646
Total litre [L]	154,000	154,000	154,000	154,000	75,100	52,018	66,852	61,701	81,173	154,000	154,000	154,000
Evaporation [L/mth]	1542	4625	10,791	28,519	43,935	47,404	55,111	48,560	28,134	13,489	4239	2312
Lake level	100%	100%	100%	100%	[illegible]	[illegible]	[illegible]	[illegible]	53%	100%	100%	100%

Figure A7. Sites 15 and 16.

Number of site	**17**	
Base area of buildings [m²]	2061	Color matches site plan
Grand total of open space [m²]	10093	Color matches site plan
Grand total of greenery [m²]	8983	
Area of main water body [m²]	659	
Capacity of main water body [m³]	263	
Capacity of extension [m³]	121	

Site plan

Irrigation rate	0%	0%	0%	100%	50%	25%	12.5%	12.5%	33%	0%	0%	0%
	January	February	March	April	May	June	July	August	September	October	November	December
Storm water inflow[L/mth]	127,020	83.849	79,377	78,223	140,249	128,318	122,548	135,571	101,694	106,949	102,910	140,641
Irrigation outtake [L/mth]	0	0	0	166,484	208,105	145,674	104,053	104,053	137,349	0	0	0
Service water inflow[L/mth]	73,152	73,152	75,590	73,152	75,590	73,152	75,590	75,590	73,152	75,590	73,152	75,590
Total litre [L]	263,000	263,000	263,000	199,152	131,803	106,588	106,490	130,612	120,029	263,000	263,000	263,000
Evaporation [L/mth]	2635	7904	18,442	48,739	75,084	81,011	94,184	82,987	48,080	23,052	7245	3952
Lake level	100%	100%	100%	76%	[illegible]	[illegible]	[illegible]	[illegible]	[illegible]	100%	100%	100%

Number of site	**18**	
Base area of buildings [m²]	2188	Color matches site plan
Grand total of open space [m²]	3131	Color matches site plan
Grand total of greenery [m²]	1691	
Area of main water body [m²]	126	
Capacity of main water body [m³]	50	
Capacity of extension [m³]	128	

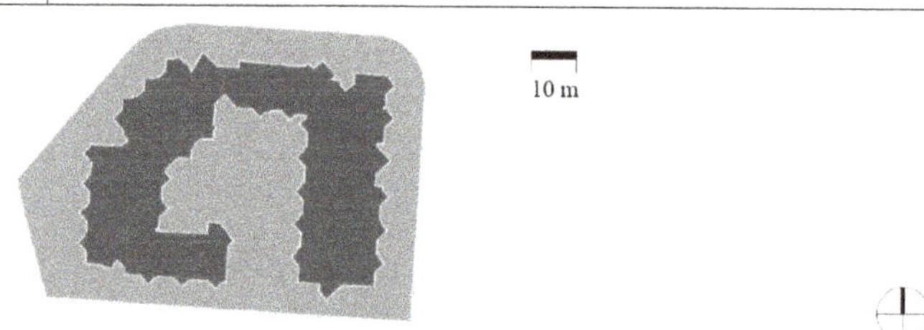

Site plan

Irrigation rate	0%	0%	0%	100%	100%	100%	100%	100%	100%	0%	0%	0%
	January	February	March	April	May	June	July	August	September	October	November	December
Storm water inflow[L/mth]	134,889	89,043	84,294	83,069	148,938	136,267	130,140	143,970	107,994	113,574	109,285	149,354
Irrigation outtake [L/mth]	0	0	0	31,294	78,234	109,528	156,468	156,468	78,234	0	0	0
Service water inflow[L/mth]	86,868	86,868	89,764	86,868	89,764	86,868	89,764	89,764	86,868	89,764	86,868	89,764
Total litre [L]	50,000	50,000	50,000	50,000	50,000	50,000	50,000	50,000	50,000	50,000	50,000	50,000
Evaporation [L/mth]	503	1510	3524	9312	14,346	15,479	17,996	15,856	9187	4405	1384	755
Lake level	100%	100%	100%	100%	100%	100%	100%	100%	100%	100%	100%	100%

Figure A8. Sites 17 and 18.

Number of site	**19**	
Base area of buildings [m²]	2917	Color matches site plan
Grand total of open space [m²]	7816	Color matches site plan
Grand total of greenery [m²]	6956	
Area of main water body [m²]	508	
Capacity of main water body [m³]	203	
Capacity of extension [m³]	171	

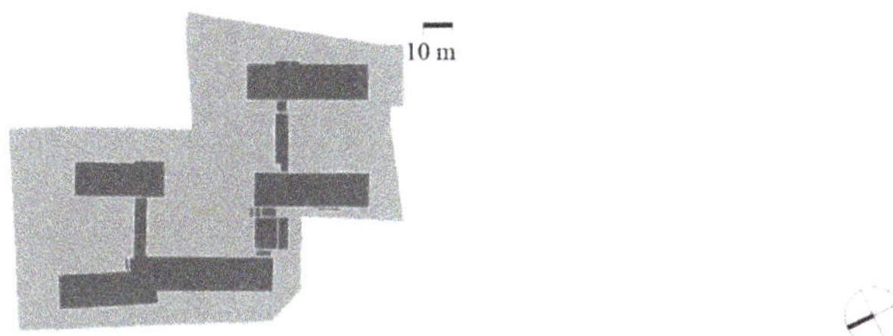

Site plan

Irrigation rate	0%	0%	0%	100%	100%	50%	33%	50%	75%	0%	0%	0%
	January	February	March	April	May	June	July	August	September	October	November	December
Storm water inflow[L/mth]	179,774	118,673	112,344	110,711	198,498	181,612	173,445	191,878	143,930	151,367	145,651	199,052
Irrigation outtake [L/mth]	0	0	0	128,954	322,385	225,670	212,774	322,385	241,789	0	0	0
Service water inflow[L/mth]	120,701	120,701	124,724	120,701	124,724	120,701	124,724	124,724	120,701	124,724	120,701	124,724
Total litre [L]	203,000	203,000	203,000	203,000	145,910	160,052	172,785	102,977	88,725	203,000	203,000	203,000
Evaporation [L/mth]	2033	6098	14,228	37,602	57,927	62,500	72,663	64,025	37,094	17,785	5589	3049
Lake level	100%	100%	100%	100%	72%	79%	85%	[illegible]		100%	100%	100%

Number of site	**20**	
Base area of buildings [m²]	1607	Color matches site plan
Grand total of open space [m²]	2322	Color matches site plan
Grand total of greenery [m²]	2067	
Area of main water body [m²]	151	
Capacity of main water body [m³]	61	
Capacity of extension [m³]	94	

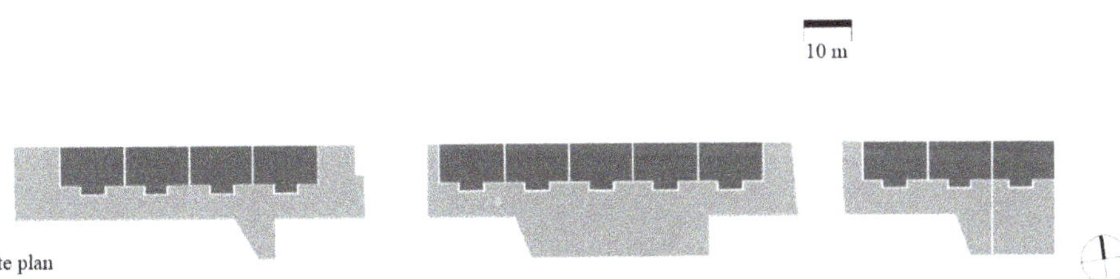

Site plan

Irrigation rate	0%	0%	0%	100%	100%	100%	100%	100%	100%	0%	0%	0%
	January	February	March	April	May	June	July	August	September	October	November	December
Storm water inflow[L/mth]	99,044	65,381	61,894	60,995	109,359	100,056	95,557	105,712	79,296	83,393	80,244	109,665
Irrigation outtake [L/mth]	0	0	0	38,310	95,774	134,084	191,549	191,549	95,774	0	0	0
Service water inflow[L/mth]	87,782	87,782	90,708	87,782	90,708	87,782	90,708	90,708	87,782	90,708	87,782	90,708
Total litre [L]	61,000	61,000	61,000	61,000	61,000	61,000	34,072	19,873	61,000	61,000	61,000	61,000
Evaporation [L/mth]	605	1816	4238	11,200	17,255	18,617	21,644	19,071	11,049	5298	1665	908
Lake level	100%	100%	100%	100%	100%	100%	56%		100%	100%	100%	100%

Figure A9. Sites 19 and 20.

Number of site	**21**	
Base area of buildings [m^2]	2580	Color matches site plan
Grand total of open space [m^2]	5268	Color matches site plan
Grand total of greenery [m^2]	4688	
Area of main water body [m^2]	343	
Capacity of main water body [m^3]	137	
Capacity of extension [m^3]	151	

Site plan

Irrigation rate	0%	0%	0%	100%	100%	100%	33%	50%	100%	0%	0%	0%
	January	February	March	April	May	June	July	August	September	October	November	December
Storm water inflow[L/mth]	159,054	104,995	99,396	97,951	175,620	160,680	153,455	169,763	127,341	133,921	128,864	176,110
Irrigation outtake [L/mth]	0	0	0	86,908	217,270	304,178	143,398	217,270	217,270	0	0	0
Service water inflow[L/mth]	94,183	94,183	97,323	94,183	97,323	94,183	97,323	97,323	94,183	97,323	94,183	97,323
Total litre [L]	137,000	137,000	137,000	137,000	137,000	45,496	103,825	110,422	89,638	137,000	137,000	137,000
Evaporation [L/mth]	1372	4116	9604	25,382	39,102	42,189	49,049	43,218	25,039	12,005	3773	2058
Lake level	100%	100%	100%	100%	100%	[illegible]	76%	81%	65%	100%	100%	100%

Number of site	**22**	
Base area of buildings [m^2]	1962	Color matches site plan
Grand total of open space [m^2]	5499	Color matches site plan
Grand total of greenery [m^2]	4894	
Area of main water body [m^2]	359	
Capacity of main water body [m^3]	144	
Capacity of extension [m^3]	115	

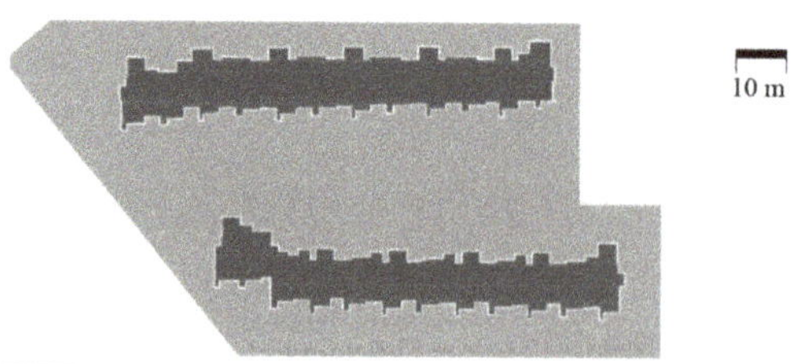

Site plan

Irrigation rate	0%	0%	0%	100%	100%	50%	33%	33%	75%	0%	0%	0%
	January	February	March	April	May	June	July	August	September	October	November	December
Storm water inflow[L/mth]	120,912	79,817	75,560	74,462	133,506	122,148	116,656	129,053	96,804	101,807	97,962	133,879
Irrigation outtake [L/mth]	0	0	0	90,699	226,749	158,724	149,654	149,654	170,062	0	0	0
Service water inflow[L/mth]	76,810	76,810	79,370	76,810	79,370	76,810	79,370	79,370	76,810	79,370	76,810	79,370
Total litre [L]	144,000	144,000	144,000	144,000	89,217	85,311	80,366	93,918	71,274	144,000	144,000	144,000
Evaporation [L/mth]	1435	4306	10,048	26,556	40,910	44,140	51,317	45,216	26,197	12,560	3947	2153
Lake level	100%	100%	100%	100%	62%	59%	[illegible]	65%	[illegible]	100%	100%	100%

Figure A10. Sites 21 and 22.

References

1. Bundesministerium für Umwelt, Naturschutz, nukleare Sicherheit und Verbraucherschutz (BMUV). *Zweiter Fortschrittsbericht zur Deutschen Anpassungsstrategie an den Klimawandel*; Federal Ministry for Environment, Nature Conservation and Nuclear Safety (Germany): Berlin, Germany, 2020.
2. World Metrological Organization (WMO). World Meteorological Organization. 2022. Available online: https://public.wmo.int/en (accessed on 25 August 2022).
3. National Oceanic and Atmospheric Administration (NOAA) and National Integrated Drought Information System (NIDIS). 2022. Available online: https://www.drought.gov/ (accessed on 25 August 2022).
4. European Centre for Medium-Range Weather Forecasts (ECMWF), ecmwf. 2022. Available online: https://www.ecmwf.int/ (accessed on 25 August 2022).
5. Statistische Bundesamt, Destatis. 2023. Available online: https://www.destatis.de/DE/Presse/Pressemitteilungen/2023/01/PD22_012_126 (accessed on 30 January 2023).
6. Raether, E. Zeit Online. 2022. Available online: https://www.zeit.de/2022/28/flutkatastrophe-ahrtal-ahr-hochwasser (accessed on 30 January 2023).
7. Intergovernmental Panel on Climate Change (IPCC). *Climate Change 2022-Impacts, Adaptation and Vulnerability*; Working Group II Contribution to the Sixth Assessment Report of the Intergovernmental Panel on Climate Change; IPCC: Geneva, Switzerland, 2022.
8. United Nations (UN). United Nations Climate Change. 2022. Available online: https://unfccc.int/process-and-meetings/the-paris-agreement/the-paris-agreement (accessed on 25 August 2022).
9. European Commission (EC), Joint Research Centre; Baranzelli, C.; Vandecasteele, I.; Aurambout, J.; Siragusa, A. (Eds.) *The Future of Cities: Opportunities, Challenges and the Way Forward*; Publications Office: Luxembourg, 2019.
10. European Enviroment Agency (EEA). Urban Adaptation to Climate Change in Europe 2016-Transforming Cities in a Changing Climate. Available online: https://op.europa.eu/en/publication-detail/-/publication/573254ad-a7f6-459d-944f-9bbfe1cb58cd/language-en/format-PDF/source-265076876 (accessed on 1 November 2023).
11. Depietri, Y. Social vulnerability assessment of the Cologne urban area (Germany) to heat waves: Links to ecosystem services. *Int. J. Disaster Risk Reduct.* **2013**, *6*, 98–117. [CrossRef]
12. Steensen, B.M.; Marelle, L.; Hodnebrog, Ø.; Myhre, G. Future urban heat island influence on precipitation. *Clim. Dyn.* **2022**, *58*, 3393–3403. [CrossRef]
13. Federal Ministry for the Environment, Nature Conservation, Building and Nuclear Safety (BMUV) (Ed.) *Recommendations for Actio Heat-Action Plans to Protect Human Health*; BMUV: Bonn, Germany, 2017.
14. Varouchakis, E.A.; Apostolakis, A.; Siaka, M.; Vasilopoulos, K.; Tasiopoulos, A. Alternatives for domestic water tariff policy in themunicipality of Chania, Greece, toward water saving using game theory. *Water Policy* **2018**, *20*, 175–188. [CrossRef]
15. He, C.; Liu, Z.; Wu, J.; Pan, X.; Fang, Z.; Li, J.; Bryan, B.A. Future global urban water scarcity and potential solutions. *Nat. Commun.* **2021**, *12*, 4667. [CrossRef] [PubMed]
16. Aboelnga, H.T.; Ribbe, L.; Frechen, F.-B.; Saghir, J. Urban Water Security: Definition and Assessment Framework. *Resources* **2019**, *8*, 178. [CrossRef]
17. United Nations (UN)-Department of Economic and Social Affairs-Population Dynamics: The World's Cities in 2018. Available online: https://www.un.org/development/desa/pd/sites/www.un.org.development.desa.pd/files/files/documents/2020/Jan/un_2018_worldcities_databooklet.pdf (accessed on 15 November 2023).
18. Radcliffe, J. History of Water Sensitive Urban Design/Low Impact Development Adoption in Australia and Internationally. In *Approaches to Water Sensitive Urban Design*; Woodhead Publishing: Sawston, UK, 2019; pp. 1–24.
19. Australian Capital Territory (ACT). *Water Sensitive Urban Design-Review Report*; ACT: Canberra, Australia, 2014.
20. Joint Steering Committee for Water Sensitive Cities (JSCWSC). *Evaluating Options for Water Sensitive Urban Design-A National Guide*; BMT WBM PTY Ltd.: Adelaide, Australia, 2009.
21. European Environment Agency (EEA). *Urban Adaptation in Europe: How Cities and Towns Respond to Climate Change*; European Environment Agency: Copenhagen, Denmark, 2020.
22. Kuller, M.; Bach, P.M.; Ramirez-Lovering, D.; Deletic, A. Framing water sensitive urban design as part of the urban form: A critical review of tools for best planning practice. *Environ. Model. Softw.* **2017**, *96*, 265–282. [CrossRef]
23. Sharma, A.K.; Pezzaniti, D.; Myers, B.; Cook, S.; Tjandraatmadja, G.; Chacko, P.; Chavoshi, S.; Kemp, D.; Leonard, R.; Koth, B.; et al. Water Sensitive Urban Design: An Investigation of Current Systems, Implementation Drivers, Community Perceptions and Potential to Supplement Urban Water Services. *Water* **2016**, *8*, 272. [CrossRef]
24. Wong, H.F. Water sensitive urban design–the journey thus far. *Australas. J. Water Resour.* **2006**, *10*, 213–222. [CrossRef]
25. U.S. Environmental Protection Agency (USEPA). *Manual: Constructed Wetlands Treatment of Municipal Wastewaters*; National Risk Management Research Laboratory: Cincinnati, OH, USA, 1999.
26. Van de Walle, A.; Kim, M.; Alam, M.K.; Wang, X.; Wu, D.; Dash, S.R.; Rabaey, K.; Kim, J. Greywater reuse as a key enabler for improving urban wastewater management. *Environ. Sci. Ecotechnology* **2023**, *16*, 100277. [CrossRef] [PubMed]

27. Juan, Y.K.; Chen, Y.; Lin, J.M. Greywater Reuse System Design and Economic Analysis for Residential Buildings in Taiwan. *Water* **2016**, *8*, 546. [CrossRef]
28. Gross, A.; Maimon, A.; Alfiya, Y.; Friedler, E. *Greywater Reuse*; CRC Press: Boca Raton, FL, USA, 2015.
29. Feng, W.; Liu, Y.; Gao, L. Stormwater treatment for reuse: Current practice and future development–A review. *J. Environ. Manag.* **2022**, *301*, 113830. [CrossRef] [PubMed]
30. Hatt, B.E.; Deletic, A.; Fletcher, T.D. Integrated treatment and recycling of stormwater: A review of Australian practice. *J. Environ. Manag.* **2006**, *79*, 102–113. [CrossRef] [PubMed]
31. Hamburg Wasser, Hamburg Water Cycle. 2024. Available online: https://www.hamburgwasser.de/umwelt/vorsorge/hamburg-water-cycle (accessed on 17 January 2023).
32. Stadt Heilbronn, Bundesgartenschau-Gelände. 2024. Available online: https://www.heilbronn.de/umwelt-mobilitaet/gruenes-heilbronn/buga-gelaende.html (accessed on 17 January 2023).
33. Amt für Wirtschaft und Stadtentwicklung (AWS)-Statistik und Stadtforschung: Zahlen in Kürze 2022. Available online: https://www.darmstadt.de/fileadmin/Bilder-Rubriken/Standort/Statistik_und_Stadtforschung/pdf/faltblaetter/ZahlenInKuerze2022_Internet.pdf (accessed on 18 November 2023).
34. Amt für Wirtschaft und Stadtentwicklung (AWS)-Statistik und Stadtforschung, Statistik und Stadtforschung. Available online: https://www.darmstadt.de/fileadmin/Bilder-Rubriken/Standort/Statistik_und_Stadtforschung/pdf/jahrbuch/04 Wirtschaft/K04-2-4.pdf (accessed on 18 November 2023).
35. Regionalverband FrankfurtRheinMain (RFRM), Metropolregion FrankfurtRheinMain, Region FrankfurtRheinMain. 2021. Available online: https://www.region-frankfurt.de/Unsere-Themen-Leistungen/Metropolregion/ (accessed on 5 September 2022).
36. Werner, P.; Enseling, A.; Koch, T.; Vaché, M.; Fahrwald, J.; Eppler, G. *Integriertes Klimaschutzkonzept für die Wissenschaftsstadt Darmstadt*; Institut für Wohnen und Umwelt (IWU): Darmstadt, Germany, 2013.
37. Umweltbundesamt (UBA). *Klimawirkungs- und Risikoanalyse 2021 für Deutschland-Kurzfassung*; UBA: Dessau-Roßlau, Germany, 2021.
38. Umweltbundesamt (UBA). *Die Risiken des Klimawandels für Deutschland-Ergebnisse der Klimawirkungs- und Risikoanalyse 2021 sowie Schlussfolgerungen der Interministeriellen Arbeitsgruppe "Anpassung an den Klimawandel"*; UBA: Dessau-Roßlau, Germany, 2022.
39. Hessisches Ministerium für Umwelt, Klimaschutz, Landwirtschaft und Verbraucherschutz (HMUKLV). *Integrierter Klimaschutzplan Hessen 2025*; HMUKLV: Wiesbaden, Germany, 2017.
40. Amt für Wirtschaft und Stadtentwicklung (AWS). *Statistik und Stadtforschung: Statistische Berichte 1/2020-Wissenschaftsstadt Darmstadt-Sonderbeitrag: Klimawandel in Darmstadt*; AWS: Darmstadt, Germany, 2020.
41. Katzschner, L.; Kupski, S. *Endbericht Klimafunktionskarte Wissenschaftsstadt Darmstadt-Gesamtstädtische Klimaanalyse mit Planungsempfehlungen und Integration der Zukünftigen Baulichen sowie Klimatischen Veränderungen*; INKEK: Lohfelden, Germany, 2016.
42. Deutscher Wetterdienst (DWD). Deutscher Wetterdienst. 2022. Available online: https://www.dwd.de/DE/Home/home_node.html (accessed on 24 October 2022).
43. DLR Rheinlandpfalz (DLRRLP). Dienstleistungszentrum Ländlicher Raum Rheinlandpfalz. Available online: https://www.wetter.rlp.de/Internet/AM/NotesAM.nsf/(Web_Stationen_Agrar)/492ce0892fa1a010c1257d710049c2e0?OpenDocument&TableRow=2.5#2 (accessed on 24 October 2022).
44. Deutsche Energie-Agentur GmbH (dena). *Der dena Gebäudereport 2016-Statistiken und Analysen zur Energieeffizienz im Gebäudebestand*; Deutsche Energie-Agentur: Berlin, Germany, 2016.
45. Bundesministerium für Wirtschaft und Energie (BMWi). *Sanierungsbedarf im Gebäudebestand-Ein Beitrag zur Energieeffizienzstrategie Gebäude*; Bundesministerium für Wirtschaft und Energie: Berlin, Germany, 2014.
46. United Nations. *Resolution adopted by the General Assembly on 6 July 2017*; United Nations: New York, NY, USA, 2017.
47. Cornel, P.; Wagner, M. *Semizentrale Ver- und Entsorgungssysteme für urbane Räume Chinas-Teilprojekt 1*; Technische Universität, Institut WAR: Darmstadt, Germany, 2006.
48. Bahr, C.; Lennerts, K. Lebens- und Nutzungsdauer von Bauteilen-Endbericht. *Aktenzeichen* **2010**, *10*, 17.
49. IP BAU. *Alterungsverhalten von Bauteilen und Unterhaltskosten-Grundlagendaten für den Unterhalt und die Erneuerung von Wohnbauten*; IP BAU: Bern, Switzerland, 1994.
50. Nützmann, G.; Moser, H. *Elemente einer analytischen Hydrologie-Prozesse-Wechselwirkungen-Modelle*; Springer: Wiesbaden, Germany, 2016.
51. Hegger, M.; Dettmar, J.; Martin, A.; Meinberg, T.; Boczek, B.; Drebes, C.; Greiner, M.; Hesse, U.; Kern, T.; Mahlke, D.; et al. *Abschlussbericht: UrbanReNet-EnEff:Stadt-Verbundprojekt Netzoptimierung-Teilprojekt: Vernetzte regenerative Energiekonzepte im Siedlungs- und Landschaftsraum*; Technical University of Darmstadt: Darmstadt, Germany, 2012.
52. Arbeitsgemeinschaft Landtechnik und Landwirtschaftliches Bauwesen in Bayern e.V. (ALB). *Bewässern nach Regeln-Wasserbedarf urbaner Grünflächen*; Beratungsblatt bef7- Ausgabe 1-5/2020; ALB Bayern e.V.: Freising, Germany, 2020.
53. Linacre, E.T. A simple formula for estimating evaporation rates in various climates, using temperature data alone. *J. Agric. Meteorol.* **1977**, *18*, 409–424. [CrossRef]

54. Feuerwehr Wiesbaden (FWW). Abteilung Vorbeugender Brandschutz, Feuerwehr Wiesbaden. Available online: https://www.wiesbaden.de/microsite/feuerwehr/medien/bindata/Ausfuehrungsbestimmungen-Feuerwehrflaechen-Wiesbaden.pdf (accessed on 5 November 2023).
55. Keith, L.; Meerow, S. *Pasreport 600 Planning for Urban Heat Resilience*; American Planning Association: Chicago, IL, USA, 2022.

water

MDPI

Review

Challenges and Innovations in Urban Drainage Systems: Sustainable Drainage Systems Focus

Anna Pia Monachese [1], María Teresa Gómez-Villarino [1,*], Jesús López-Santiago [1], Ernesto Sanz [2,3], Andrés F. Almeida-Ñauñay [2,3] and Sergio Zubelzu [1,2]

[1] School of Agricultural, Food and Biosystems Engineering, Universidad Politécnica de Madrid, 28040 Madrid, Spain; monacheseannapia@gmail.com (A.P.M.); jesus.lopez.santiago@upm.es (J.L.-S.); sergio.zubelzu@upm.es (S.Z.)

[2] Centro de Estudios e Investigación para la Gestión de Riesgos Agrarios y Medioambientales, CEIGRAM, Universidad Politécnica de Madrid, Senda del Rey, 13, 28040 Madrid, Spain; ernesto.sanz@upm.es (E.S.); af.almeida@upm.es (A.F.A.-Ñ.)

[3] Grupo de Sistemas Complejos, Universidad Politécnica de Madrid, Avda. Puerta de Hierro, n° 2-4, 28040 Madrid, Spain

* Correspondence: teresa.gomez.villarino@upm.es

Abstract: Sustainable Urban Drainage Systems (SUDS) represent a paradigm shift in stormwater management, offering holistic solutions to urban water challenges. This review examines SUDS principles, design strategies, effectiveness and barriers to implementation. SUDS prioritize infiltration and mimic natural hydrological processes to reduce flood risk, improve water quality and support ecosystems in urban environments. Effective SUDS design integrates different components such as permeable pavements, green roofs, and rain gardens, tailored to the local context. Evidence suggests that well-designed SUDS can mitigate peak flows, reduce runoff volumes, and purify water. However, barriers to widespread adoption include lack of awareness, upfront costs, and regulatory complexity. Overcoming these will require collaborative stakeholder action to prioritize education, policy support, and funding opportunities. Future research should focus on optimizing SUDS design, assessing long-term performance, and quantifying socio-economic benefits. By integrating SUDS into urban landscapes, cities can strengthen hydrological resilience, promote sustainability, and enrich urban life.

Keywords: SUDS evolution; stormwater management; urban water challenges; hydrological resilience; sponge city

1. Introduction

The rapid pace of urbanisation, coupled with the escalating impacts of climate change, has created an urgent need for transformative approaches to urban water management [1]. As cities expand and populations increase, the challenges of flooding and water pollution become more pronounced. Urbanisation contributes to the degradation of land structure through the proliferation of hardened surfaces. This phenomenon often leads to an increased incidence of urban flooding during rainfall events [2–5]. Furthermore, the exacerbation of these challenges by climate change, characterised by more frequent and intense rainfall events, underlines the critical importance of adopting resilient and sustainable stormwater management strategies.

SUDS emerged several decades ago as pioneering solutions that revolutionised traditional approaches to stormwater management and ecological harmony within urban landscapes. A pivotal moment came in 1992 with the publication of the "Scope for Control of Urban Runoff" guidelines [6], which provided technical guidance on the management of urban runoff through innovative solutions. SUDS represent a paradigm shift towards emulating and respecting the environment while addressing the multiple challenges posed by urbanisation and climate change. Looking at the historical development of urban drainage, it is clear that early engineering feats laid the foundations for modern infrastructure development [7]. However, as urbanisation accelerated in the 19th and 20th centuries, the focus shifted to grey infrastructure solutions to meet the growing demand for water, sanitation, transport, and energy services in burgeoning cities [8]. The limitations of traditional grey infrastructure have become increasingly apparent in the face of growing urban challenges and the increasing impacts of climate change. Initially referred to by various terms such as Best Management Practices (BMPs), Green Infrastructure (GI), Low Impact Development (LID), and Water Sensitive Urban Design (WSUD), among others, the concept of SUDS has evolved over time to encompass a holistic and integrated approach to urban drainage [9–11]. Scientific work demonstrates that SUDS can play a relevant role not only in managing the urban water cycle, but also in providing multiple benefits to society from different (complementary) functions such as aesthetics, environment, or amenity.

As we will show in this manuscript, an exceptional number of studies have been presented that demonstrate the effectiveness of SUDS from different perspectives, purposes, and insights. From flood protection to the provision of amenity or ecosystem services, either by installing SUDS in isolation or by integrating them into spatial planning processes, using hydrological, social, or economic tools, researchers have addressed almost every potential topic related to SUDS. Despite this enormous effort, we strongly believe that some relevant issues remain unresolved and new insights are needed to develop effective strategies for SUDS provision. Both scientific research and field implementation are progressing rapidly, but in a somewhat atomised and disjointed manner. In this context, an attempt to summarise and review the research to date in order to determine the future direction of development may help to clarify some relevant gaps.

In this manuscript, we aim to review the state of the art in order to critically present the challenges and opportunities for SUDS provision. Our main objective is to provide new insights that can help guide future efforts by critically reviewing previous work and existing initiatives.

2. Methodology

This review adopts a comprehensive approach to examine the evolution of SUDS and conventional stormwater management practices and the interplay between both dimensions as well. The methodology encompasses the following key steps:

Literature Search Strategy:

1. Identification of Relevant Keywords: Key terms related to SUDS, urban drainage, stormwater management, climate change impacts, and sustainable infrastructure were identified. In particular, we have searched for the set of concepts considered in the following list:
    (a) SUDS planning
    (b) sponge city
    (c) stormwater management
    (d) urban water challenges
    (e) hydrological resilience
2. Database Selection: Academic databases such as PubMed, Google Scholar, Web of Science, and Scopus were utilized, along with relevant government reports and institutional publications.

Inclusion and Exclusion Criteria:

1. Inclusion Criteria: Peer-reviewed journal articles, conference papers, books, reports, and other scholarly publications focusing on SUDS, conventional stormwater management, urban hydrology, and related topics were included.
2. Exclusion Criteria: Literature unrelated to the review scope or lacking empirical evidence, such as opinion pieces and non-peer-reviewed sources, was excluded. Publications not available in English were also excluded.

Synthesis and Interpretation:

1. Thematic Analysis: Extracted data were thematically categorized based on key themes and concepts identified in the literature, such as challenges of conventional stormwater management and the evolution of SUDS. The bibliometric analysis was performed using the Bibliometrix tool for R developed by [12]. We used Tidy Version [13] for efficient data management and plotting, and the igraph library [14] for co-occurrence analysis and graph generation. The database consisted of 3176 articles with the bibliometrics DB object containing relevant information such as authors, title, keywords, abstract and cited references. For further details, it is recommended to consult the documentation of the Bibliometrix package (version 4.2.3).
2. Synthesis of Findings: Data synthesized from selected literature were analyzed to identify overarching themes, relationships between variables, and knowledge gaps. Comparative analysis was conducted to contrast findings from different studies.
3. Interpretation and Conclusion: Synthesized findings were interpreted in the context of research objectives and existing theoretical frameworks. Implications for practice, policy, and future research were discussed.

Limitations:

1. Scope Limitations: The review focused primarily on English-language literature published within the past 2 decades, potentially excluding relevant studies in other languages or older publications.
2. Publication Bias: The review may be subject to publication bias, as it primarily relied on peer-reviewed literature accessible through academic databases.
3. Generalizability: Findings synthesized from selected literature may not be generalizable to all geographical regions or contexts.

3. Results

3.1. Sustainable Urban Drainage Development

An analysis of conventional sewerage systems and SUDS highlights key developments in urban water management and the challenges they pose. Conventional systems focus on the rapid removal of wastewater to prevent flooding, prioritizing volume management, often at the expense of environmental sustainability. These older systems, many of which were designed in the 19th century, are now under significant pressure from urbanization and climate change, which are increasing runoff, causing network overload, and degrading water quality.

SUDS, introduced in the late 1980s, represent a shift towards sustainable stormwater management that aims to restore natural hydrological cycles [11].

With SUDS solutions such as green roofs, permeable pavements, and rain gardens, SUDS reduce runoff volume, improve water quality, and provide additional benefits such as improved urban aesthetics, recreational space, and biodiversity. Figure 1 shows that 3176 papers containing the term "SUDS" were identified, with China and the USA leading the way. in terms of the number of publications. For a more detailed investigation, we included an analysis of the main journals publishing research on this topic (Table 1).

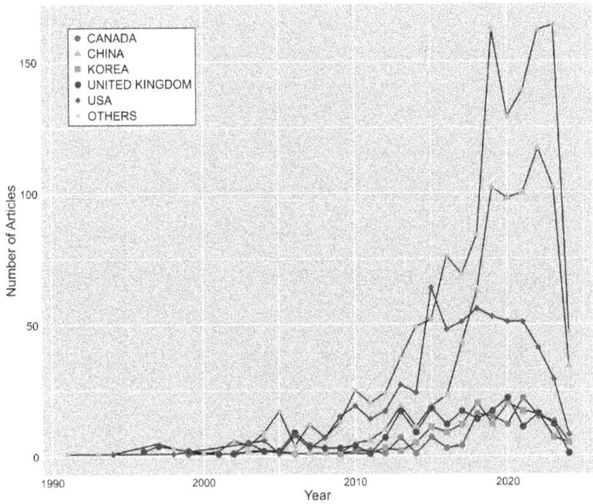

Figure 1. The review of 3176 papers containing the term "Sustainable Urban Drainage Systems" indicates that the primary contributors are China and the USA. China (761), USA (600), United Kingdom (216), Korea (141), and Canada (131) were selected as they represent the top five publishing countries on the SUDS theme over the last 30 years.

Table 1. The most important journals contributing to the topic of SUDS and the number of articles counted.

Journals	Number of Articles
WATER	277
SUSTAINABILITY	137
WATER SCIENCE AND TECHNOLOGY	132
JOURNAL OF HYDROLOGY	130
JOURNAL OF ENVIRONMENTAL MANAGEMENT	114
SCIENCE OF THE TOTAL ENVIRONMENT	107
JOURNAL OF CLEANER PRODUCTION	73
URBAN WATER JOURNAL	65
JOURNAL OF HYDROLOGIC ENGINEERING	63
DESALINATION AND WATER TREATMENT	50
WATER RESOURCES MANAGEMENT	47
LID: IT WORKS IN ALL CLIMATES AND SOILS	44
WATER RESEARCH	41
ENVIRONMENTAL SCIENCE AND POLLUTION RESEARCH	38
JOURNAL OF ENVIRONMENTAL ENGINEERING	38
JOURNAL OF IRRIGATION AND DRAINAGE ENGINEERING	38
JOURNAL OF SUSTAINABLE WATER IN THE BUILT ENVIRONMENT	38
ECOLOGICAL ENGINEERING	33
NEW TRENDS IN URBAN DRAINAGE MODELLING, UDM 2018	33
SUSTAINABLE CITIES AND SOCIETY	30

From the pioneering works in early 1990s [11], SUDS philosophy has matured. At present time, authors mostly assume SUDS provide society with not only hydrology-related goods, but also with environmental, recreational, and aesthetic services overcoming their initial role related to sustainable rainwater management (Figure 2).

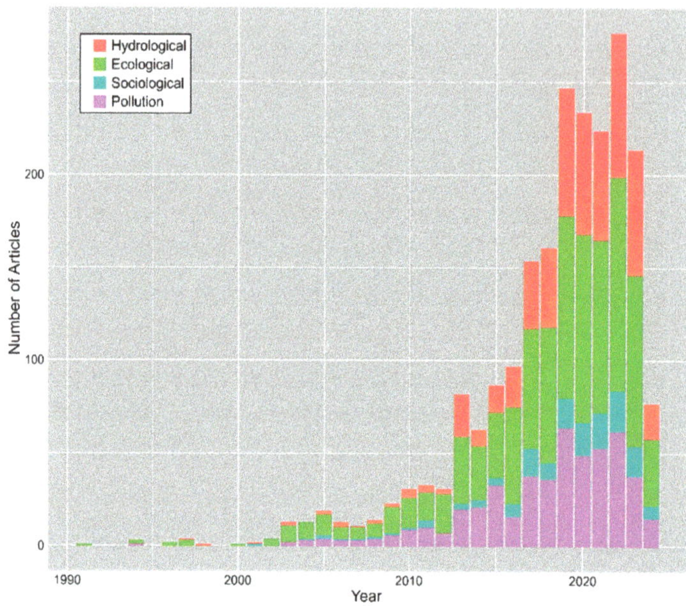

Figure 2. The main utilities of SUDS include hydrological, ecological, sociological, and health-related benefits, particularly concerning pollution management. Each category aggregate the number of articles containing the most relevant keywords for its theme, hydrological ("Flooding", Runoff", "Hydrology", "Flow", "Drainage", "Flood", "Inundation", "Hydrologic", "Hydrological", "Detention", "Peak", "Volume"), social ("Aesthetic", "Recreational", "Social"), ecological ("Environment", "Environmental", "Ecosystem", "Wildlife"), and pollution ("Pollutant", "Pollution", "Quality", "Treatment", "Purification", "Bioretention", "Removal", "Nitrogen", "Phosphorus", "Waste-water", "Nitrate", "Microplastic", "Metals").

Recently, there has been a notable increase in research focusing on the challenges of implementing SUDS, as shown in Figure 3. These challenges include regulatory barriers, financial constraints, and public perception issues. This growing body of literature highlights the urgent need to address these barriers in order to facilitate the effective integration of SUDS into urban environments, particularly in light of public concerns about water quality. In contrast to the limited adaptability of traditional sewerage systems, SUDS offer a more resilient solution when combined with existing infrastructure [15,16], a point widely recognized in the literature, as illustrated in Figure 4. In addition, the construction of SUDS is significantly less invasive than the renovation of current conventional sewer networks. This characteristic allows cities to more effectively adapt their overall stormwater management strategies, improving their ability to respond to urban challenges.

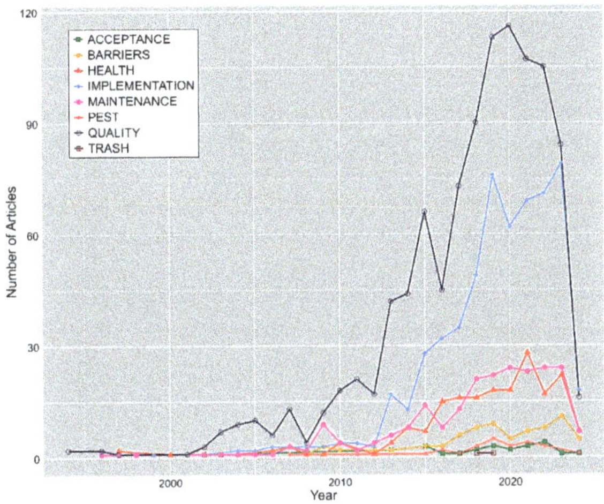

Figure 3. Several factors contribute to the barriers affecting the acceptability of SUDS. The following terms, identified as barriers to SUDS implementation, were searched in the abstracts, titles, or keywords of all related SUDS articles.

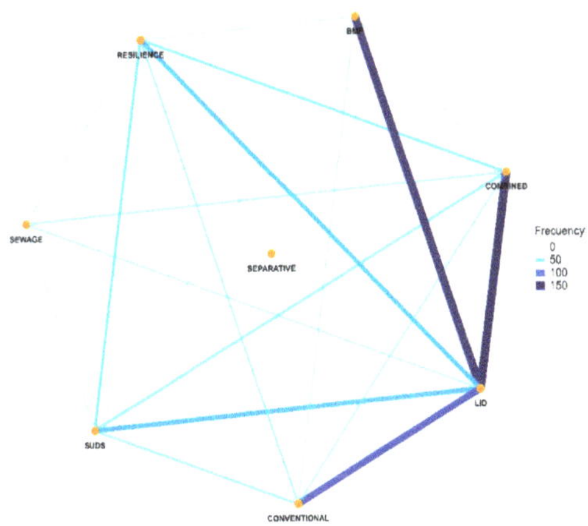

Figure 4. Keywords co-occurrence of SUDS, LID, BMP, Conventional (sewerage), Combined sewerage) and Separative (sewerage), Sewage and Resilience. Each term was checked in the abstract, title, and keywords of all SUDS articles. The main cited words are: LID and BMP and Conventional and Combined (sewerage). All pairwise combinations between variables were counted, with the width of the connecting lines and the intensity of their color represents the strength of co-occurrence between terms.

3.2. Climate Change Adaptation Through Sustainable Urban Drainage Systems

Climate change poses significant challenges to urban water management, with increased variability in precipitation and more frequent extreme weather events. SUDS offer a proactive approach to adapting to these challenges by mimicking natural hydrological processes.

SUDS are designed to effectively manage stormwater, reduce runoff, and mitigate flood risk. By incorporating features such as green roofs, permeable pavements, and rain gardens, SUDS enhance urban resilience. These systems not only capture and absorb rainwater, but also improve water quality by filtering pollutants, thereby reducing the burden on conventional drainage systems. SUDS also contribute to urban biodiversity and create green spaces that enhance the aesthetic appeal of cities. They provide recreational areas for residents and help regulate local temperatures, contributing to the overall climate resilience of cities. In summary, the integration of SUDS into urban planning is a critical component of climate change adaptation. SUDS not only allow cities to manage stormwater more effectively, but also reduce the risk of flooding, protect vital infrastructure, and improve the overall quality of life for urban residents (Figure 5). While listing all the benefits of SUDS, it is important to note that many design plans focus on projections of population growth, water demand, and land use change, assuming that future climate conditions will be similar to those of the past. This assumption leaves water systems vulnerable to the very conditions they are likely to face and for which they should be designed [17]. Therefore, stormwater engineers and decision-makers need to focus on three key aspects: understanding how current Stormwater Management Systems (SMS) Stormwater Control Measures (SCM) designs will perform under future climate conditions, determining the necessary design changes at local or regional scales to ensure resilience and effectiveness, and identifying the critical factors to consider in relation to climate change [18]. By reducing surface water runoff and promoting natural water infiltration, SUDS help to minimize the environmental impact of drainage systems and contribute to the sustainability of urban environments. Recent literature underlines a growing recognition of the multiple benefits of SUDS, highlighting their effectiveness in attenuating peak flows, controlling localized flooding, managing runoff volumes, and improving water quality. They also play a key role in restoring natural hydrological cycles and reducing the urban heat island effect, further emphasizing their importance in building resilient, climate-smart cities.

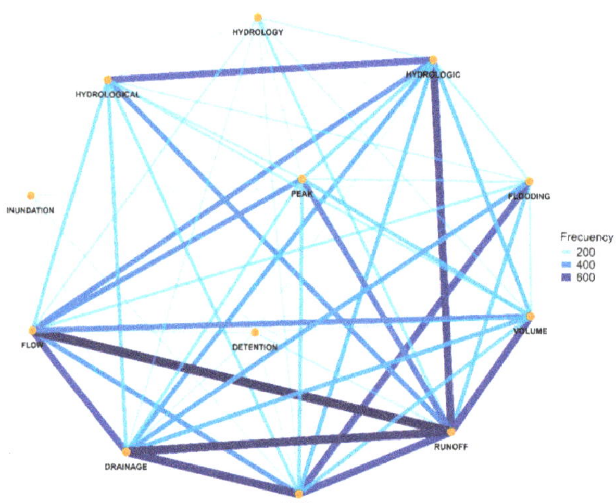

Figure 5. Co-occurrence of hydrological challenges adressed over the past years using SUDS. Each term was checked in the abstract, title, and keywords of all SUDS articles. All pairwise combinations between variables were counted, with the width of the connecting lines and the intensity of their color represents the strength of co-occurrence between terms.

SUDS are generally classified into two main groups, each of which has subcategories:

1. Infiltration-Based SUDS focus on allowing stormwater to infiltrate directly into the ground at its source. Examples include:
    - Swales: Shallow, vegetated channels designed to convey and filter stormwater, facilitating infiltration into the soil.
    - Infiltration Trenches: Excavated trenches filled with porous materials like gravel to capture and promote the infiltration of stormwater.

- Permeable Pavement: Materials used for roads, sidewalks, or parking lots that allow water to infiltrate, thereby reducing runoff and enhancing groundwater recharge.

2. Storage-Based SUDS involve the temporary storage of water, enabling its gradual release. Examples include:
    - Retention Ponds: Artificial ponds or basins designed to capture and temporarily store stormwater, thereby reducing peak flows and attenuating runoff. The stored water is released gradually to mitigate downstream flooding.
    - Detention Basins: Similar to retention ponds, detention basins temporarily hold stormwater runoff, primarily aimed at controlling peak flow rates rather than retaining water permanently.

And as previously mentioned, all SUDS technologies can function simultaneously to address hydrological, ecological, social, and health-related issues (Figure 6).

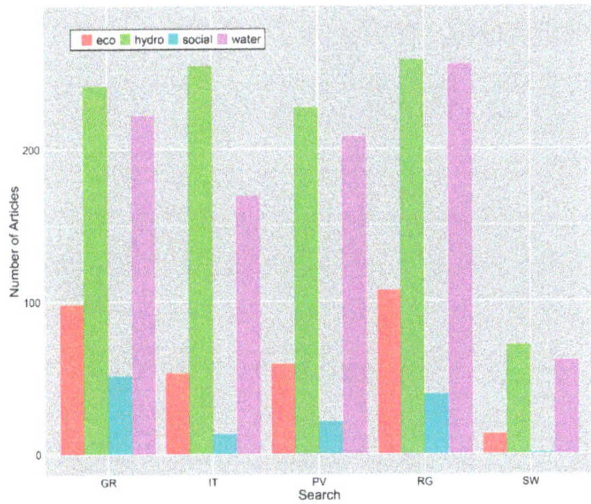

Figure 6. All SUDS technologies rules compared for each category. The green roofs (GR), rain gardens (RG), infiltration trenches (IT), permeable pavements (PV), and swales (SW) and number of manuscripts by SUDS' functions. Each category aggregate the number of articles containing the most relevant keywords for its theme, hydro ("Flooding", Runoff", "Hydrology", "Flow", "Drainage", "Flood", "Inundation", "Hydrologic", "Hydrological", "Detention", "Peak", "Volume"), social ("Aesthetic", "Recreational", "Social"), eco ("Environment", "Environmental", "Ecosystem", "Wildlife"), and water ("Pollutant", "Pollution", "Quality", "Treatment", "Purification", "Bioretention", "Removal", "Nitrogen", "Phosphorus", "Waste-water", "Nitrate", "Microplastic", "Metals").

3.3. *Water Quality and Water Reuse in SUDS*

SUDS are not only effective in managing stormwater but also play a key role in improving water quality and facilitating water reuse. By mimicking natural processes, SUDS reduce pollution in runoff by filtering pollutants such as heavy metals, sediments, and organic compounds through structures such as permeable pavements, rain gardens, and swales. These methods tackle urban pollution at its source, minimizing harmful discharges into water bodies.

SUDS also contribute to water reuse by capturing and treating rainwater and making it available for non-potable uses such as irrigation, toilet flushing, and groundwater recharge. This provides cities with sustainable water resource management options, reducing reliance on conventional systems and mitigating the effects of urbanization and climate change [19]. Originally designed to restore the natural hydrological balance of catchments, SUDS promote the local use of non-infiltrated rainfall, particularly for aquifer recharge and irrigation of green spaces. Figure 7 illustrates the increasing number of studies focusing on the potential of SUDS to enhance these processes.

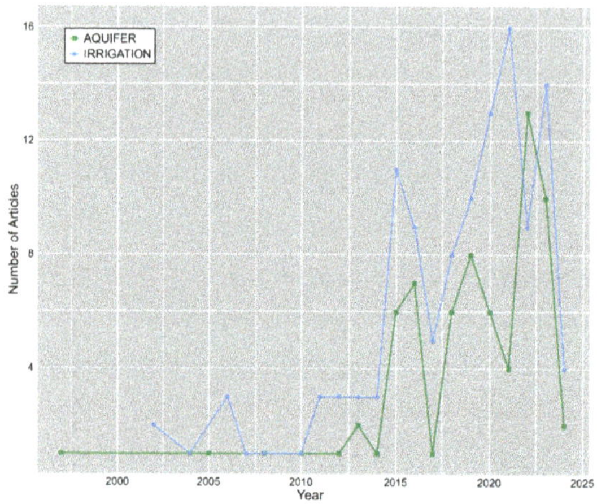

Figure 7. Number of manuscripts (temporal evolution and distribution by country) related to runoff management and SUDS. The terms related to runoff management ("Aquifer" and "Irrigation") were searched in the abstracts, titles, or keywords of all relevant SUDS articles in recent years.

To manage pollution (Figure 8), SUDS use green infrastructure to simulate natural filtration. Permeable pavements allow water to infiltrate through layers of soil, removing pollutants before they reach aquifers or rivers [20]. Vegetated areas such as swales and rain gardens also capture stormwater and promote the degradation of pollutants through plant and soil absorption [21].

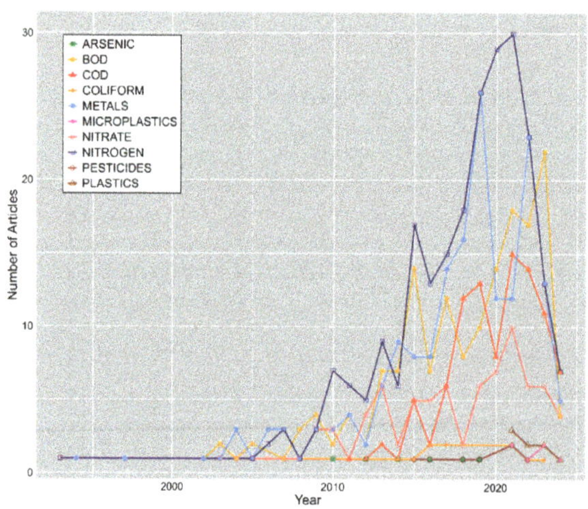

Figure 8. Temporal evolution of pollutants included in manuscripts related to SUDS. Each contaminant was checked in the abstract, title, and keywords of all SUDS articles in the database.

A major benefit of SUDS is their ability to trap heavy metals such as copper (Cu) and zinc (Zn), which are common in stormwater runoff from heavily trafficked and industrial areas. By filtering these pollutants, SUDS reduce their harmful concentrations before they reach water bodies [19].

SUDS also help to prevent pollution from combined sewer overflows. During heavy rainfall, conventional drainage systems are often overwhelmed, resulting in the release of untreated sewage into water bodies. SUDS mitigate this by capturing and storing excess rainwater, reducing the load on sewer systems and preventing such overflows [11,22].

In addition to chemical pollutants, SUDS control the movement of sediments and organic matter that can otherwise clog waterways and reduce oxygen levels in aquatic ecosystems. Through natural filtration, sedimentation, and absorption processes, SUDS improve water quality, benefiting both the environment and urban populations.

4. Discussion

4.1. Clarifying Research Gaps: The Balance Between the Traditional Sewage and the SUDS

The core design of modern water supply and wastewater treatment infrastructure dates back more than a century. Since their inception, these systems have been progressively expanded to meet the demands of growing urban populations while adapting to evolving public health and environmental standards. Characterized by long operating lives and significant upfront investments, these infrastructures exhibit a strong technological path dependency. However, recent discussions have increasingly focused on alternative approaches, with SUDS emerging as a key innovation within this evolving urban water management landscape. The challenges faced by urban drainage networks are largely due to urbanization and climate change, resulting in increased runoff exceeding sewer capacity [23,24]. These challenges include increased runoff volumes and higher peak flows, resulting in environmental damage, network overloads, river overflows, lowered groundwater levels, erosion, and habitat degradation [25,26]. Climate change is exacerbating these challenges by altering precipitation patterns and increasing the frequency of extreme weather events that stress urban water systems [27]. In addition to threatening the physical capacity of drainage networks, urban stormwater is emerging as a significant source of pollutants, negatively impacting water quality and ecosystem health [28]. A wide range of pollutants can be found in stormwater runoff, including heavy metals, xenobiotic organic compounds such as polycyclic aromatic hydrocarbons, pesticides, flame retardants, detergents, antioxidants, plasticizers, and more. These pollutants are present at levels that have the potential to adversely affect the quality of surface waters [29]. Despite the obvious ecological consequences of water pollution, effluents from conventional drainage systems affect landscape aesthetics, ecology, and beneficial uses of water, interfering with recreational activities and drinking water supplies [7,30].

The management of conventional urban drainage infrastructure is currently facing some relevant issues, as recently highlighted by [31–33]. Aging infrastructure is a major concern in many urban areas, with decades-old pipes, treatment plants, and distribution networks at risk of failure and service disruption [34].

Dealing with ageing infrastructure requires significant investment in maintenance and replacement, which places a financial burden on local authorities [34,35]. The significant cost and time required to install or upgrade these drainage networks is considerable [36], especially considering that most of them are located under fully developed, populated urban areas. As a result, changes to the conventional system are often incremental and lack the flexibility to adapt to changing circumstances [37]. Given the challenges of climate change and urbanization, simply extending the underground pipe system may not be consistent with sustainability goals, as discussed by Chocat et al. [38].

Conceived in the UK in the late 1980s [11], SUDS seek to restore the harmony of natural hydrological processes that existed before the onset of urbanization [39]. In contrast to traditional networks, green infrastructure systems such as green roofs, rain gardens, and permeable pavements are intended to help capture, absorb, and filter rainwater, thereby reducing excess runoff and associated pollution risks. Scientific interest has grown rapidly since the early 1990s, as shown in Figure 1, which reports on the number of manuscripts containing any concept related to SUDS, particularly in China and the USA, the countries where the largest number of scientific manuscripts have been published. Envisioning a new paradigm for managing stormwater in urban environments (Figure 2), the concept of sustainable stormwater management emerged decades ago, which aims to effectively manage stormwater runoff in urban areas while minimizing negative impacts on the environment, public health, and infrastructure, and promoting long-term resilience and sustainability [40].

Urban water infrastructure, such as water supply and sewerage network, includes structures designed to both provide people with potable water and to manage excess water [41,42]. As part of conventional sewerage systems, urban drainage networks (mainly) involve gravity flow from all end points over the entire drainage area to either treatment works or natural watercourses. Stormwater management in cities has traditionally relied on centralized systems designed to remove runoff as quickly as possible (Wang et al. [43]) to prevent flooding, following design criteria established in the early 1800s [44]. These systems typically prioritize water quantity management over concerns related to what we now understand as sustainable water management.

Recently, there has been a notable increase in research focusing on the challenges of implementing SUDS, as shown in Figure 3. These challenges include regulatory barriers, financial constraints, and issues of public perception and preventive maintenance [45]. This growing body of literature highlights the urgent need to address these barriers in order to facilitate the effective integration of SUDS in urban areas, particularly in light of public concerns about water quality; for example, a great potential for obtaining filtered rainwater from porous pavements has been observed, particularly for flushing toilets and lavatories in Brazil, with a filtration capacity of 70.1% and 80.0% for two different porous pavements [46].

In contrast to the limited adaptive capacity of traditional sewer systems, as highlighted in the recent study by Rodriguez et al., green infrastructure, while increasing resilience, may not fully counteract the changes expected under future scenarios [15]. Even when green infrastructure strategies are implemented to their maximum spatial extent, the study finds that they fall short of achieving 100% adaptive capacity for all indicators assessed. The highest observed capacity is only 48%, highlighting the complex challenges of improving the resilience of urban drainage systems.

Recognizing both the limitations and benefits of SUDS, they are now increasingly being designed to complement traditional drainage systems. This integrated approach improves urban water networks by improving water management and reducing the risk of flooding, while minimizing negative impacts on the surrounding environment. By strategically combining SUDS with conventional infrastructure (Figure 4), cities can create a more resilient and efficient water management system that supports sustainability goals and preserves natural ecosystems.

4.2. Integrating SUDS for Urban Flood Resilience

Urban drainage networks are critical for preventing damage from extreme weather events, especially as rainfall patterns change [47,48]. Increasingly frequent flood events are causing significant socioeconomic losses and environmental impacts worldwide [49,50]. Urban areas are at particularly high risk of flooding due to variations in land surface and changes in hydrological processes caused by human activities [51,52]. Intense rainfall events often overwhelm the capacity of these drainage systems, leading to flooding when the inflow exceeds the normal network capacity [53,54].

To mitigate such risks (Figure 5), SUDS are designed to manage runoff and protect communities and infrastructure from extreme weather events. The literature is increasingly highlighting the importance of SUDS, with many references emphasizing their role.

Typically, designers determine peak flows during rain events to size SUDS. This process involves the analysis of historical data, rainfall intensity-duration-frequency curves, and hydraulic models [55]. Many studies, including those by Viavattene & Ellis [56], Ferrans et al. [57,58], and Karami et al. [59], illustrate the application of these design criteria.

Research indicates that SUDS can meet hydrological objectives, although some studies raise concerns. SUDS also play an important role in reducing urban flooding, as shown by D'Ambrosio et al. [60] and Guptha et al. [61]. According to Almaaitah et al., SUDS help to reduce the hydraulic stress on conventional sewer systems, making them essential for adaptation to climate change [62]. For example, Qin et al. demonstrated how SUDS features such as swales, permeable pavements, and green roofs effectively reduced urban flooding in Shenzhen, China, during various rainfall events [63]. Similarly, Tao et al. found that SUDS were effective in managing peak flows during low-intensity rainfall events, although their effectiveness decreased during higher-intensity events [64].

Studies comparing grey infrastructure and SUDS, such as that by Dong et al., found that while grey infrastructure can effectively reduce overflow frequency, SUDS are more efficient at managing overflow volume [65]. Similarly, research by Macdonald et al. found that initial runoff losses were lower when SUDS were implemented compared to conventional systems [66]. However, it has been observed that while SUDS techniques have an impact on water flows, their ability to reduce water volumes during extreme events is limited and highly sensitive to local conditions, including rainfall magnitude and duration, and soil material and texture. For example, clay soils, which cover much of the northern hemisphere and were deposited by glaciers as ice-contact sediments, present particular challenges due to their low permeability. Bockhorn et al. emphasize that in such contexts, gravel-filled infiltration trenches or green spaces (such as parks and gardens) are better solutions for stormwater management under low permeability conditions [67]. However, Zhou argues that the most effective approach is to integrate SUDS with traditional drainage systems to create a synergistic design capable of addressing the complexities of extreme rainfall events, soil variability, and urban infrastructure needs [68].

For example, Raimondi et al. found that converting just 2% of paved surfaces in Bresso, Milan, to permeable pavement could reduce peak stormwater runoff by 6.5% [69]. In Cincinnati, Fu et al. found that the installation of green infrastructure alone did not completely eliminate overflows, highlighting the need for additional detention infrastructure to achieve the desired runoff reductions [70].

Studies evaluating different types of SUDS show varying effectiveness. The Puengpapat study reported that permeable pavement reduced runoff volume by 50% and peak flow by 60%, making it the most effective option [71]. Green roofs reduced both metrics by 30%, while other solutions such as bioretention cells and rain gardens showed reductions of 24% to 20%.

Some agencies have documented the implementation of SUDS in various cities, including New York, London, Copenhagen, Portland, and Sheffield [72–77]. However, the effectiveness of SUDS in relation to various conventional sewerage systems has yet to be thoroughly investigated.

4.3. The Multifaceted Role of SUDS in Urban Water Management

The role of SUDS extends well beyond flood risk management, positioning them as adaptable solutions that integrate with natural processes to tackle the multifaceted challenges of urban water management. The Committee of the Regions of the European Union emphasizes that effective water policy must incorporate three core elements—storage, containment, and drainage—each essential for mitigating peaks in the water cycle [78]. This allows for the natural discharge of surplus water while ensuring that adequate reserves are available during dry periods. Sustainable stormwater management practices, as highlighted by the EU Water Framework Directive, empower local controls to manage both the quality and quantity of runoff at its source or in close proximity [79]. This insight elevates the significance of SUDS in managing rainfall resources, advocating for a comprehensive approach that surpasses traditional methods focused solely on damage prevention. The foundational definition of SUDS makes it clear that they are pivotal in ensuring the sustainable management of hydrological systems, striving to replicate pre-development hydrology [80]. This restoration involves restoring the original water balance in terms of infiltration, runoff, and recharge to natural watercourses and aquifers (Figure 7). In particular, Zubelzu et al. is one of the few studies that has examined the full balance of water resources at the catchment scale, emphasising the importance of maintaining minimum recharge rates to aquifers alongside contributions to natural water bodies [81]. According to the original concept of SUDS articulated by Chang et al., these systems should facilitate the local use of non-infiltrated rainfall, thereby improving both aquifer recharge and water storage for green space irrigation [82]. For example, research by El Bouzidi et al. demonstrated a significant breakthrough in water conservation at the University of Settat in Morocco, where rainwater harvesting systems reduced potable water consumption by over 50%, providing a sustainable supply in the face of increasing scarcity [83]. Similarly, Archer et al. investigated rainfall infiltration and groundwater levels within a permeable SUDS pavement and found that infiltration rates were significantly higher (1925 mm/h) compared to an adjacent green area (56 mm/h) [84]. This system effectively channeled rainfall into the underlying aquifer, resulting in a rapid rise in groundwater levels within an hour of a significant rainfall event (32.8 mm/h). The reduction of pollutants carried by surface runoff has also traditionally been valued as a function of SUDS and research. As can be seen from Figure 8, scientists have mostly investigated the role of SUDS in relation to nitrogen and metals.

Kuruppu et al. investigated the effectiveness of permeable pavement systems in reducing heavy metal runoff. Rainfall intensity affects the ability of permeable pavements to retain heavy metals such as Al, Cr, Cu, Mo, and Sr, with heavy rainfall events causing Cr desorption and reducing Mo attenuation [85]. Rainfall duration also affects metal retention, particularly for Cr, Sr, and Mo. Evidence suggests that untreated heavy metals from traditional pavements may leach into nearby water sources. Physical treatment processes depend on material properties such as aggregate selection and stratification, while chemical processes involve the interaction of metals with surface materials. PPSs with specific design elements can enhance heavy metal biosorption, thereby contributing to pollutant attenuation. These findings inform local authorities in setting PPS design parameters to meet stormwater guidelines. Studies in different geographical locations by Liu et al. and Koc et al. [86,87] show significant pollutant reductions. The results of Liu et al. showed that the removal rates of GR for TSS, COD, and TN reached 31.60%, 25.10%, and 37.8%, respectively, while Koc et al. showed 41.15%, 39.73%, and 29.58%, respectively. Taken together, these findings underscore the transformative potential of SUDS in urban environments, highlighting their role not only in stormwater management, but also in advancing broader sustainability goals.

5. Conclusions

SUDS offer innovative solutions to urban water challenges and have significant strengths, as well as some limitations. SUDS effectively manage stormwater by reducing runoff volumes and attenuating peak flows, as demonstrated in studies where converting just 2% of paved areas to permeable pavement reduced peak stormwater runoff by 6.5%. Infiltration-based systems, such as swales and permeable pavements, improve groundwater recharge and have been shown to infiltrate rainfall into aquifers. In addition, SUDS play a vital role in improving water quality by trapping pollutants such as zinc and copper from stormwater runoff, thereby reducing environmental risks. The environmental and social benefits of SUDS are equally remarkable, as green infrastructure such as rain gardens not only improve urban biodiversity but also provide recreational spaces that enhance the quality of urban life. In addition, SUDS contribute to climate change adaptation by mimicking natural hydrological processes to increase resilience, as demonstrated in Shenzhen, China, where green roofs and rain gardens have effectively mitigated urban flooding during various rainfall events. SUDS also integrate seamlessly with existing infrastructure, offering less invasive and more adaptable alternatives to conventional drainage systems. However, SUDS face significant challenges. High initial costs and regulatory hurdles often prevent their widespread adoption, while a lack of regular maintenance can compromise their long-term performance. Although SUDS are effective in managing low to moderate rainfall events, their capacity diminishes during extreme weather events, as shown in studies where high-intensity rainfall significantly reduced their effectiveness. Their performance is also highly dependent on site-specific conditions such as soil permeability, with clay-rich areas requiring bespoke solutions like gravel-filled trenches. Public awareness is another critical barrier, as misconceptions about the safety and benefits of SUDS limit their adoption and integration into urban planning. Despite these challenges, SUDS represent a transformative approach to urban water management, balancing hydrological efficiency, environmental benefits, and social value. Addressing these limitations through improved policy frameworks, targeted public education, and innovative designs will unlock their full potential to create sustainable and resilient urban environments. Furthermore, integrating SUDS with traditional drainage systems offers a synergistic design capable of addressing the complexities of extreme rainfall events, soil variability, and urban infrastructure needs. By combining these approaches, cities can achieve greater resilience and sustainability in managing urban water systems.

Author Contributions: Conceptualization, S.Z., A.P.M. and M.T.G.-V.; Methodology, S.Z., A.P.M. and M.T.G.-V.; Validation, S.Z., A.P.M., J.L.-S., E.S., A.F.A.-Ñ. and M.T.G.-V.; Formal analysis, S.Z., A.P.M., J.L.-S., E.S., A.F.A.-Ñ. and M.T.G.-V.; Investigation, S.Z., A.P.M. and M.T.G.-V.; Data curation, S.Z., A.P.M., J.L.-S., E.S., A.F.A.-Ñ. and M.T.G.-V.; Writing—original draft preparation, S.Z., A.P.M. and M.T.G.-V.; Writing—review and editing, S.Z., A.P.M. and M.T.G.-V.; Visualization, S.Z., A.P.M. and M.T.G.-V.; Supervision, S.Z., A.P.M. and M.T.G.-V. All authors have read and agreed to the published version of the manuscript.

Funding: This research Project has been funded by The Comunidad de Madrid through the call Research Grants for Young Investigators from Universidad Politécnica de Madrid.

Data Availability Statement: Data are available upon request to the corresponding author.

Conflicts of Interest: The authors declare no conflicts of interest.

References

1. Srivastava, R.K. *Managing Urbanization, Climate Change and Disasters in South Asia*; Springer Nature: Singapore, 2020.
2. Wu, W. The impact of urbanization on rainfall-runoff response in a rapidly urbanizing catchment. *J. Hydrol.* **2014**, *517*, 574–584.
3. Ren, P.; Li, J.; Feng, P.; Guo, Y.; Ma, Q. Evaluation of multiple satellite precipitation products and their use in hydrological modelling over the Luanhe River basin, China. *Water* **2018**, *10*, 677. [CrossRef]
4. He, J.; Mu, Y.; Ren, C.; Wu, S. Urbanization and urban flooding: An integrated system dynamics approach. *Sustainability* **2019**, *11*, 2749.

5. Liu, D.; Jiang, W.; Mu, L.; Wang, S. Streamflow prediction using deep learning neural network: Case study of Yangtze River. *IEEE Access* **2020**, *8*, 90069–90086. [CrossRef]
6. CIRIA. *Sustainable Urban Drainage Systems—Best Practice Manual for England, Scotland, Wales and Northern Ireland*; CIRIA Report No. CR086A; CIRIA: London, UK, 2001.
7. Charlesworth, S.M. A review of the adaptation and mitigation of global climate change using sustainable drainage in cities. *J. Water Clim. Chang.* **2010**, *1*, 165–180. [CrossRef]
8. Durrans, S.R.; Burian, S.J.; Pitt, R. Enhancement of precipitation data for small storm hydrologic prediction. *J. Hydrol.* **2004**, *299*, 180–185. [CrossRef]
9. Wong, T.H. *Towards Sustainable Urban Drainage Systems: An Examination of Institutional Barriers and Opportunities in Australia*; Cooperative Research Centre for Catchment Hydrology: Canberra, Australia, 2006.
10. Benedict, M.A.; McMahon, E.T. *Green Infrastructure: Linking Landscapes and Communities*; Island Press: Washington, DC, USA, 2012.
11. Fletcher, T.D.; Shuster, W.; Hunt, W.F.; Ashley, R.; Butler, D.; Arthur, S.; Trowsdale, S.; Barraud, S.; Semadeni-Davies, A.; Bertrand-Krajewski, J.-L.; et al. SUDS, LID, BMPs, WSUD and more—The evolution and application of terminology surrounding urban drainage. *Urban Water J.* **2015**, *12*, 525–542. [CrossRef]
12. Aria, M.; Cuccurullo, C. Bibliometrix : An R-Tool for Comprehensive Science Mapping Analysis. *J. Informetr.* **2017**, *11*, 959–975. [CrossRef]
13. Wickham, H.; Averick, M.; Bryan, J.; Chang, W.; McGowan, L.; François, R.; Grolemund, G.; Hayes, A.; Henry, L.; Hester, J.; et al. Welcome to the Tidyverse. *J. Open Source Softw.* **2019**, *4*, 1686. [CrossRef]
14. Csárdi, G.; Nepusz, T.; Traag, V.; Horvát, S.; Zanini, F.; Noom, D.; Müller, K. *Igraph: Network Analysis and Visualization*, R package version 1(0.9002). 2023.
15. Rodriguez, M.; Fu, G.; Butler, D.; Yuan, Z.; Cook, L. The effect of green infrastructure on resilience performance in combined sewer systems under climate change. *J. Environ. Manag.* **2024**, *353*, 120229. [CrossRef] [PubMed]
16. Rodriguez, M.; Fu, G.; Butler, D.; Yuan, Z.; Sharma, K. Exploring the spatial impact of green infrastructure on urban drainage resilience. *Water* **2021**, *13*, 1789. [CrossRef]
17. Sowby, R.B.; Jones, D.R.; George, G.A. Policy Options to Support Climate-Conscious Urban Water Planning. *Earth* **2024**, *5*, 896–903. [CrossRef]
18. Hathaway, J.M.; Bean, E.Z.; Bernagros, J.T.; Christian, D.P.; Davani, H.; Ebrahimian, A.; Fairbaugh, C.M.; Gulliver, J.S.; McPhillips, L.E.; Palino, G.; et al. A Synthesis of Climate Change Impacts on Stormwater Management Systems: Designing for Resiliency and Future Challenges. *J. Sustain. Water Built Environ.* **2024**, *10*, 04023014. [CrossRef]
19. Chang, N.; Lu, J.; May Chui, T.F.; Hartshon, N. Global Policy of low impact 442 development for storm water management in urban regions. *Land Use Policy* **2018**, *70*, 368–383. [CrossRef]
20. Göbel, P.; Stubbe, H.; Weinert, M.; Zimmermann, J.; Fach, S.; Dierkes, C.; Kories, H.; Messer, J.; Mertsch, V.; Geiger, W.F.; et al. Near-natural stormwater management and its effects on the water budget and groundwater surface in urban areas taking account of the hydrogeological conditions. *J. Hydrol.* **2004**, *299*, 267–283. [CrossRef]
21. Ballard, B.W.; Wilson, S.; Udale-Clarke, H.; Illman, S.; Scott, T.; Ashley, R.; Kellagher, R. *The SUDS Manual*; CIRIA: London, UK, 2015.
22. Butler, D.; Digman, C.J.; Makropoulos, C.; Davies, J.W. *Urban Drainage*; CRC Press: Boca Raton, FL, USA, 2018.
23. Hobart City Council. *Water Sensitive Urban Design Site Development Guidelines and Practice Notes*; Hobart City Council: Hobart, Australia, 2006.
24. Butler, D.M.; Ranells, N.N.; Franklin, D.H.; Poore, M.H.; Green, J.T., Jr. Runoff water quality from manured riparian grasslands with contrasting drainage and simulated grazing pressure. *Agric. Ecosyst. Environ.* **2008**, *126*, 250–260. [CrossRef]
25. Marsalek, J. Overview of urban drainage impacts on aquatic habitat. In *Integrated Urban Water Resources Management*, 1st ed.; Hlavinek, P., Kukharchyk, T., Marsalek, J., Mahrikova, I., Eds.; NATO Security through Science Series; Springer: Dordrecht, The Netherlands, 2006; pp. 181–190.
26. Schiel, D.R.; Gerrity, S.; Orchard, S.; Alestra, T.; Dunmore, R.A.; Falconer, T.; Thomsen, M.S.; Tait, L.W. Cataclysmic disturbances to an intertidal ecosystem: Loss of ecological infrastructure slows recovery of biogenic habitats and diversity. *Front. Ecol. Evol.* **2021**, *9*, 767548. [CrossRef]
27. Gasperi, J.; Gromaire, M.C.; Kafi, M.; Moilleron, R.; Chebbo, G. Contributions of wastewater, runoff and sewer deposit erosion to wet weather pollutant loads in combined sewer systems. *Water Res.* **2010**, *44*, 5875–5886. [CrossRef]
28. Malaviya, P.; Sharma, R.; Sharma, P.K. Rain gardens as stormwater management tool. In *Sustainable Green Technologies for Environmental Management*; Springer: Singapore, 2019; pp. 141–166.
29. Eriksson, E.; Ledin, A.; Baun, A.; Lützhøft, H.-C.H.; Mikkelsen, P.S. Stormwater Priority Pollutants Versus Surface Water Quality Criteria. In *Vandkvalitet i Grudvand/Overfladevand—Hvordan Griber vi Det an?* ATV Jord og Grundvand: Kongens Lyngby, Denmark, 2011; pp. 33–44.

30. Thévenot, D.R.; Förster, M.; Deutsch, J.C.; Geldof, G.; Svensson, G.; Mikkelsen, P.S.; Revitt, D.M.; Aftias, E.; Metelka, T.; Sieker, H.; et al. DayWater: The challenge to develop an Adaptive Decision Support System for urban stormwater source control. In *DayWater: An Adaptive Decision Support System for Urban Stormwater Management*; IWA Publishing: London, UK, 2008; Chapter 1.
31. Smith, J.; Brown, K.; Miller, R. Water scarcity in the face of uncertainty: An analysis of UK household water consumption practices. *J. Clean. Prod.* **2018**, *197*, 202–211.
32. Green, D.; O'Donnell, E.; Johnson, M.; Slater, L.; Thorne, C.; Zheng, S.; Stirling, R.; Chan, F.K.S.; Li, L.; Boothroyd, R.J. Green infrastructure: The future of urban flood risk management? *Wiley Interdiscip. Rev. Water* **2021**, *8*, e1560. [CrossRef]
33. Lee, H.; Song, K.; Kim, G.; Chon, J. Flood-adaptive green infrastructure planning for urban resilience. *Landsc. Ecol. Eng.* **2021**, *17*, 427–437. [CrossRef]
34. Wong, T.H.; Brown, R.R. The water sensitive city: Principles for practice. *Water Sci. Technol.* **2009**, *60*, 673–682. [CrossRef]
35. Teng, J.; Jakeman, A.J.; Vaze, J.; Croke, B.F.; Dutta, D.; Kim, S.J.E.M. Flood inundation modelling: A review of methods, recent advances and uncertainty analysis. *Environ. Model. Softw.* **2017**, *90*, 201–216. [CrossRef]
36. Wilderer, P.A. Integration of the environmental engineering into the urban water management, concepts, constraints, case studies. *Water Sci. Technol.* **2004**, *49*, 11–18.
37. Sieker, H.; Helm, B.; Krebs, P.; Schlottmann, P.; Tränker, J. Flexibility—A planning criterion for stormwater management. In Proceedings of the 11th International Conference on Urban Drainage, Edingburgh, UK, 31 August–5 September 2008.
38. Chocat, B.; Deroubaix, J.F.; Lesturgeon, Y. Integrated urban drainage modelling: Historical development and recent advances. *J. Hydroinform.* **2004**, *6*, 163–182.
39. Zhu, Z.; Chen, Z.; Chen, X.; Yu, G. An assessment of the hydrologic effectiveness of low impact development (LID) practices for managing runoff with different objectives. *J. Environ. Manag.* **2019**, *231*, 504–514. [CrossRef]
40. Pitt, R.; Clark, S.E. Integrated storm-water management for watershed sustainability. *J. Irrig. Drain. Eng.* **2008**, *134*, 548–555. [CrossRef]
41. Blumensaat, F.; Wolfram, M.; Krebs, P. Sewer model development under minimum data requirements. *Environ. Earth Sci.* **2012**, *65*, 1427–1437. [CrossRef]
42. Mair, M.; Zischg, J.; Rauch, W.; Sitzenfrei, R. Where to find water pipes and sewers?—On the correlation of infrastructure networks in the urban environment. *Water* **2017**, *9*, 146. [CrossRef]
43. Wang, M.; Zhang, Y.; Zhang, D.; Zheng, Y.; Li, S.; Tan, S.K. Life-cycle cost analysis and resilience consideration for coupled grey infrastructure and low-impact development practices. *Sustain. Cities Soc.* **2021**, *75*, 103358. [CrossRef]
44. Delleur, J.W. The evolution of urban hydrology: Past, present, and future. *J. Hydraul. Eng.* **2003**, *129*, 563–573. [CrossRef]
45. DelGrosso, Z.L.; Hodges, C.C.; Dymond, R.L. Identifying key factors for implementation and maintenance of green stormwater infrastructure. *J. Sustain. Water Built Environ.* **2019**, *5*, 05019002. [CrossRef]
46. Hammes, G.; Thives, L.P.; Ghisi, E. Application of stormwater collected from porous asphalt pavements for non-potable uses in buildings. *J. Environ. Manag.* **2018**, *222*, 338–347. [CrossRef] [PubMed]
47. IPCC. Climate change 2013: The physical science basis. In *Contribution of Working Group I to the Fifth Assessment Report of the Intergovernmental Panel on Climate Change*; IPCC: Geneva, Switzerland, 2013; p. 1535.
48. UNISDR (United Nations International Strategy for Disaster Reduction). Suggested elements for the post-2015 framework for disaster risk reduction. June 2014. In Proceedings of the Third United Nations World Conference on Disaster Risk Reduction, Preparatory Committee, First session, Geneva, Switzerland, 14–15 July 2014.
49. Barredo, J.I. Normalised flood losses in Europe: 1970–2006. *Nat. Hazards Earth Syst. Sci.* **2009**, *9*, 97–104. [CrossRef]
50. Barredo, J.I.; Saurí, D.; Llasat, M.C. Assessing trends in insured losses from floods in Spain 1971–2008. *Nat. Hazards Earth Syst. Sci.* **2012**, *12*, 1723–1729. [CrossRef]
51. Sillanpää, N.; Harjanne, A.; Malve, O. Impacts of urbanisation and urban structure on flooding: A case study in Finland. *J. Hydrol.* **2015**, *525*, 294–304.
52. Guan, M.; Carrivick, J.L.; Wright, N.G.; Sleigh, P.A.; Staines, K.E. Quantifying the combined effects of multiple extreme floods on river channel geometry and on flood hazards. *J. Hydrol.* **2016**, *538*, 256–268. [CrossRef]
53. Falconer, R.H.; Cobby, D.; Smyth, P.; Astle, G.; Dent, J.; Golding, B. Rainfall Flooding: New Approaches to Alerting, Mapping, and Managing Flood Risk. *J. Flood Risk Manag.* **2009**, *2*, 198–208. [CrossRef]
54. Chen, A.S.; Ye, S.; Yang, J.; Chen, Y. Risk analysis of urban storm water flooding in China: A review. *Water* **2015**, *7*, 352–370.
55. Viessman Jr, W. Integrated water management. *J. Contemp. Water Res. Educ.* **1996**, *106*, 2.
56. Viavattene, C.; Ellis, J.B. The management of urban surface water flood risks: SUDS performance in flood reduction from extreme events. *Water Sci. Technol.* **2013**, *67*, 99–108. [CrossRef] [PubMed]
57. Ferrans, P.; Torres, M.N.; Temprano, J.; Sánchez, J.P.R. Sustainable Urban Drainage System (SUDS) modeling supporting decision-making: A systematic quantitative review. *Sci. Total Environ.* **2022**, *806*, 150447. [CrossRef]
58. Ferrans, P.; Reyes-Silva, J.D.; Krebs, P.; Temprano, J. Flood Management with SUDS: A Simulation–Optimization Framework. *Water* **2023**, *15*, 426. [CrossRef]

59. Karami, M.; Behzadian, K.; Ardeshir, A.; Hosseinzadeh, A.; Kapelan, Z. A multi-criteria risk-based approach for optimal planning of SuDS solutions in urban flood management. *Urban Water J.* **2022**, *19*, 1066–1079. [CrossRef]
60. D'Ambrosio, R.; Balbo, A.; Longobardi, A.; Rizzo, A. Re-think urban drainage following a SuDS retrofitting approach against urban flooding: A modelling investigation for an Italian case study. *Urban For. Urban Green.* **2022**, *70*, 127518. [CrossRef]
61. Guptha, G.C.; Swain, S.; Al-Ansari, N.; Taloor, A.K.; Dayal, D. Assessing the role of SuDS in resilience enhancement of urban drainage system: A case study of Gurugram City, India. *Urban Clim.* **2022**, *41*, 101075. [CrossRef]
62. Almaaitah, T.; Appleby, M.; Rosenblat, H.; Drake, J.; Joksimovic, D. The potential of Blue-Green infrastructure as a climate change adaptation strategy: A systematic literature review. *Blue-Green Syst.* **2021**, *3*, 223–248. [CrossRef]
63. Qin, Y.; Lu, J.; Du, P.; Chen, Z.; Ren, Y.; Wu, T.; Miller, J.T.; Wen, J.; Miller, D.J.; Zhang, Z.; et al. In situ fabrication of porous-carbon-supported α-MnO_2 nanorods at room temperature: Application for rechargeable Li–O_2 batteries. *Energy Environ. Sci.* **2013**, *6*, 519–531. [CrossRef]
64. Tao, X.; Kongjian, Y.; Dihua, L.; Miao, W. Assessment and Impact Factor Analysis on Stormwater Regulation and Storage Capacity of Urban Green Space in China and Abroad. *China City Plan. Rev.* **2023**, *32*, 6–16.
65. Dong, X.; Guo, H.; Zeng, S. Enhancing future resilience in urban drainage system: Green versus grey infrastructure. *Water Res.* **2017**, *124*, 280–289. [CrossRef] [PubMed]
66. Macdonald, K.; Jefferies, C. Performance and design details of SUDS. In Proceedings of the National Hydrology Seminar, Tullamore, Ireland, 11 November 2003; pp. 93–102.
67. Bockhorn, B.; Klint, K.E.S.; Jensen, M.B.; Møller, I. Use of geological mapping tools to improve the hydraulic performance of SuDS. *Water Sci. Technol.* **2015**, *71*, 1492–1499. [CrossRef]
68. Zhou, Q. A review of sustainable urban drainage systems considering the climate change and urbanization impacts. *Water* **2014**, *6*, 976–992. [CrossRef]
69. Raimondi, F.; Marchioni, M.L.; Dresti, C.; Kian, D.; Mambretti, S.; Becciu, G. Urban flood risk management: Impact of combined strategies. *Int. J. Environ. Impacts* **2021**, *4*, 219–230. [CrossRef]
70. Fu, X.; Goddard, H.; Wang, X.; Hopton, M.E. Development of a scenario-based stormwater management planning support system for reducing combined sewer overflows (CSOs). *J. Environ. Manag.* **2019**, *236*, 571–580. [CrossRef]
71. Puengpapat, P. Hydrologic Performance of Green Infrastructure Across Climate Regions. Master's Thesis, University of Illinois at Chicago, Chicago, IL, USA, 2023.
72. McPhearson, T.; Hamstead, Z.A.; Kremer, P. Urban ecosystem services for resilience planning and management in New York City. *Ambio* **2014**, *43*, 502–515. [CrossRef]
73. Fairbrass, A.J.; Jones, K.; McIntosh, A.L.S.; Yao, Z.; Malki-Epshtein, L.; Bell, S.J. *Green Infrastructure for London: A Review of the Evidence*; UCL, The Engineering Exchange: London, UK, 2018.
74. Amorim, J.H.; Engardt, M.; Johansson, C.; Ribeiro, I.; Sannebro, M. Regulating and cultural ecosystem services of urban green infrastructure in the nordic countries: A systematic review. *Int. J. Environ. Res. Public Health* **2021**, *18*, 1219. [CrossRef]
75. Suteerasan, S. Blue-Green Infrastructure on the Move: How Resilience Concepts Travel Between Cities. Master's Thesis, Kungliga Tekniska högskolan, Stockholm, Sweden, 2021.
76. Booth, C.A.; Charlesworth, S.M. An overture of sustainable surface water management. In *Sustainable Surface Water Management: A Handbook for SUDS*; John Wiley & Sons, Ltd.: Hoboken, NJ, USA, 2016; pp. 1–9.
77. Melville-Shreeve, P.; Cotterill, S.; Grant, L.; Arahuetes, A.; Stovin, V.; Farmani, R.; Butler, D. State of SuDS delivery in the United Kingdom. *Water Environ. J.* **2018**, *32*, 9–16. [CrossRef]
78. Committee of The Regions of the European Union. The role of regional and local authorities in promoting sustainable water management. In Proceedings of the 91st Plenary Session, Brussels, Belgium, 30 June–1 July 2011; Doc. CDR5-2011-FIN-AC_EN.
79. *Directive 2000/60/EC of the European Parliament and of the Council of 23 October 2000 Establishing a Framework for Community Action in the Field of Water Policy*; European Union: Brussels, Belgium, 2000.
80. Pappalardo, V.; La Rosa, D.; Campisano, A.; La Greca, P. The potential of green infrastructure application in urban runoff control for land use planning: A preliminary evaluation from a southern Italy case study. *Ecosyst. Serv.* **2017**, *26*, 345–354. [CrossRef]
81. Zubelzu, S.; Rodríguez-Sinobas, L.; Sordo-Ward, A.; Pérez-Durán, A.; Cisneros-Almazán, R. Multi-objective approach for determining optimal sustainable urban drainage systems combination at city scale. The case of San Luis Potosí (México). *Water* **2020**, *12*, 835. [CrossRef]
82. Chang, C.H.; Huang, C.S.; Yeh, H.D. Analysis of three-dimensional unsaturated–saturated flow induced by localized recharge in unconfined aquifers. *Hydrol. Earth Syst. Sci.* **2018**, *22*, 3951–3963. [CrossRef]
83. El Bouzidi, A.; Anouar, A.; Bouzziri, M. Management and valuation of rainwater by alternative techniques, case of the university of Settat, in Morocco. *Water Cycle* **2024**, *5*, 109–120. [CrossRef]
84. Archer, N.A.; Bell, R.A.; Butcher, A.S.; Bricker, S.H. Infiltration efficiency and subsurface water processes of a sustainable drainage system and consequences to flood management. *J. Flood Risk Manag.* **2020**, *13*, e12629. [CrossRef]

85. Kuruppu, U.; Rahman, A. Evaluation of Permeable Pavement Systems for Removing Heavy Metals from Stormwater. *Water* **2023**, *15*, 1573. [CrossRef]
86. Liu, Y.; Bralts, V.F.; Engel, B.A. Evaluating the effectiveness of management practices on hydrology and water quality at watershed scale with a rainfall-runoff model. *Sci. Total Environ.* **2015**, *511*, 298–308. [CrossRef]
87. Koc, K.; Ekmekcioğlu, Ö.; Özger, M. An integrated framework for the comprehensive evaluation of low impact development strategies. *J. Environ. Manag.* **2021**, *294*, 113023. [CrossRef] [PubMed]

water

MDPI

Article

Designing a High-Performance Oil–Water Filtration System: Surface-Enhanced Quartz with Hydrophilic Nanoparticles for Sustainable Water Reuse and Global Water Scarcity Solutions

Nthabiseng Ramanamane * and Mothibeli Pita

Department of Mechanical Engineering, Bioresources, and Biomedical Engineering, College of Science, Engineering and Technology, University of South Africa, Florida 1710, South Africa; pitam@unisa.ac.za
* Correspondence: ramannj@unisa.ac.za

Abstract: The increasing demand for freshwater resources, coupled with industrial pollution, necessitates improved water treatment technologies. This study investigates the potential of quartz-based filtration systems enhanced with hydrophilic nanoparticles for efficient oil-water separation. The quartz material, abundant and cost-effective, was processed and modified through sequential coatings to enhance its hydrophilicity and separation efficiency. Comprehensive characterization techniques, including SEM, XRD, and Raman spectroscopy, were employed to evaluate surface morphology, chemical composition, and structural integrity at different stages of coating. The findings demonstrated that the first coating achieved the most uniform nanoparticle distribution, significantly improving hydrophilicity and separation efficiency, reducing oil content in filtrates to 17.3 mg/L. Subsequent coatings resulted in agglomeration and pore clogging, leading to diminished performance. Validation through mathematical models corroborated experimental observations, confirming the first coating's superior balance of nanoparticle integration, permeability, and separation efficiency. This research highlights the potential of surface-engineered quartz as a scalable, cost-effective solution for sustainable water reuse. Future work will focus on optimizing coating techniques, scaling up, and integrating the system with complementary technologies to enhance water treatment processes.

Keywords: oil–water separation; hydrophilic nanoparticles; quartz filtration system; water reuse; membrane technology; surface modification

1. Introduction

The availability of clean water is increasingly becoming a global concern, with growing demand driven by population expansion, industrialization, and climate change [1]. A significant challenge in maintaining water quality is the contamination of freshwater resources with oily wastewater, which originates from petrochemical industries, manufacturing processes, and transportation sectors [2,3]. The presence of oil in water disrupts natural ecosystems, reduces water reuse potential, and complicates wastewater treatment efforts, necessitating the development of effective oil–water separation technologies [4].

Traditional oil–water separation methods, including gravity separation, skimming, and chemical coagulation, often face limitations in handling stable emulsions, achieving high separation efficiency, and ensuring cost-effectiveness in large-scale applications [5,6]. Membrane-based filtration systems, particularly those utilizing polymeric and ceramic membranes, offer enhanced separation efficiency but are often constrained by high fabrication costs, susceptibility to fouling, and energy-intensive operation [7,8]. As a result,

https://doi.org/10.3390/w17040501

research efforts are increasingly focused on developing alternative filtration materials that are both efficient and economically viable.

Quartz, a naturally abundant and mechanically robust material, has emerged as a potential low-cost alternative for oil–water separation due to its chemical stability and ease of processing [9,10]. However, its inherent hydrophobic nature limits its effectiveness in water purification applications. To enhance its separation capability, surface modification techniques have been explored to improve hydrophilicity and oil repellency, thereby increasing the efficiency of quartz-based filtration systems.

A critical component of designing advanced filtration systems is the integration of mathematical modeling. Mathematical models provide a systematic framework for understanding and optimizing key parameters, including the distribution of nanoparticles, the impact of surface modifications on filtration performance, and the interaction dynamics between oil droplets and the filtration medium. These models enable researchers to simulate operational conditions, predict performance outcomes, and refine system designs before implementation. Incorporating modeling into the development process ensures that the system is not only theoretically sound but also practically viable for real-world applications.

This study centers on the development of a high-performance oil–water filtration system utilizing quartz particles enhanced with hydrophilic nanoparticles. The research aims to balance efficiency, scalability, and environmental sustainability, positioning the system as a practical solution to combat global water scarcity. By incorporating advanced surface engineering techniques alongside mathematical modeling, the study offers a comprehensive approach to enhancing oil–water separation technologies. The outcomes align with the broader objective of fostering water reuse and advancing sustainable resource management in the face of growing global water challenges.

2. Materials and Methods

This study utilized dry-graded quartz particles (SiO_2: 98%, Fe_2O_3: 0.18%, particle size: 0.8–1.8 mm) as the primary filtration medium for oil–water separation. Quartz was selected due to its abundance, chemical stability, and cost-effectiveness, particularly in South Africa, where it is readily available and suitable for water treatment applications. The dry-graded quartz particles used on this study was obtained from Sallies Silica Mine, Brits, South Africa. To ensure the quartz particles met the required specifications, they underwent a structured preparation process involving several stages. Initially, the quartz was mined and transported to the processing facility. It was then subjected to size reduction through a two-stage crushing process, where a jaw crusher was used for primary crushing, followed by further refinement using a cone crusher to achieve the desired particle size. The crushed quartz was then washed with purified water to remove surface contaminants, dust, and residual impurities. After washing, the quartz was air-dried for three days under controlled conditions before undergoing a heating process to eliminate residual moisture, enhancing its stability for further modification.

To improve the filtration performance, the quartz surface was modified using hydrophilic nanoparticles. The incorporation of nanoparticles was aimed at increasing the hydrophilicity and oil-repelling properties of the filtration medium. A high-pressure spray coating technique was used for nanoparticle deposition, ensuring a controlled and uniform surface modification process. The quartz samples were divided into six groups: raw quartz, washed quartz, quartz with one nanoparticle coating, quartz with two coatings, quartz with three coatings, and quartz with four coatings. The coating process was standardized by maintaining a fixed nozzle-to-quartz distance of 5 cm, a 90-degree spray angle, and precise timing and pressure settings to ensure uniform nanoparticle dispersion. Each coated sample was air-dried at room temperature (approximately 25 °C) for 48 h to allow for nanoparticle stabilization and adhesion.

To assess the effectiveness of the surface modification, the coated quartz samples underwent detailed characterization using multiple analytical techniques. Scanning Electron Microscopy (SEM) was employed to examine the nanoparticle size, morphology, and distribution on the quartz surface. The analysis confirmed that the nanoparticles ranged between 23 and 78 nm, ensuring stable and well-dispersed surface modification. Raman Spectroscopy and X-Ray Diffraction (XRD) were used to verify the chemical composition and structural stability of the coated quartz, ensuring that the modification process did not alter the material's fundamental integrity. Additionally, mathematical modeling was applied to evaluate the coating uniformity and nanoparticle distribution patterns, optimizing the surface modification process for enhanced filtration performance.

2.1. Performance Evaluation for Oil–Water Separation

Figure 1 illustrates the custom-designed test rig developed to evaluate the oil–water separation performance of quartz particles. This system was engineered as a cost-effective and scalable solution, utilizing filtration beds as the primary separation mechanism. To enhance efficiency, integrated cartridges were incorporated into the design, facilitating multi-stage filtration to ensure the production of high-quality permeate. The resulting treated water meets the necessary standards for reuse, making this system a promising approach to addressing global water scarcity and sustainability challenges.

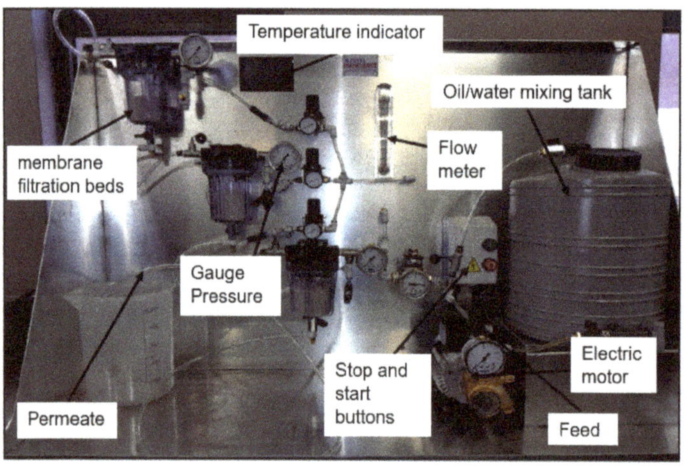

Figure 1. Illustrates the filtration system used for oil and water separation.

Filtration experiments were conducted using a three-stage integrated cartridge system, developed to simulate real-world filtration conditions. The experimental setup was assessed at varying flow rates (50–200 mL/min), cross-flow velocities (0.1–0.5 m/s), and pressure differentials (0.05–0.35 bar) to examine the influence of nanoparticle coatings on permeability and separation efficiency.

For the first nanoparticle coating, the highest flux was recorded at 75.3 L/m^2·h, demonstrating optimal permeability and efficient oil rejection. However, as additional coatings were applied, hydraulic resistance increased, leading to a decline in flux to 12.8 L/m^2·h by the fourth coating. This suggests that excessive nanoparticle deposition can impede water transport due to pore blockage and agglomeration.

The first coating also significantly enhanced hydrophilicity, reducing the contact angle of quartz surfaces and promoting stronger water affinity while simultaneously reducing oil adhesion. This modification resulted in an 83.7% oil removal efficiency within the first 30 min of filtration. However, with increased nanoparticle layering, oil detachment was hindered, reducing efficiency to 47.2% by the fourth coating.

Overall, the highest separation efficiency was achieved with a single nanoparticle coating, whereas additional layers led to clogging and reduced performance due to excessive deposition. These findings emphasize the need for precise nanoparticle integration to maintain an optimal balance between permeability, separation efficiency, and long-term operational stability in oil–water filtration systems.

2.1.1. Comparative Analysis of Quartz-Based Filtration with Conventional Technologies

To contextualize the performance of the quartz-based filtration system, it is essential to compare its efficiency, cost-effectiveness, and operational feasibility against established oil–water separation technologies. Various filtration methods, including polymeric membranes, ceramic membranes, activated carbon adsorption, and electrocoagulation, have been widely used for oil–water separation. Each technology presents distinct advantages and limitations based on separation efficiency, fouling resistance, operational complexity, and cost sustainability.

Comparison of Filtration Technologies and Advantages of Quartz-Based Filtration

As presented in Table 1, quartz-based filtration system developed in this study offers several advantages over conventional methods:

Table 1. Comparison of Filtration Technologies.

Filtration Technology	Separation Efficiency (%)	Flux (L/m^2·h)	Fouling Resistance	Energy Consumption	Cost Considerations	References
Polymeric Membranes	85–99% (varies with material)	10–120	Moderate (prone to fouling)	High (pressure-driven)	High (frequent replacement)	[11,12]
Ceramic Membranes	90–99%	50–300	High (resistant to fouling)	Moderate	Very High (fabrication costs)	[13,14]
Activated Carbon Adsorption	60–90%	N/A	Moderate (clogging issues)	Low	Medium (replacement needed)	[15,16]
Electrocoagulation	75–95%	N/A	High (self-cleaning ability)	High	Medium to High	[17,18]
Quartz-Based System (This Study)	83.7% (first coating)	75.3 (first coating)	High (hydrophilic nanoparticle surface)	Low (operates at 0.05–0.35 bar)	Low (cost-effective quartz material)	Current study

High Permeability and Moderate Oil Removal Efficiency: The system achieves a flux of 75.3 L/m^2·h, surpassing most polymeric membranes while maintaining a comparable oil removal efficiency.

Lower Energy Requirements: Unlike pressure-driven polymeric and ceramic membranes, this system operates at low pressure (0.05–0.35 bar), significantly reducing energy consumption and making it more suitable for off-grid applications.

Superior Fouling Resistance: The hydrophilic nanoparticle-modified quartz surface resists oil adhesion, reducing membrane clogging and prolonging operational lifespan—an issue commonly encountered in polymeric membranes.

Cost-Effective and Scalable: Unlike ceramic membranes, which involve high fabrication costs, quartz is an abundant, low-cost material that can be easily processed for large-scale applications. Overall, the quartz-based filtration system presents a cost-effective, energy-efficient, and scalable alternative for oil–water separation, particularly in resource-limited settings. Compared to conventional polymeric membranes, it offers higher permeability and lower operational costs, while its fouling resistance and low-energy operation make it a viable choice for sustainable water reuse applications.

2.1.2. Scalability and Practical Implementation of the Quartz-Based Filtration System

Table 2 highlights the key scalability factors that determine the real-world applicability of filtration technologies, particularly in industrial wastewater treatment, oil spill remediation, and sustainable water reuse initiatives. The successful deployment of any filtration system at a large scale depends on its material availability, fabrication complexity, operational efficiency, energy consumption, maintenance requirements, cost implications, environmental impact, and industrial adaptability.

Table 2. Scalability Assessment of Quartz-Based Filtration System.

Scalability Factor	Key Considerations
Material Availability	Quartz is naturally abundant and cost-effective, making large-scale implementation feasible.
Fabrication Complexity	Simple mechanical processing and nanoparticle coating allow for scalable manufacturing.
Operational Efficiency	Modular design enables easy expansion for high-throughput applications.
Energy Consumption	Operates at low pressure (0.05–0.35 bar), significantly reducing energy demands compared to polymeric membranes.
Maintenance Requirements	Exhibits low fouling rates, requiring minimal maintenance and longer operational lifespan.
Cost Considerations	Lower initial and operational costs than polymeric and ceramic membranes; potential for oil recovery.
Environmental Impact	Supports sustainable water reuse, reduces secondary pollution, and minimizes waste generation.
Industrial Application Potential	Suitable for wastewater treatment plants, petroleum industries, and decentralized water purification systems.

The quartz-based filtration system developed in this study offers high separation efficiency and low operational costs, making it a promising candidate for large-scale implementation. However, its scalability must be critically assessed in terms of resource availability, manufacturing feasibility, energy efficiency, and long-term economic sustainability to ensure its viability in commercial and industrial applications. The insights provided in Table 2 establish a structured evaluation of these factors, positioning the quartz-based system within the broader landscape of filtration technologies.

2.2. *Development of Mathematical Models for Oil–Water Filtration Systems*

To design advanced filtration systems that integrate surface-enhanced quartz with hydrophilic nanoparticles, mathematical modeling is a critical tool. Below are developed mathematical models that address key parameters in the filtration process.

Equation (1) represents the Nanoparticle Distribution Model. This model captures the uniformity and density of hydrophilic nanoparticles on the quartz surface, optimizing their coverage for maximum water affinity, where $D_{np}(x,y)$ is nanoparticle density at coordinates (x,y), A is the total quartz surface area, N is the number of nanoparticles distributed, (x_i, y_i) is the coordinates of individual nanoparticles, and σ is the standard deviation controlling nanoparticle spread:

$$D_{np}(x,y) = \frac{1}{A}\sum_{i=1}^{N}\left[exp\left(\frac{(x-x_i)^2+(y-y_i)^2}{2\sigma^2}\right)\right]. \quad (1)$$

Equation (2) represents the Surface Modification Efficiency Model. This model quantifies the improvement in surface wettability due to hydrophilic nanoparticle coatings, where (E_s) is the surface modification efficiency, $\Delta\theta$ is the reduction in contact angle $(\theta_r - \theta_c)$, θ_r is initial contact angle of raw quartz, θ_c is the contact angle after coating, D_{vac} being the void area coefficient (uncoated regions), and D_{max} is the maximum area available for coating:

$$E_s = \frac{\Delta\theta}{\theta_r} \times \left(1 - \frac{D_{vac}}{D_{max}}\right). \quad (2)$$

Equation (3) indicates the Oil–water Interaction Dynamics Model. This model examines the interaction between oil droplets and the filtration surface, focusing on the adhesion force and removal efficiency, where F_{adh} is the adhesion force between oil droplet and surface, γ_{ow} is the interfacial tension between oil and water, θ_c being the contact angle of the modified quartz surface, k_e is the empirical constant for surface-oil interaction, d is the distance between oil droplet and surface, and β being the decay rate of interaction force with distance:

$$F_{adh} = \gamma_{ow} cos(\theta_c) - \frac{k_e}{d} \cdot exp(-\beta d). \tag{3}$$

Equation (4) represents the Permeability and Flux Model. This model calculates the water flux through the filtration system and evaluates the permeability of the quartz-based medium, where J_w, is the water flux, ΔP being the pressure drop across the medium, k is the permeability constant of the quartz-nanoparticle composite, μ is the viscosity of water, and L being the thickness of the filtration medium:

$$J_w = \frac{\Delta P \cdot k}{\mu \cdot L}. \tag{4}$$

Equation (5) represents the Separation Efficiency Model. This model evaluates the overall efficiency of the system in separating oil from water, where η_s is the separation efficiency, C_{in} is the oil concentration in the feed water, and C_{out} being the oil concentration in the permeate water:

$$\eta_s = \frac{(C_{in} - C_{out})}{C_{in}} \times 100. \tag{5}$$

The proposed mathematical models are interlinked, forming a comprehensive framework for the design and optimization of the filtration system. The models allow for precise customization, ensuring that the system can be tailored to diverse water contamination scenarios. This adaptability makes the filtration system a valuable tool for addressing global water scarcity challenges, particularly in regions where water reuse is critical for sustaining communities and industries.

3. Results

The experimental outcomes represent the culmination of detailed characterization and rigorous performance testing to evaluate the effectiveness of the developed filtration system. A suite of advanced analytical methods and practical assessments was applied to gain a thorough understanding of the material's properties and separation performance. Scanning Electron Microscopy (SEM) was utilized to analyze the surface morphology of the quartz samples, both in their untreated form and after modification with hydrophilic nanoparticles. Raman spectroscopy was employed to identify chemical bonds, and functional groups present on the quartz surface, while X-Ray Diffraction (XRD) provided critical insights into the crystalline structure of the material before and after nanoparticle integration. Additionally, oil and grease analysis quantified the residual oil content in the permeate, offering a precise measure of the system's separation efficiency.

3.1. Scanning Electron Microscopy (SEM) Quartz Surface Analysis

The following SEM images provide a detailed representation of the quartz surface at various stages of preparation and coating. The sequence begins with the untreated raw quartz surface, showcasing its natural morphology. Next, the surface is displayed after being thoroughly washed with a descaling chemical to remove impurities and enhance its cleanliness. The progression continues with the application of the first nanoparticle coating, revealing the initial layer of modification. Subsequent images capture the quartz surface after the addition of the second, third, and fourth coatings, illustrating the incremental buildup of material layers and changes in surface structure as each coating is applied.

The Scanning Electron Microscopy (SEM) image of raw quartz in Figure 2 reveals a heterogeneous surface morphology, characterized by irregularities, sharp edges, micro-cracks, and layered formations. The presence of these features suggests that the quartz has undergone mechanical stress, likely during the mining and crushing process [19]. The rough surface lacks significant porosity, which may influence its interaction with water and oil droplets during filtration. Additionally, the layered structure observed in the image is characteristic of silica-rich minerals, which can play a role in nanoparticle adhesion when the quartz is subjected to surface modification [20]. The high surface roughness could also impact oil–water separation efficiency, affecting how water permeates through the system.

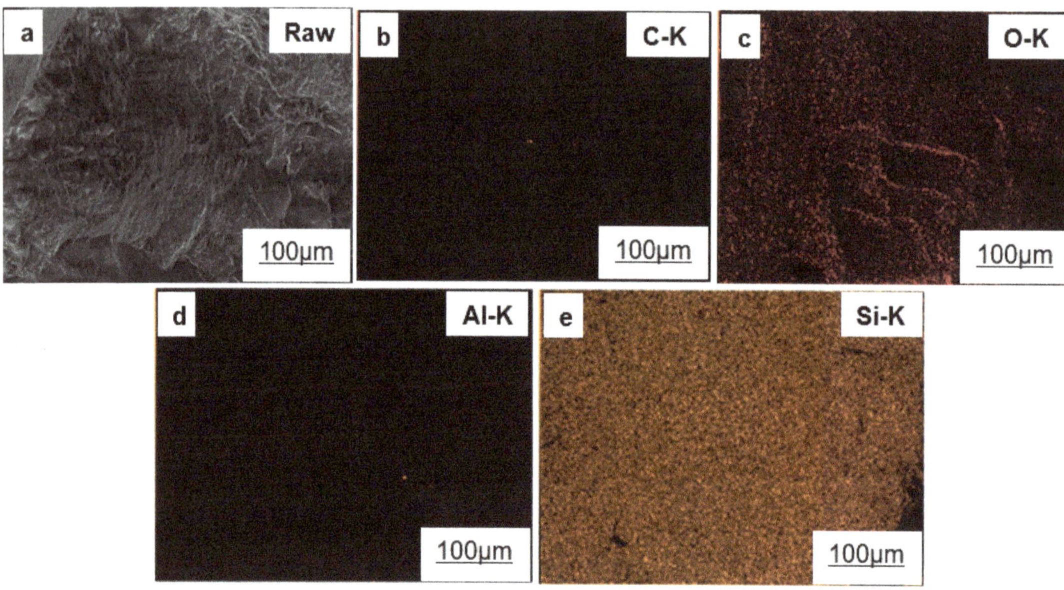

Figure 2. SEM image of raw quartz illustrating the distribution of nanoparticles. (**a**) SEM Image of Raw Quartz reveals a rough and irregular texture, characteristic of natural quartz, which influences adsorption properties and membrane performance. (**b**) Carbon (C-K) Distribution indicates negligible carbon content, confirming minimal organic contamination, which is beneficial for reducing fouling in membrane applications. (**c**) Oxygen (O-K) Distribution shows a uniform presence of oxygen, consistent with SiO_2 (silicon dioxide), the primary component of quartz, ensuring structural stability. (**d**) Aluminium (Al-K) Distribution detects trace amounts of aluminium impurities, which may contribute to mechanical reinforcement but remain a minor component. (**e**) Silicon (Si-K) Distribution confirms the predominant presence of SiO_{25}, with high intensity and uniform dispersion, indicating chemical stability and making the quartz suitable for membrane fabrication.

The elemental mapping analysis provides insights into the chemical composition and distribution of key elements present in the raw quartz. The carbon (C-K) mapping indicates a low presence of carbonaceous material, suggesting that the quartz is relatively free from organic contaminants or hydrocarbon deposits. This is beneficial for filtration applications, as it minimizes unwanted fouling. The oxygen (O-K) mapping shows a high and uniform distribution, which aligns with the expected silicon dioxide (SiO_2) composition of quartz. The strong oxygen signal confirms the oxidized state of the material, indicating no significant elemental impurities affecting its chemical stability. Additionally, aluminum (Al-K) mapping reveals minor traces of alumina (Al_2O_3), which are common in natural quartz deposits. The presence of aluminum could slightly influence the hydrophilicity of the quartz, as alumina-based compounds tend to alter surface wettability. Finally, the silicon (Si-K) mapping demonstrates a dominant and homogenous presence of SiO_2, verifying that the quartz is structurally stable and chemically pure, with no noticeable non-silica-based inclusions.

From a filtration perspective, the observed morphological and chemical properties have several implications. The rough and uneven surface of the raw quartz provides an advantageous platform for nanoparticle adhesion, which is crucial for hydrophilic modification in oil–water separation systems [21]. However, excessive surface roughness could lead to non-uniform coating distribution, necessitating precise nanoparticle deposition techniques. The low carbon presence ensures that the quartz is free from organic fouling, reducing the likelihood of oil absorption during filtration. Additionally, the layered microstructure suggests that the material is mechanically durable, although prolonged high-pressure flow conditions could contribute to surface erosion over extended operational cycles [22]. The minor presence of alumina (Al_2O_3) may introduce surface charge variations, which could influence the hydrophilicity and water permeability of the quartz material.

The scanning electron microscopy (SEM) image of washed quartz in Figure 3 presents a smoother and more refined surface morphology compared to raw quartz, indicating the removal of surface impurities, residual debris, and fine dust through the washing process. The surface appears to have undergone structural rearrangement, exhibiting reduced microfractures and fewer surface irregularities. However, while the roughness has diminished, some surface undulations and minor imperfections remain, which may influence nanoparticle adhesion and coating uniformity in later modification steps [23]. The reduction in surface irregularities suggests that the washing procedure effectively eliminated weakly bound particles, thereby improving the structural stability of the quartz material. This enhanced smoothness is expected to facilitate more uniform hydrophilic nanoparticle coating, which is critical for optimizing oil–water separation performance in filtration applications.

Elemental mapping analysis through energy-dispersive X-ray spectroscopy (EDS) reveals key chemical composition characteristics of the washed quartz. The carbon (C-K) signal appears more prominent than in raw quartz, possibly due to adsorption of trace organic residues from the washing solution. However, the carbon presence remains minimal, indicating that the quartz surface is still largely free from hydrocarbon contamination, which is beneficial for maintaining high water affinity. The oxygen (O-K) mapping displays an intensified and homogeneously distributed signal, consistent with the silicon dioxide (SiO_2) composition of quartz. This suggests that surface silica groups are more exposed after washing, enhancing reactivity and potential bonding sites for further nanoparticle deposition.

The presence of aluminium (Al-K) remains detectable, though its distribution is relatively sparse, indicating that alumina (Al_2O_3) impurities persist despite washing. These residual aluminium traces could subtly influence surface charge properties, potentially affecting water permeability and oil repellence. Meanwhile, the silicon (Si-K) signal exhibits a well-defined, structured distribution, confirming a high-purity quartz framework with no significant phase separations or structural degradation. The phosphorus (P-K) mapping introduces a new element that was absent in raw quartz, likely resulting from descaling agents or washing additives. The phosphorus appears to be concentrated in localized regions, suggesting possible surface interactions with quartz that may influence its hydrophilicity and long-term chemical stability.

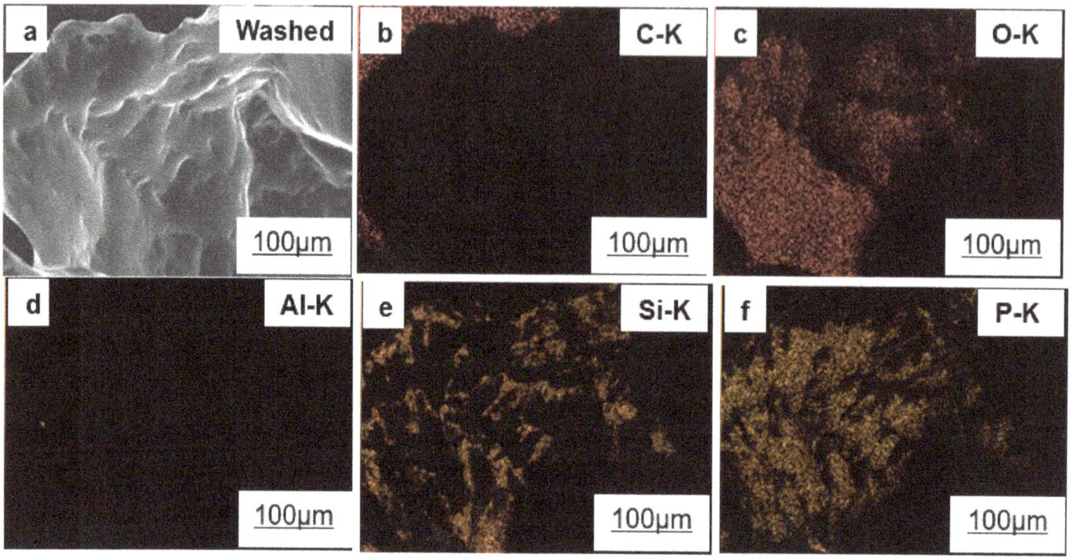

Figure 3. SEM image of washed quartz illustrating the distribution of nanoparticles. (**a**) SEM Image of Washed Quartz shows a smoother surface morphology compared to the raw quartz, indicating the removal of impurities and surface irregularities, which may enhance wettability and membrane performance. (**b**) Carbon (C-K) Distribution reveals an increased presence of carbonaceous deposits, possibly due to environmental exposure or residual organic contaminants from the washing process. (**c**) Oxygen (O-K) Distribution demonstrates a well-distributed oxygen presence, confirming the dominance of SiO_2 (silicon dioxide) while ensuring structural stability. (**d**) Aluminum (Al-K) Distribution shows minor traces of aluminum, suggesting that the washing process did not significantly alter its presence, meaning it remains as an inherent impurity in the quartz matrix. (**e**) Silicon (Si-K) Distribution highlights a higher intensity of silicon, indicating better exposure of the quartz framework, which can influence hydrophilicity and surface reactivity. (**f**) Phosphorus (P-K) Distribution confirms the presence of phosphorus, which plays a role in enhancing the hydrophilic properties of the quartz, making it more suitable for oil-water separation applications.

The changes observed in surface morphology and elemental composition have significant implications for the filtration performance of the quartz material [24]. The smoother and cleaner surface provides a more uniform substrate for nanoparticle adhesion, ensuring consistent hydrophilic modifications that are essential for oil–water separation applications. However, the persistence of aluminium impurities may introduce variations in wettability, while the presence of phosphorus suggests potential secondary surface interactions that require further investigation. The washing process appears to have successfully enhanced the surface purity of quartz, but minor structural imperfections and elemental variations may still influence filtration efficiency and long-term stability.

The scanning electron microscopy (SEM) image of the first-coated quartz in Figure 4 demonstrates a well-distributed nanoparticle layer, making it the optimal coating stage for filtration applications. The surface appears significantly smoother and more structured than uncoated quartz, indicating that the nanoparticles have adhered effectively without excessive agglomeration. This uniform distribution enhances the hydrophilicity of the surface, ensuring efficient water permeability while repelling oil droplets, which is a critical factor in oil–water separation systems. The moderate surface roughness observed in the micrograph promotes enhanced interaction with water molecules, preventing oil accumulation and fouling, a common limitation in traditional filtration systems.

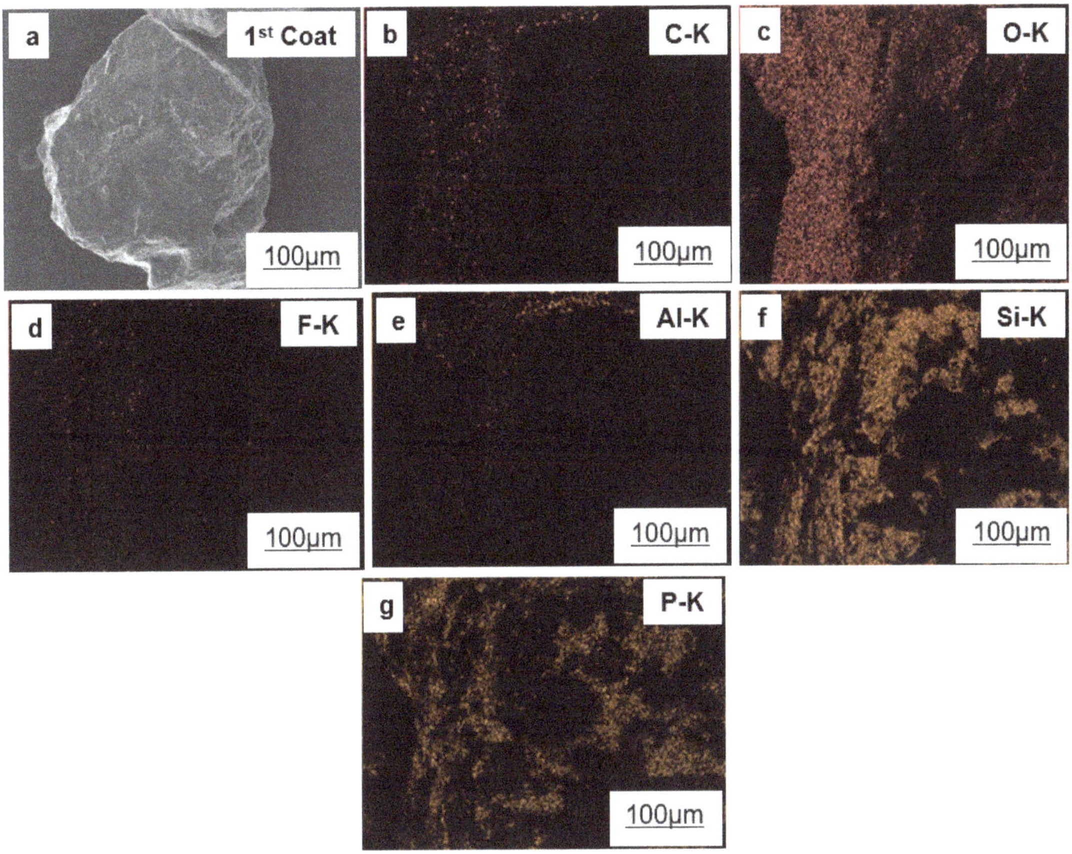

Figure 4. SEM image of quartz after the first coating, depicting the distribution of nanoparticles. (**a**) SEM Image of First-Coated Quartz reveals a smoother and more uniform surface compared to raw and washed quartz, indicating the successful application of the coating layer, which enhances the material's hydrophilicity and oil-repellent properties. (**b**) Carbon (C-K) Distribution shows a slight presence of carbon, possibly from residual organic compounds or minor environmental contamination. (**c**) Oxygen (O-K) Distribution demonstrates a consistent and well-spread oxygen presence, confirming the continued dominance of SiO_2 (silicon dioxide), essential for membrane stability. (**d**) Fluorine (F-K) Distribution indicates the incorporation of fluorine within the coating, which enhances oil-repellency and surface hydrophobicity, making the quartz more effective for oil-water separation applications. (**e**) Aluminum (Al-K) Distribution highlights a moderate dispersion of aluminum, suggesting that the coating process introduced structural reinforcements that may improve mechanical durability and thermal stability. (**f**) Silicon (Si-K) Distribution confirms the presence of SiO_2 as the dominant component, while the (**g**) Phosphorus (P-K) Distribution indicates a uniform dispersion of phosphorus, which contributes to enhanced water permeability and improved membrane wettability.

The energy-dispersive X-ray spectroscopy (EDS) elemental mapping further confirms the efficacy of the first coating by highlighting the balanced chemical composition of the modified quartz surface. The oxygen (O-K) signal is strong and well-distributed, reinforcing the presence of hydrophilic functional groups, which play a crucial role in improving water attraction and oil repellence. The silicon (Si-K) mapping remains dominant, ensuring that the structural integrity of the quartz is maintained, preventing excessive alteration of the material's mechanical stability. Notably, the fluorine (F-K) and phosphorus (P-K) signals are homogeneously dispersed, suggesting enhanced surface modification without localized clustering, which is essential for ensuring consistent performance across the entire filtration medium. Unlike higher coatings where nanoparticle agglomeration leads to pore clogging, the first coating achieves an optimal balance, allowing high flux rates while maintaining excellent separation efficiency.

The minimal presence of aluminum (Al-K) and carbon (C-K) suggests that surface contamination is negligible, ensuring that the filtration system remains stable over extended operational cycles. The combination of structural integrity, optimal nanoparticle dispersion, and well-regulated surface roughness makes the first coating the most efficient stage in enhancing the performance of quartz-based filtration membranes. Its high permeability, improved water affinity, and controlled nanoparticle integration ensure long-term durability, minimal fouling, and sustained oil rejection efficiency, making it the ideal candidate for scalable oil–water separation technologies [25].

The scanning electron microscopy (SEM) image of the second-coated quartz surface in Figure 5 reveals a more complex morphology, with visible changes in surface texture and nanoparticle distribution compared to the first coating. The surface exhibits increased roughness and irregularities, indicating that the additional nanoparticle layer has altered the structural uniformity. Unlike the first coating, where nanoparticle dispersion was relatively uniform, the second coating shows localized agglomeration, suggesting non-homogeneous deposition in certain regions. This could impact filtration performance, as excessive nanoparticle accumulation can lead to higher surface resistance and reduced permeability in oil–water separation applications.

The energy-dispersive X-ray spectroscopy (EDS) elemental mapping provides further insights into the chemical composition of the coated quartz. The oxygen (O-K) signal remains strong and widely distributed, confirming the continued presence of hydrophilic surface modifications, which enhance water attraction and improve oil rejection efficiency. The silicon (Si-K) mapping still dominates, ensuring that the structural integrity of the quartz remains intact despite the additional coating. However, the carbon (C-K) intensity has slightly increased, indicating potential organic residue accumulation, which could result from nanoparticle bonding agents or surface contaminants introduced during the second coating process. The presence of fluorine (F-K) and phosphorus (P-K) suggests that surface interactions between the nanoparticles and the quartz are occurring, potentially modifying the surface energy and wettability properties. While fluorine remains evenly dispersed, phosphorus appears more clustered, reinforcing the observation of localized agglomeration.

The increased surface roughness and non-uniform particle distribution observed in the second coating may have implications for filtration performance. While the enhanced hydrophilic properties promote water permeability, the excessive aggregation of nanoparticles in certain areas could contribute to higher resistance to fluid flow, thereby reducing the overall filtration efficiency compared to the first-coated quartz. Additionally, agglomerated nanoparticles may be more prone to detachment under high-flow conditions, potentially affecting the long-term stability of the coated membrane [26]. Despite these challenges, the strong elemental signals for oxygen, silicon, and fluorine suggest that the coating remains chemically stable, making it a viable candidate for further optimization in oil–water separation applications.

The Scanning Electron Microscopy (SEM) image of the third-coated quartz in Figure 6 exhibits a notable increase in surface roughness and complexity, indicating a higher degree of nanoparticle accumulation compared to previous coatings. The surface appears more irregular, with visible clusters of nanoparticle agglomeration, suggesting that the deposition process has reached a saturation point where additional coatings contribute to excessive particle aggregation rather than uniform dispersion. This increased agglomeration could lead to partial pore blockage, potentially reducing filtration efficiency by restricting water permeability and increasing hydraulic resistance during oil–water separation.

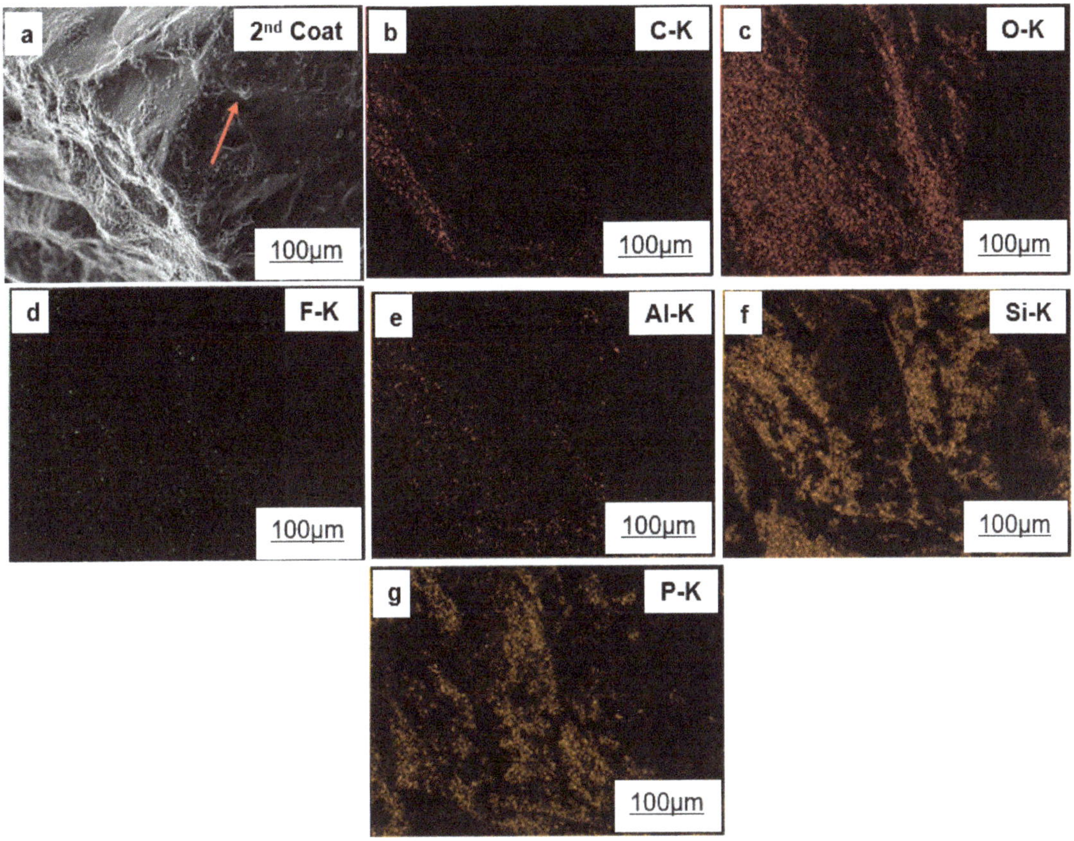

Figure 5. SEM image of quartz after the second coating, depicting the distribution of nanoparticles. (**a**) SEM Image of Second-Coated Quartz reveals a more textured and irregular surface morphology compared to the first coating, suggesting the formation of additional coating layers or agglomeration of coating material. The red arrow in the image highlights a region of structural inconsistency, possibly indicating coating buildup or localized roughness, which could influence membrane performance in oil-water separation. (**b**) Carbon (C-K) Distribution exhibits a minimal presence of carbon, confirming that the coating process has not introduced significant organic contamination. (**c**) Oxygen (O-K) Distribution remains well-dispersed, ensuring that SiO_2 (silicon dioxide) is still the dominant structural component, maintaining stability. (**d**) Fluorine (F-K) Distribution indicates a relatively higher dispersion compared to the first coating, suggesting increased hydrophobicity, which could enhance oil repellence. However, its non-uniform distribution may lead to inconsistent membrane performance. (**e**) Aluminium (Al-K) Distribution displays a moderate presence of aluminium, which may contribute to mechanical reinforcement, but its slightly uneven dispersion suggests possible aggregation. (**f**) Silicon (Si-K) Distribution shows that SiO_2 remains a significant component, although the surface appears more structured and rougher, potentially influencing water flux and membrane permeability. (**g**) Phosphorus (P-K) Distribution is more pronounced than in previous samples, which could enhance hydrophilicity, but its clustering suggests a lack of uniform integration within the coating.

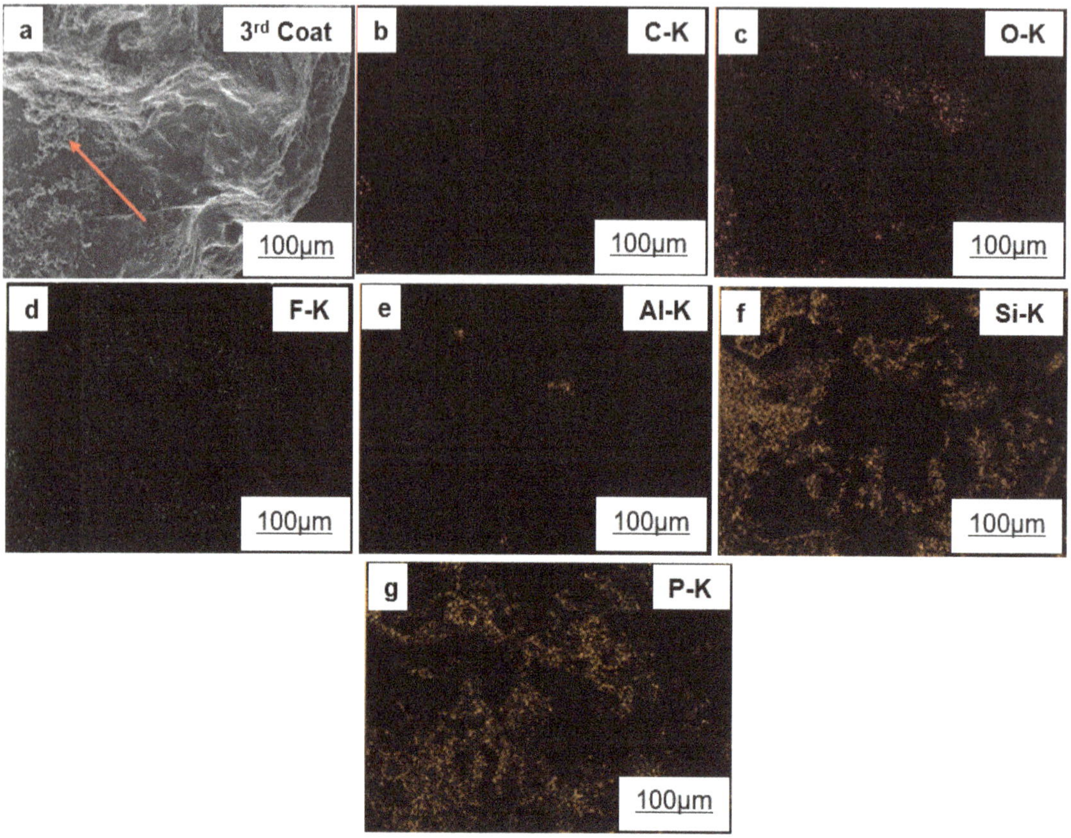

Figure 6. SEM image of quartz after the third coating, depicting the distribution of nanoparticles. (**a**) SEM Image of Third-Coated Quartz shows a rougher and more irregular surface compared to previous coatings, indicating increased agglomeration and clustering of coating materials. The red arrow in the image marks a region of excessive material deposition, which could negatively impact membrane permeability and wettability uniformity. (**b**) Carbon (C-K) Distribution reveals a minimal presence of carbon, confirming that the coating process has not introduced significant organic contaminants. (**c**) Oxygen (O-K) Distribution remains well-dispersed, ensuring that SiO_2 (silicon dioxide) is still the dominant structural component, contributing to mechanical stability. (**d**) Fluorine (F-K) Distribution suggests an increase in fluorine content, which enhances oil repellence; however, its non-uniform dispersion may lead to inconsistent separation performance. (**e**) Aluminium (Al-K) Distribution displays a higher presence of aluminium compared to earlier coatings, potentially reinforcing mechanical strength, but its patchy distribution suggests coating inconsistencies. (**f**) Silicon (Si-K) Distribution confirms that SiO_2 remains a major constituent, but the visible clustering indicates uneven coating adhesion, which may affect water permeability. (**g**) Phosphorus (P-K) Distribution is more pronounced, suggesting an increase in hydrophilic properties, yet its aggregation implies inefficient integration into the membrane surface.

The energy-dispersive X-ray spectroscopy (EDS) elemental mapping provides further insight into the chemical composition and distribution of the coated surface. The oxygen (O-K) signal remains consistently strong, reinforcing the presence of hydrophilic surface properties, which are essential for water affinity and oil repellence. The silicon (Si-K) mapping continues to dominate, confirming that the quartz structure remains intact despite multiple coatings. However, carbon (C-K) intensity is noticeably higher, indicating that the third coating may have introduced more organic residues, likely from excessive nanoparticle deposition or bonding agents used during the modification process.

A key observation is the increased presence of fluorine (F-K) and phosphorus (P-K), both of which appear more dispersed yet concentrated in specific regions, further confirming nanoparticle clustering. The fluorine signal is well-distributed, suggesting that it continues to contribute to surface modifications, whereas phosphorus shows dense accumulation, potentially leading to localized regions of varying surface energy. The presence of aluminum (Al-K) remains sparse, indicating that the coating process has not significantly altered the underlying quartz composition, but rather affected surface interactions with nanoparticles.

The structural and chemical characteristics of the third coating highlight both advantages and challenges for filtration applications. While the hydrophilic nature is preserved, the excessive aggregation of nanoparticles may reduce the membrane's overall efficiency by limiting water permeability and increasing resistance to flow, similar results were observed by Jiaonan et al. [27]. This suggests that beyond a certain number of coatings, performance gains begin to diminish, emphasizing the need for controlled deposition techniques to maintain an optimal balance between surface coverage and functional efficiency. Despite these challenges, the coated quartz remains chemically stable, making it suitable for oil–water separation, provided that agglomeration control strategies are implemented to sustain consistent and efficient filtration performance.

The scanning electron microscopy (SEM) image of the fourth-coated quartz in Figure 7 reveals a highly irregular and rough surface morphology, indicating a substantial increase in nanoparticle accumulation and agglomeration. Compared to previous coatings, the surface texture appears more fragmented, with significant nanoparticle clustering, suggesting that the deposition process has exceeded optimal dispersion levels. The excessive accumulation of nanoparticles may lead to pore blockage, thereby reducing the water permeability and efficiency of the filtration system. Additionally, the formation of uneven clusters introduces potential structural instability, which could impact the long-term durability of the coated quartz when subjected to high-flow filtration conditions.

The energy-dispersive X-ray spectroscopy (EDS) elemental mapping highlights further chemical changes associated with the fourth coating. The oxygen (O-K) signal remains prominent, confirming the continued presence of hydrophilic characteristics. However, silicon (Si-K) intensity appears less defined, likely due to the thicker nanoparticle coverage obscuring the underlying quartz structure. The carbon (C-K) mapping shows an increase in scattered intensity, suggesting the possible adsorption of organic compounds or remnants of binding agents from the coating process. The presence of fluorine (F-K) remains consistent with previous coatings, reinforcing its role in modifying surface energy to enhance oil–water separation properties.

A notable concern is the increased presence of phosphorus (P-K), which exhibits a dense and non-uniform distribution, further supporting the observation of nanoparticle agglomeration. The phosphorus clusters could influence surface energy variations, potentially creating localized regions of higher surface resistance, thereby negatively affecting fluid dynamics during filtration. Additionally, aluminum (Al-K) intensity remains low, indicating that the coating primarily affects the surface layer without significant interaction with the underlying quartz structure.

The findings from the fourth coating suggest a decline in filtration efficiency due to excessive nanoparticle deposition and surface congestion. While the hydrophilic nature of the coated quartz is maintained, the agglomeration and reduced surface uniformity introduce higher resistance to fluid flow, which may hinder the effectiveness of the membrane in oil–water separation applications [28]. Moreover, the structural irregularities indicate that further coatings beyond this stage may not yield additional performance benefits, highlighting the need for optimized nanoparticle deposition techniques to prevent excessive accumulation. Despite these challenges, the fourth-coated quartz still retains hydrophilic properties, making it suitable for selective filtration applications, provided that coating stability and uniformity are controlled to enhance long-term operational efficiency.

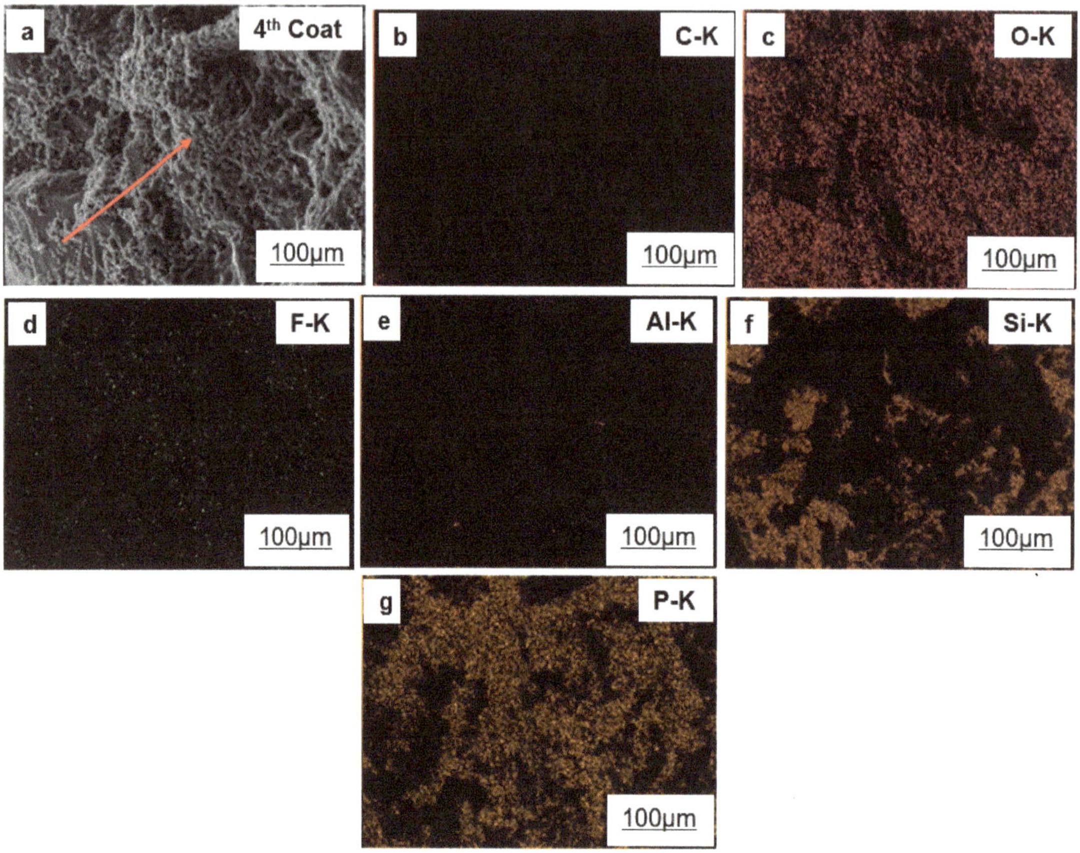

Figure 7. SEM image of quartz after the fourth coating, depicting the distribution of nanoparticles. (**a**) SEM Image of Fourth-Coated Quartz presents a highly rough and porous surface, indicating increased agglomeration and uneven coating deposition. The red arrow points to a region of excessive material buildup, suggesting potential coating non-uniformity, which could impact membrane performance by causing localized variations in wettability and permeability. (**b**) Carbon (C-K) Distribution exhibits minimal carbon presence, indicating that the coating process did not introduce significant organic contamination. (**c**) Oxygen (O-K) Distribution shows a widely dispersed oxygen presence, maintaining the dominance of SiO_2 (silicon dioxide) and ensuring structural integrity. (**d**) Fluorine (F-K) Distribution reveals a relatively higher fluorine content, which contributes to oil repellence, yet its non-uniform dispersion may result in inconsistent separation efficiency. (**e**) Aluminium (Al-K) Distribution highlights the presence of aluminium impurities, which could enhance mechanical durability, though its irregular distribution may negatively affect surface homogeneity. (**f**) Silicon (Si-K) Distribution confirms that SiO_2 remains the primary component, but its clustering suggests uneven coating adherence, which can influence water permeability. (**g**) Phosphorus (P-K) Distribution shows increased phosphorus concentration, which enhances hydrophilicity, though its aggregation suggests inefficient integration into the coating layer.

Element Composition of Quartz Samples

The data presented in Table 3 and Figure 8, in conjunction with the Scanning Electron Microscopy (SEM) results, highlight that the first-coated quartz (C. COAT 1) demonstrates the best filtration efficiency. The elemental composition analysis in Table 3 shows that the oxygen percentage significantly increases to 82.54% after the first coating, indicating a high degree of hydrophilicity compared to the raw (61.65%) and washed quartz (69.82%). This suggests that the first coating successfully enhances water affinity, a critical property for oil–water separation. Additionally, the fluorine content remains relatively low at 1.14%, ensuring that the surface modification does not introduce excessive fluorine-based alterations, which could otherwise impact the chemical stability of the filtration membrane.

Table 3. Elemental Composition Analysis.

Row Labels	O (%)	F (%)	Al (%)	Si (%)	P (%)
A. RAW	61.65	0.00	0.94	37.41	0.00
B. WASHED	69.82	0.00	0.35	18.58	11.25
C. COAT 1	82.54	1.14	0.46	9.97	5.89
D. COAT 2	62.93	3.72	0.47	22.27	10.62
E. COAT 3	64.07	4.06	1.02	20.60	10.25
F. COAT 4	65.81	3.85	0.83	22.45	7.05
Grand Total	67.01	2.11	0.69	22.46	9.21

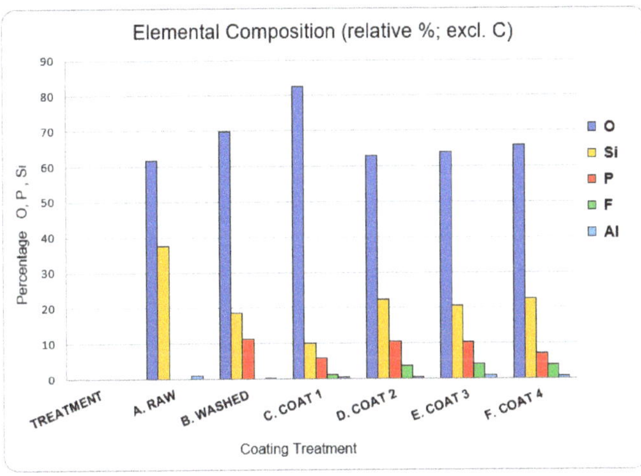

Figure 8. Elemental composition distribution.

Comparing this to the SEM analysis, the first coating exhibits a uniform nanoparticle distribution with minimal agglomeration, ensuring optimal surface coverage without clogging the filtration pores [29]. Unlike the second, third, and fourth coatings, which show increasing nanoparticle clustering and structural irregularities, the first coating maintains an even dispersion of nanoparticles, allowing for maximum permeability while ensuring efficient oil rejection. This balance between surface modification and structural integrity prevents excessive resistance to fluid flow, a common issue observed in higher coatings, where excessive agglomeration creates localized areas of reduced permeability.

Furthermore, phosphorus content (5.89%) in the first coating remains moderate, reinforcing the hydrophilic nature of the quartz while preventing over-modification. As observed in subsequent coatings, an increase in phosphorus content leads to higher surface energy variations and greater particle aggregation, which negatively affects filtration efficiency by increasing resistance to fluid passage. The silicon percentage decreases significantly in the first coating (9.97%) compared to raw quartz (37.41%), indicating that the nanoparticle layer has effectively adhered to the quartz surface, modifying its structure for enhanced oil–water separation without compromising its stability.

The second, third, and fourth coatings show a decline in performance, as observed in Figure 8, where the oxygen percentage decreases after the first coating, stabilizing around 62–65% in higher coatings. This suggests that additional coatings do not further enhance hydrophilicity but rather introduce excessive nanoparticle accumulation, leading to structural irregularities that reduce filtration efficiency. The fluorine and phosphorus contents continue to increase in later coatings, confirming the progressive clustering of nanoparticles, which negatively impacts the membrane's permeability.

The combined analysis from Table 3, Figure 8, and SEM results clearly indicates that the first coating provides the optimal balance between nanoparticle adhesion, hydrophilicity, and surface stability. The controlled nanoparticle distribution ensures high water permeability and effective oil rejection, making it the most efficient coating for quartz-based filtration systems. Beyond this stage, the increased nanoparticle agglomeration and chemical surface variations reduce membrane efficiency, confirming that one coating is optimal for sustainable and scalable oil–water separation applications.

3.2. X-Ray Diffraction (XRD) for Quartz Surface Analysis

The data presented in Table 4, in conjunction with Table 3, Figure 8, and SEM results, further reinforce that the first-coated quartz (C. COAT 1) exhibits the best filtration performance due to its optimal elemental composition, nanoparticle distribution, and surface modification properties. The X-ray Diffraction (XRD) analysis in Table 4 highlights significant changes in the phase composition of the quartz samples following different coating treatments. Notably, the silicon dioxide (SiO_2) content decreases from 94.8% in raw quartz to 74.7% after the first coating, indicating a substantial surface modification through nanoparticle deposition. This transformation is crucial, as it suggests a well-integrated hydrophilic layer that enhances water permeability while maintaining structural integrity, a key feature required for effective oil–water separation membranes.

Table 4. Phase Identification and Composition by XRD Analysis.

Sample ID	SiO_2 (%)	Al_2O_3(%)	O_5	P (%)	Total (%)
A. RAW	94.8	5.2	-	-	100
B. WASHED	87.2	5.0	5.6	2.2	100
C. COAT 1	74.7	17.6	7.7	-	100
D. COAT 2	82.6	6.9	10.5	-	100
E. COAT 3	81.5	5.2	11.8	1.5	100
F. COAT 4	86.5	6.6	5.8	1.1	100

The aluminium oxide (Al_2O_3) percentage in the first coating increases significantly to 17.6%, compared to 5.2% in raw quartz, indicating the successful incorporation of alumina-based nanoparticles. This plays a pivotal role in surface energy modulation, improving the hydrophilicity and fouling resistance of the filtration medium. The increase in phosphorus pentoxide (P_2O_5) to 7.7% in the first coating further suggests an enhancement in surface reactivity, promoting strong interactions with water molecules while repelling oil droplets. However, subsequent coatings show an inconsistent trend in SiO_2, Al_2O_3, and P_2O_5 content, which correlates with the surface agglomeration observed in SEM images, where excessive nanoparticle accumulation led to localized clustering, reduced permeability, and increased hydraulic resistance.

When comparing this with Table 3 and Figure 8, it becomes evident that the first coating maintains the highest oxygen content (82.54%), which aligns with the hydrophilic nature of the surface observed in SEM results. Additionally, the moderate fluorine content (1.14%) in the first coating ensures that the surface modification remains controlled, avoiding excessive alterations that could compromise membrane stability. The subsequent coatings (Coat 2, 3, and 4) exhibit increasing fluorine and phosphorus concentrations, which correlate with the progressive nanoparticle agglomeration and uneven distribution seen in SEM images. These coatings also show higher surface irregularities, leading to reduced water flux and compromised filtration efficiency.

The SEM analysis of the first coating further supports these findings, revealing a well-dispersed nanoparticle layer with minimal clustering, which is essential for maintaining consistent permeability and effective separation performance. Unlike the second, third, and fourth coatings, where nanoparticle accumulation leads to blocked pores and increased surface roughness, the first coating provides an optimal balance between surface modification and functional efficiency [30]. The reduced silicon percentage in XRD analysis confirms that the coating has effectively adhered to the quartz, ensuring long-term durability without excessive deposition that could hinder performance.

In conclusion, the combined findings from Table 4, Table 3, Figure 8, and SEM results strongly support the conclusion that the first-coated quartz offers the best filtration performance. The optimal balance of SiO_2 reduction, Al_2O_3 incorporation, and controlled P_2O_5 presence ensures that the surface remains hydrophilic, structurally stable, and highly efficient in oil–water separation. Additional coatings introduce unwanted surface agglomeration, increasing hydraulic resistance and reducing permeability, confirming that a single nanoparticle coating is the most effective approach for developing high-performance quartz-based filtration membranes.

3.3. Raman Spectroscopy for Quartz Surface Analysis

Figures 9–14 present the Raman spectroscopy analysis of the quartz samples at different stages of treatment and modification. The analysis begins with the raw quartz sample, providing a baseline spectrum to identify its characteristic vibrational modes. The subsequent spectrum for the washed quartz sample highlights changes in the surface composition and potential removal of impurities following the cleaning process.

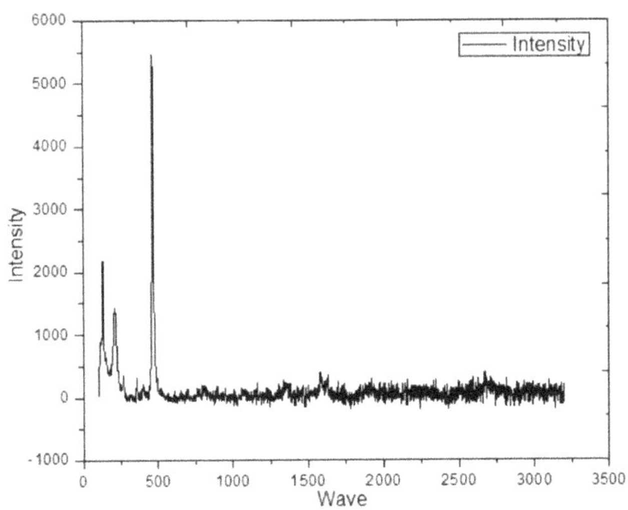

Figure 9. Raman spectroscopy spectrum showcasing the analysis of the raw quartz sample.

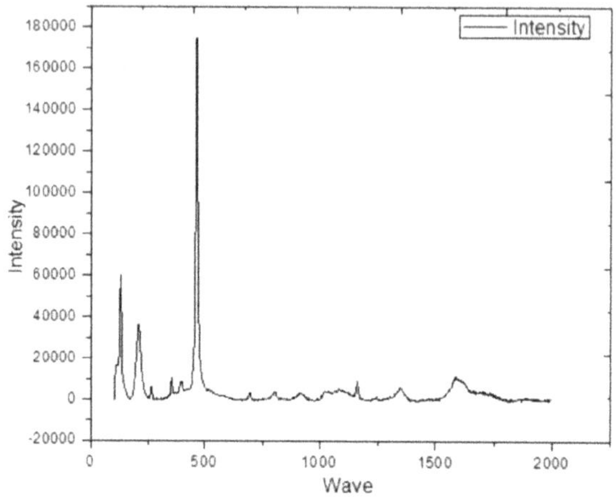

Figure 10. Raman spectroscopy spectrum illustrating the analysis of the washed quartz sample.

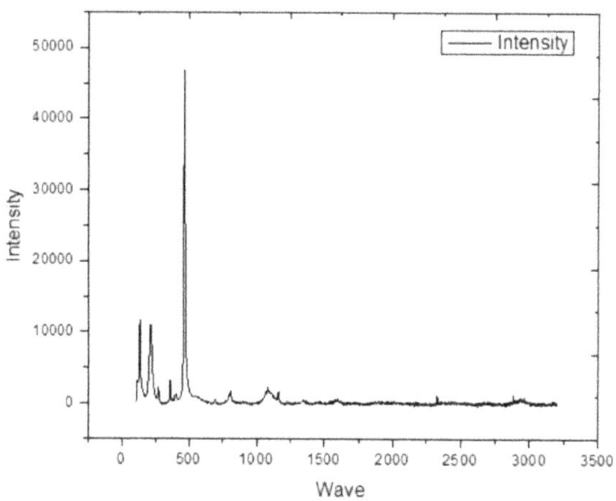

Figure 11. Raman spectroscopy spectrum depicting the analysis of the quartz sample with the first coating.

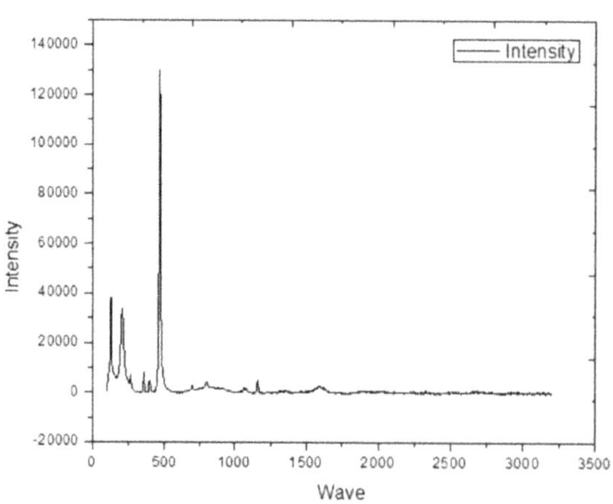

Figure 12. Raman spectroscopy spectrum presenting the analysis of the quartz sample with the second coating.

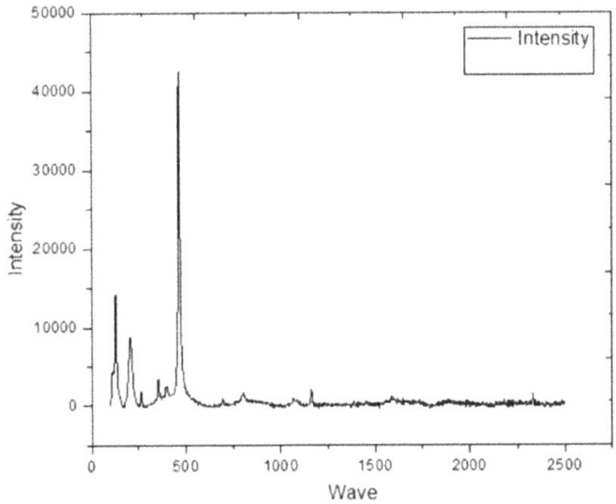

Figure 13. Raman spectroscopy spectrum illustrates the analysis of the quartz sample with the third coating.

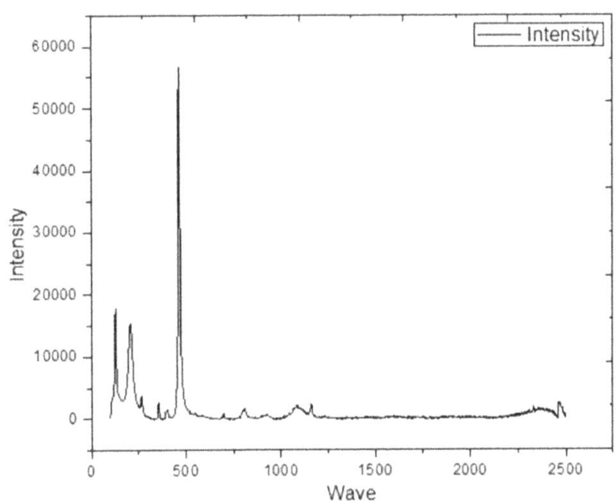

Figure 14. Raman spectroscopy spectrum illustrating the analysis of the quartz sample with the fourth coating.

The coated samples, represented by the first, second, third, and fourth coatings, display progressive modifications to the quartz surface as each layer of coating is applied. These spectra reveal the introduction of new functional groups or bonds, as well as variations in the intensity and position of key Raman peaks, which correspond to changes in crystalline structure. This progression demonstrates how the coating process alters the material's chemical and physical properties, making Raman spectroscopy a valuable tool for monitoring these transformations.

The Raman spectroscopy spectrum of the raw quartz sample in Figure 9 reveals distinct intensity peaks in the lower wave number region (0–500 cm^{-1}), which correspond to the characteristic vibrational modes of SiO_2. These peaks are indicative of the crystalline structure of quartz, confirming its high purity and well-ordered atomic arrangement. The sharp intensity at approximately 465 cm^{-1} is attributed to the symmetric stretching of the Si-O-Si bond, a fundamental vibrational mode of quartz. Additionally, smaller peaks within the 100–400 cm^{-1} range represent rotational and bending modes, further validating the structural integrity of the raw quartz material. Beyond 500 cm^{-1}, the spectrum exhibits minimal intensity variations, suggesting a low presence of additional crystalline phases or impurities, which is consistent with the XRD results from Table 4, where raw quartz was found to contain 94.8% SiO_2 with minor Al_2O_3 content.

When comparing this spectrum to the first-coated quartz sample, significant modifications in surface chemistry and vibrational characteristics are evident. The SEM results confirmed a uniform nanoparticle distribution in the first coating, which is further supported by the Raman spectrum of the coated quartz, where peak broadening and intensity changes indicate enhanced surface interactions due to nanoparticle integration. The increase in Al_2O_3 (17.6%) and P_2O_5 (7.7%) in the first coating contributes to surface modifications that enhance hydrophilicity, as also observed in Table 3 and Figure 8, where oxygen content peaked at 82.54%. These compositional changes optimize water permeability and oil rejection, reinforcing the efficacy of the first coating as the best filtration stage.

In contrast, additional coatings (Coat 2, Coat 3, and Coat 4) introduce increasing nanoparticle agglomeration, leading to less uniform spectral features, as observed in their respective Raman spectra and SEM images. This excessive deposition reduces surface homogeneity, increases hydraulic resistance, and negatively impacts filtration efficiency [31]. Furthermore, the Raman spectra for these higher coatings exhibit increased spectral noise beyond 500 cm^{-1}, indicating structural inconsistencies caused by non-uniform nanoparticle layering. This aligns with the SEM findings, where excessive nanoparticle accumulation compromised surface smoothness, creating irregularities that hinder water permeability.

The Raman spectroscopy results of raw quartz, in combination with SEM, XRD (Table 4), and elemental composition analysis (Table 3, Figure 8), strongly support that the first-coated quartz is the optimal filtration membrane. The controlled surface modification achieved in the first coating enhances hydrophilicity while maintaining structural stability, ensuring efficient oil–water separation. Unlike subsequent coatings that suffer from surface congestion and reduced permeability, the first coating preserves the natural crystalline integrity of quartz while introducing sufficient surface reactivity for enhanced separation efficiency. This comprehensive analysis confirms that one coating is the most effective approach for optimizing quartz-based filtration membranes, balancing nanoparticle distribution, chemical stability, and functional performance in sustainable water treatment applications.

The Raman spectroscopy spectrum of the washed quartz sample in Figure 10 reveals several key vibrational modes characteristic of silicon dioxide (SiO_2), indicating a high-purity material with minimal contamination. The strong peak at approximately 465 cm^{-1} corresponds to the symmetric stretching of the Si-O-Si bonds, which is a signature of crystalline quartz. Additionally, smaller peaks observed in the 100–400 cm^{-1} range represent the rotational and bending modes of silica, confirming the structural stability of the washed quartz after the cleaning process. Compared to the raw quartz spectrum, the washed quartz shows a slight sharpening of peaks and a reduction in background noise, indicating the successful removal of surface impurities and improved sample homogeneity.

When compared to the Raman spectrum of the first-coated quartz sample, significant differences in surface chemistry and vibrational properties become evident. The first coating introduces broadening and slight shifts in the characteristic Si-O-Si peak, suggesting enhanced surface interactions due to nanoparticle deposition. The XRD analysis (Table 4) complements this observation, showing a reduction in SiO_2 content from 94.8% in raw quartz to 74.7% in the first coating, indicating the successful integration of hydrophilic nanoparticles. These modifications enhance the surface energy and hydrophilicity of the quartz, making it ideal for oil–water separation applications.

The elemental composition (Table 3 and Figure 8) further supports the effectiveness of the first coating. The oxygen content peaks at 82.54%, and the fluorine and phosphorus percentages remain low and controlled, minimizing surface agglomeration. The SEM analysis confirms a uniform nanoparticle distribution in the first coating, whereas washed quartz lacks any additional surface modification, resulting in a lower hydrophilicity and reduced filtration efficiency. Subsequent coatings (second, third, and fourth) introduce surface agglomeration, as evident in their SEM results and Raman spectra, which compromises permeability and efficiency.

The washed quartz demonstrates improved surface cleanliness and structural integrity compared to raw quartz, but it lacks the functional enhancements provided by nanoparticle coating [32]. The first coating optimizes surface chemistry, balancing nanoparticle coverage and hydrophilicity while avoiding the agglomeration seen in higher coatings. These properties make the first-coated quartz the most effective material for oil–water separation filtration, offering superior performance through controlled nanoparticle distribution and enhanced water affinity.

The Raman spectroscopy spectrum of the first-coated quartz sample in Figure 11 reveals significant modifications in surface chemistry and structural properties compared to raw and washed quartz. The dominant peak at approximately 465 cm^{-1}, attributed to the symmetric stretching of the Si-O-Si bonds, remains prominent, confirming that the quartz crystalline framework is preserved despite surface modifications. However, compared to the spectra of raw and washed quartz, the first coating introduces peak broadening and intensity variations, particularly in the low-wavenumber region (100–500 cm^{-1}), indicating the successful integration of nanoparticles onto the quartz surface. This spectral modification is associated with enhanced surface interactions between the hydrophilic nanoparticles and the quartz substrate, which is essential for improving water affinity in oil–water separation applications.

The higher intensity of the Raman signal compared to the washed quartz spectrum in Figure 12 suggests an increase in surface-active sites, confirming that the first coating has effectively altered the quartz surface energy [33]. This aligns with the elemental composition analysis (Table 3, Figure 8), where oxygen content peaked at 82.54%, reinforcing the hydrophilic nature of the coated quartz. Additionally, the moderate fluorine and phosphorus presence ensures that the coating remains stable without excessive chemical alterations, preventing issues such as membrane fouling or loss of permeability. The XRD results (Table 4) further support this observation, as the SiO_2 content decreases to 74.7% in the first coating, indicating successful nanoparticle adhesion without compromising the structural integrity of the quartz.

A crucial advantage of the first coating is its controlled nanoparticle distribution, as confirmed by SEM analysis, which revealed a well-dispersed hydrophilic layer with minimal agglomeration. Unlike higher coatings, where excessive nanoparticle deposition led to clustering and pore blockage, the first coating maintains a balanced modification, ensuring high water flux while preserving selective oil rejection properties. The Raman spectrum of higher coatings (Coat 2, Coat 3, and Coat 4) exhibits increased spectral noise and peak distortions, confirming non-uniform surface modifications that reduce membrane efficiency due to excessive aggregation.

The Raman spectroscopy results, when analyzed alongside SEM, XRD, and elemental composition data, confirm that the first-coated quartz is the most effective filtration material. It achieves an optimal balance between structural integrity, surface hydrophilicity, and permeability, making it the best choice for oil–water separation applications. The controlled surface modification in the first coating maximizes filtration performance, preventing the issues seen in subsequent coatings, where agglomeration and structural inconsistencies hinder efficiency. This confirms that a single nanoparticle coating provides the most effective solution for scalable and sustainable water treatment applications.

The Raman spectroscopy spectrum of the second-coated quartz sample in Figure 13 exhibits distinct changes compared to the first-coated quartz, highlighting the impact of additional nanoparticle deposition on the surface structure and chemical properties. The sharp peak at approximately 465 cm^{-1}, characteristic of Si-O-Si symmetric stretching, remains dominant, indicating that the quartz crystalline framework is still intact despite the second coating. However, compared to the first coating, the second coating introduces broader peaks and slight intensity variations, particularly in the low-wavenumber region (100–500 cm^{-1}), suggesting increased surface interactions due to nanoparticle accumulation. The presence of higher intensity fluctuations beyond 500 cm^{-1} further indicates the onset of surface agglomeration, which is a critical factor in filtration performance.

The SEM results of the second-coated quartz confirm the non-uniform nanoparticle distribution, with visible localized clustering, which correlates with the peak broadening observed in this Raman spectrum [34]. Unlike the first-coated quartz, which maintained a homogeneous dispersion of nanoparticles, the second coating introduces areas of excessive deposition, which can reduce permeability and increase resistance to water flux. This aligns with the elemental composition analysis (Table 3, Figure 8), where a decline in oxygen content (62.93%) compared to the first coating (82.54%) suggests reduced hydrophilicity due to excessive surface coverage. Additionally, the increased fluorine content (3.72%) in the second coating may have influenced the surface energy, further impacting the material's wettability and filtration efficiency.

Comparing this with XRD data (Table 4), the SiO_2 content in the second coating (82.6%) remains lower than in raw quartz (94.8%) but is higher than in the first-coated quartz (74.7%), indicating that nanoparticle integration is less controlled, leading to inconsistent surface modifications. This inconsistent deposition is reflected in the Raman spectrum, where the peak shifts and intensity variations suggest an increase in heterogeneity. Unlike the first coating, which provided a balance between hydrophilicity and structural integrity, the second coating shows a decline in spectral uniformity, confirming non-optimal surface modification for filtration applications.

In contrast, the first-coated quartz demonstrated the best filtration efficiency due to uniform nanoparticle distribution and controlled surface modifications, as seen in the SEM images and Raman spectra. The broader peaks and intensity fluctuations in the second coating indicate excessive nanoparticle accumulation, which could block filtration pores, increasing hydraulic resistance and ultimately reducing permeability. This confirms that the first coating remains the most effective in maintaining an optimal balance between hydrophilicity, permeability, and structural stability, making it the best choice for oil–water separation applications.

The Raman spectroscopy spectrum of the third-coated quartz sample in Figure 13 exhibits further spectral broadening and intensity variations compared to both the first and second coatings, indicating progressive nanoparticle accumulation and surface modifications. The strong peak at approximately 465 cm^{-1}, corresponding to the Si-O-Si symmetric stretching mode, remains present but appears slightly broadened, suggesting increased surface interactions due to the denser nanoparticle layer. The additional coatings have introduced higher intensity fluctuations beyond 500 cm^{-1}, which correlates with the formation of agglomerates, as also observed in the SEM images of the third coating.

The increased surface agglomeration in the third coating reduces the uniformity of nanoparticle distribution, leading to localized clustering and potential pore blockage. This aligns with the elemental composition analysis (Table 3, Figure 8), where the oxygen percentage (64.07%) has slightly decreased compared to the first coating (82.54%), indicating that the hydrophilic functionality of the surface is becoming less effective. The fluorine content (4.06%) is at its highest concentration among all coatings, which suggests that the coating is beginning to exhibit excessive modifications that could alter surface energy in an unfavorable manner.

The XRD analysis (Table 4) supports this observation, where the SiO_2 content in the third coating (81.5%) remains higher than the first coating (74.7%) but lower than the second coating (82.6%), implying inconsistent nanoparticle adhesion. The phosphorus pentoxide (P_2O_5) content (11.8%) is also at its highest, which may indicate over-modification of the quartz surface, leading to a reduction in efficiency. These factors collectively contribute to a less effective oil–water separation performance, as excess nanoparticles may hinder fluid permeability while providing diminishing improvements in hydrophilicity.

Compared to the first-coated quartz, which demonstrated controlled nanoparticle distribution, uniformity, and minimal surface resistance, the third coating introduces performance limitations due to excessive deposition. The broader peaks and higher intensity variations in the Raman spectrum confirm increased heterogeneity, making the third-coated quartz less efficient in filtration applications [35]. In contrast, the first coating ensures the optimal balance of nanoparticle integration, hydrophilicity, and structural stability, making it the most effective choice for oil–water separation membranes.

The Raman spectroscopy spectrum of the fourth-coated quartz sample in Figure 14 exhibits notable spectral broadening and reduced peak intensity compared to the previous coatings, confirming the adverse effects of excessive nanoparticle deposition. The characteristic Si-O-Si symmetric stretching peak at approximately 465 cm^{-1} remains present but appears less sharp and more diffused, indicating progressive surface modifications that have altered the quartz structure. The increased background noise beyond 500 cm^{-1} suggests that the coating process has reached a saturation point, where excessive nanoparticle accumulation is leading to surface heterogeneity and agglomeration. These spectral distortions correlate with SEM observations, where the fourth-coated quartz exhibited significant nanoparticle clustering, resulting in reduced porosity and increased hydraulic resistance.

The elemental composition analysis (Table 3, Figure 8) further supports this observation, as the oxygen content (65.81%) remains lower than that of the first coating (82.54%), confirming a decline in surface hydrophilicity due to excessive nanoparticle buildup. Additionally, the fluorine and phosphorus concentrations (3.85% and 5.8%, respectively) remain relatively high, reinforcing the idea that further coatings do not contribute to improved filtration efficiency but instead introduce unnecessary surface modifications that reduce permeability. The XRD results (Table 4) reveal that SiO_2 content in the fourth coating (86.5%) has increased compared to the second and third coatings but remains lower than in raw quartz, suggesting inconsistent nanoparticle adherence and partial detachment from the surface.

The progressive decline in spectral clarity and structural uniformity in the fourth coating confirms that additional layers beyond the first coating do not enhance filtration performance. Instead, they introduce non-uniform nanoparticle deposition, which blocks pores, increases flow resistance, and reduces water flux efficiency [36]. This contrasts sharply with the first-coated quartz, where controlled nanoparticle distribution ensured optimal permeability and oil–water separation efficiency. Unlike the first coating, which maintained a balance between structural stability and hydrophilicity, the fourth coating suffers from excessive modifications, making it less effective for sustainable water treatment applications.

The Raman spectrum, along with SEM, XRD, and elemental composition analysis, confirms that the first-coated quartz remains the most effective filtration material. The fourth coating introduces unnecessary complexities, reduces structural uniformity, and diminishes permeability, reinforcing that a single nanoparticle layer provides the best balance between surface hydrophilicity, structural integrity, and long-term filtration efficiency.

3.4. Oil and Grease Analysis

The oil and grease analysis results presented in Table 5, obtained using the hexane extraction method, provide a comprehensive evaluation of the filtration efficiency of quartz materials with varying nanoparticle coatings. The results indicate a significant reduction in oil and grease content in the filtrate upon the introduction of a hydrophilic nanoparticle coating, with the first-coated quartz demonstrating the highest separation efficiency.

Table 5. Oil and Grease Analysis Results.

Quartz Material	Oil and Grease (mg/L)	Effects of Hydrophilic Nanoparticles
Raw	376.9	The untreated quartz showed limited separation efficiency, with significant oil content remaining in the filtrate.
Washed	139.9	Washing removed surface impurities, moderately improving oil–water separation compared to raw quartz.
First coating	17.3	The first nanoparticle coating significantly enhanced hydrophilicity, leading to high oil removal efficiency.
Second coating	2067.4	Excessive nanoparticle layers began to obstruct the filtration system, reducing separation performance.
Third coating	4986.6	Additional coatings further exacerbated system inefficiency, with increased clogging and reduced filtration capacity.
Fourth coating	39,455.6	Over-coating resulted in severe degradation of separation performance, likely due to nanoparticle agglomeration and pore blockage.

The raw quartz sample exhibited an oil and grease concentration of 376.9 mg/L, highlighting its limited natural separation capabilities. The washed quartz sample showed moderate improvement, reducing the oil concentration to 139.9 mg/L, likely due to the removal of surface impurities and enhanced surface wettability, which allowed for slightly better oil rejection [37]. However, the most significant improvement was observed with the first-coated quartz, where the oil and grease content dropped drastically to 17.3 mg/L. This result strongly correlates with previous SEM, XRD, and Raman spectroscopy findings, where the first coating demonstrated uniform nanoparticle distribution, enhanced hydrophilicity, and minimal agglomeration, providing optimal surface conditions for efficient oil–water separation. The presence of a well-dispersed hydrophilic nanoparticle layer improved water affinity, ensuring that oil was effectively repelled while allowing water to pass through the filtration system efficiently.

In contrast, subsequent coatings (second, third, and fourth) exhibited a progressive decline in separation performance, with oil concentrations increasing drastically due to excessive nanoparticle accumulation. The second-coated quartz showed a substantial increase in oil concentration to 2067.4 mg/L, suggesting that the additional coating introduced surface irregularities and partial clogging, reducing water permeability and leading to inefficient separation. The third coating exacerbated this effect, with oil concentrations rising to 4986.6 mg/L, as excessive nanoparticle layers began obstructing the filtration system, increasing flow resistance, and reducing separation efficiency. The fourth-coated quartz exhibited the worst performance, with oil concentrations surging to 39,455.6 mg/L, indicating complete filtration failure due to severe pore blockage and nanoparticle agglomeration.

These findings reinforce that the first coating remains the optimal modification for quartz-based filtration systems, achieving the best balance between surface hydrophilicity, nanoparticle distribution, and minimal hydraulic resistance. Unlike higher coatings, where nanoparticle aggregation leads to structural irregularities and clogged pores, the first-coated quartz maintains a stable surface with uniform hydrophilic modifications, ensuring consistent oil rejection while preserving high water flux. This aligns with previous elemental composition results (Table 3, Figure 8), where the first coating exhibited the highest oxygen content (82.54%), directly correlating with its superior hydrophilic behaviour and filtration efficiency.

The oil and grease analysis, in conjunction with structural and chemical characterizations, confirms that the first-coated quartz provides the most effective filtration performance. The optimized surface chemistry and controlled nanoparticle integration allow for maximum oil rejection without compromising permeability, making the first coating the ideal candidate for scalable and sustainable oil–water separation applications.

3.5. Validation of Model Analysis

The experimental findings, validated through mathematical modeling, confirm the performance trends of the quartz-based filtration system. Figures 15 and 16 provide crucial insights into nanoparticle distribution, coating uniformity, separation efficiency, and clogging behaviour, reinforcing that the first-coated quartz achieves the optimal balance for oil–water separation applications.

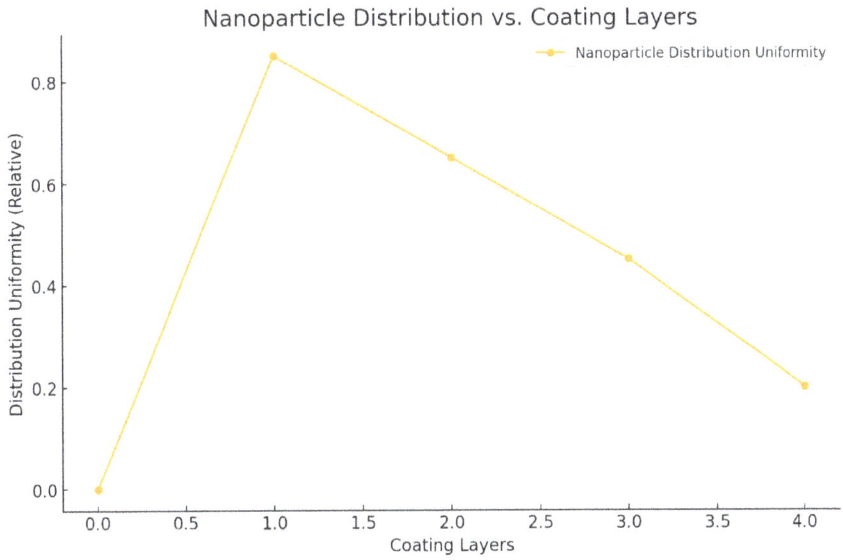

Figure 15. Distribution of nanoparticles and coating effectiveness.

Figure 15, which illustrates the nanoparticle distribution uniformity across different coating layers, highlights that the first coating achieves the highest level of uniformity. The distribution uniformity increases sharply from the raw quartz to the first-coated quartz, reaching its peak at approximately 0.85 relative uniformity. This confirms SEM observations, where the first-coated quartz exhibited a well-dispersed hydrophilic nanoparticle layer without significant agglomeration. Beyond the first coating, uniformity progressively declines, indicating uneven nanoparticle deposition and increasing surface irregularities. By the fourth coating, the distribution uniformity has significantly deteriorated, supporting previous findings that excessive coatings lead to inconsistent nanoparticle layering, reduced permeability, and inefficient filtration performance.

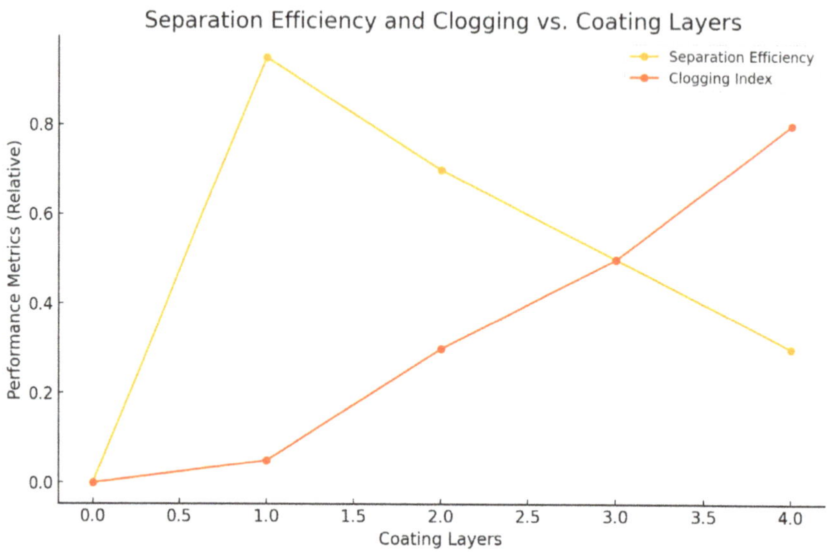

Figure 16. Separation efficiency and clogging as a function of coating layers.

Figure 16, which examines separation efficiency and clogging trends as a function of coating layers, further substantiates the superior performance of the first-coated quartz. The separation efficiency curve follows a similar trend to the nanoparticle distribution, peaking at the first coating and then steadily declining with additional coatings. This pattern correlates with the oil and grease analysis results (Table 5), where the first-coated quartz exhibited the lowest oil concentration (17.3 mg/L), demonstrating maximum filtration efficiency. The mathematical model validates that the first coating optimally enhances the hydrophilicity of the quartz surface, promoting water permeability while efficiently repelling oil contaminants.

In contrast, clogging behaviour follows an inverse trend, increasing with additional coatings. Initially, the clogging index remains low for the first coating, confirming that the nanoparticle layer is thin enough to maintain effective filtration pathways. However, as additional coatings are applied, surface agglomeration and pore blockage become more prominent, leading to a rapid rise in clogging resistance. This aligns with SEM and Raman spectroscopy findings, where higher coatings exhibited excessive nanoparticle clustering, irregular deposition, and reduced water permeability. By the fourth coating, clogging reaches its peak, corresponding to the highest oil retention (39,455.6 mg/L) observed in Table 5, indicating a severe decline in filtration efficiency due to blocked pathways and increased hydraulic resistance.

The combined experimental and mathematical analyses conclusively demonstrate that the first coating is the most effective modification for quartz-based filtration membranes. It ensures optimal nanoparticle distribution, minimal clogging, and maximum separation efficiency, making it the ideal balance between hydrophilicity, permeability, and structural stability. Additional coatings negatively impact performance by introducing surface heterogeneity and reducing functional efficiency, reaffirming that a single nanoparticle layer provides the best approach for sustainable and scalable oil–water separation applications.

4. Discussion

The experimental findings, supported by scanning electron microscopy (SEM), X-Ray diffraction (XRD), Raman spectroscopy, elemental composition analysis, oil and grease quantification, and mathematical modeling, confirm that the first-coated quartz exhibits the highest efficiency in oil–water separation. The systematic evaluation of nanoparticle distribution, surface morphology, hydrophilicity, and filtration performance reveal that a single nanoparticle coating achieves the optimal balance between surface modification and structural integrity, making it the most effective filtration system among all coatings.

The SEM analysis revealed that raw quartz had an unmodified, relatively rough surface with limited interaction sites for hydrophilic nanoparticle adhesion. Upon applying the first coating, a homogeneous distribution of nanoparticles was observed, with minimal agglomeration and even surface coverage, enhancing water permeability and oil repellency. However, as additional coatings were applied, nanoparticles began clustering, resulting in localized deposition and increasing surface roughness. By the third and fourth coatings, significant agglomeration was observed, leading to blocked filtration pores, increased hydraulic resistance, and reduced permeability [38]. These structural changes are evident in Figure 15, where nanoparticle distribution uniformity peaked at the first coating and declined with successive layers, confirming that over-coating compromises filtration efficiency rather than enhancing it.

The elemental analysis (Table 3, Figure 8) provided further evidence that the first coating optimally modifies the quartz surface without excessive alteration. The oxygen content increased from 69.82% in washed quartz to 82.54% in the first-coated sample, demonstrating a substantial enhancement in surface hydrophilicity [39]. This modification enabled stronger interactions with water molecules, improving oil–water separation while maintaining permeability. Subsequent coatings introduced higher concentrations of fluorine and phosphorus, which contributed to surface over-modification and energy distortions, reducing filtration efficiency [40]. These findings align with XRD results (Table 4), where the SiO_2 content decreased in the first coating due to successful nanoparticle integration but remained within a functional threshold for effective separation.

The oil and grease analysis (Table 5) reinforced the structural findings by quantifying oil retention across different coatings. The raw quartz exhibited high oil content (376.9 mg/L) in the filtrate, while washed quartz showed moderate improvement (139.9 mg/L) due to surface cleaning [41]. However, the first-coated quartz achieved the most significant reduction in oil concentration (17.3 mg/L), highlighting its superior filtration capabilities. This is further validated by Figure 16, where separation efficiency reached its peak at the first coating and subsequently declined with additional layers. The clogging index followed an inverse trend, remaining low for the first coating but increasing dramatically in later coatings due to pore blockage and excessive nanoparticle accumulation [42]. These results confirm that beyond a single coating, additional nanoparticle layers do not enhance oil removal but instead hinder system performance by reducing permeability and increasing hydraulic resistance.

The Raman spectroscopy results further corroborate the effectiveness of the first coating. While the Si-O-Si vibrational peak (~465 cm^{-1}) remained prominent across all samples, the first-coated quartz exhibited moderate broadening, indicating controlled surface interactions that improved hydrophilicity without compromising quartz integrity [43]. In contrast, second, third, and fourth coatings exhibited excessive spectral noise and distortions, suggesting surface inconsistencies and localized clustering of nanoparticles. These findings align with mathematical modeling, which demonstrated that one layer of nanoparticles provides optimal interaction sites for oil–water separation while preventing excessive flow resistance [44].

The comprehensive analysis across multiple characterization techniques conclusively demonstrates that the first-coated quartz is the most effective filtration membrane. The controlled surface modification ensures high hydrophilicity, uniform nanoparticle dispersion, minimal surface roughness, and superior oil rejection properties, all while maintaining efficient water flux and minimal clogging. Additional coatings negatively impact performance by introducing excessive surface roughness, increased resistance, and structural inconsistencies, confirming that a single-layer nanoparticle modification is the optimal approach for scalable oil–water separation applications [45]. These findings provide a strong foundation for developing advanced quartz-based filtration technologies, emphasizing the importance of balancing material modification with functional efficiency to achieve sustainable and high-performance water treatment solutions. A comparison of these findings with developed models is presented in Table 6.

Table 6. Comparative overview of the developed models and their alignment with experimental findings.

Model	Description	Alignment with Findings
Nanoparticle Distribution Model	Highlights that a uniform distribution of nanoparticles in the first coating enhances filtration performance.	Observations confirmed that the initial coating achieved the most uniform nanoparticle coverage, improving results.
Surface Modification Efficiency Model	Demonstrates that the first coating enhances wettability and hydrophilicity of the surface.	Experimental results showed improved hydrophilicity and wettability after applying the initial coating.
Permeability and Flux Model	Explains how additional coatings increase resistance, leading to reduced system performance.	Data confirmed that subsequent coatings reduced permeability due to increased resistance, consistent with the model.
Separation Efficiency Model	Validates that the first coating provides the best performance for oil–water separation.	Tests revealed that the first coating achieved maximum oil separation efficiency, aligning with model projections.

5. Conclusions

This study demonstrates that a single-layer hydrophilic nanoparticle coating on quartz filtration membranes provides the optimal balance between permeability, oil–water separation efficiency, and structural stability. The first-coated quartz exhibited the most uniform nanoparticle distribution, minimizing surface roughness and preventing agglomeration, which maintained open filtration pathways and enhanced water permeability. Elemental composition analysis confirmed that the first coating significantly improved surface hydrophilicity (oxygen content: 82.54%), leading to efficient oil rejection. Raman spectroscopy results indicated that the structural integrity of quartz remained intact with the first coating, while additional coatings introduced surface irregularities and excessive nanoparticle clustering. Oil and grease analysis further validated this, showing that oil concentration in the filtrate dropped to 17.3 mg/L for the first coating, while additional coatings led to clogging and performance degradation.

Mathematical modeling aligned with experimental findings, confirming that the first coating maximized separation efficiency while maintaining permeability, whereas subsequent coatings increased resistance and reduced effectiveness. Overall, this study establishes that a single nanoparticle coating is the most effective modification for quartz filtration membranes, ensuring sustainable and high-performance oil–water separation while avoiding the inefficiencies caused by excessive coatings. These insights contribute to the development of scalable and efficient filtration technologies for industrial wastewater treatment and environmental applications. Future studies could focus on investigating advanced nanoparticle deposition techniques to improve uniformity and reduce agglomeration during multi-layer coating applications.

Author Contributions: Conceptualization, M.P. and N.R.; methodology, N.R.; software, N.R.; validation, M.P and N.R.; formal analysis, N.R.; investigation, N.R.; resources, M.P.; data curation, N.R.; writing—original preparation, N.R.; writing—review and editing, M.P. and N.R.; visualization, N.R.; supervision, M.P.; project administration, M.P. All authors have read and agreed to the published version of the manuscript.

Funding: This research received no external funding.

Data Availability Statement: The data that support the findings of this study are not publicly available due to ethical restrictions. However, summaries and processed results can be made available upon reasonable request to the corresponding author.

Acknowledgments: The authors would like to acknowledge the financial support provided by the University of South Africa (UNISA).

Conflicts of Interest: The authors declare no conflicts of interest.

References

1. Waheed, A.; Baig, U.; Abussaud, B.; Aljundi, I.H. Fabrication of supported carbide-derived-carbon membrane by two phases of interfacial polymerization for oil/water separation. *Ceram. Int.* **2022**, *48*, 8125–8135. [CrossRef]
2. Venzke, C.D.; Giacobbo, A.; Klauck, C.R.; Viegas, C.; Hansen, E.; Monteiro de Aquim, P.; Rodrigues, M.A.S.; Bernardes, A.M. Integrated Membrane Processes (EDR-RO) for Water Reuse in the Petrochemical Industry. *J. Membr. Sci. Res.* **2018**, *4*, 218–226. [CrossRef]
3. Bhattacharyya, A.; Liu, L.; Walsh, M.; Lee, K. Membrane technology for treating decanted oily wastewater from marine oil spill operations: Comparison between membrane filtration and membrane bioreactor. *Mar. Pollut. Bull.* **2023**, *194*, 115397. [CrossRef] [PubMed]
4. Martirosyan, A.; Ilyushin, Y.; Afanaseva, O.; Kukharova, T.; Asadulagi, M.; Khloponina, V. Development of an Oil Field's Conceptual Model. *Int. J. Eng. Trans. B Appl.* **2025**, *38*, 381–388. [CrossRef]
5. Wang, K.; Liu, X.; Dong, Y.; Zhang, S.; Li, J. A biomimetic Janus delignified wood membrane with asymmetric wettability prepared by thiol-ol chemistry for unidirectional water transport and selective oil/water separation. *Colloids Surf. A Physicochem. Eng. Asp.* **2022**, *652*, 129793. [CrossRef]
6. Yang, Y.; Jiang, X.; Goh, K.; Wang, K. The separation of oily water using low-cost natural materials: Review and development. *Chemosphere* **2021**, *285*, 131398. [CrossRef]
7. Bolto, B.; Zhang, J.; Wu, X.; Xie, Z. A Review on Current Development of Membranes for Oil Removal from Wastewaters. *Membranes* **2020**, *10*, 65. [CrossRef] [PubMed]
8. Hubadillah, S.K.; Othman, M.H.D.; Matsuura, T.; Ismail, A.F.; Rahman, M.A.; Harun, Z.; Jaafar, J.; Nomura, M. Fabrications and applications of low cost ceramic membrane from kaolin: A comprehensive review. *Ceram. Int.* **2018**, *44*, 4538–4560. [CrossRef]
9. Peng, B.; Zhao, Y.; Nie, D.; Yi, R.; Chen, L.; Zhang, L. Selective transport properties of graphene oxide membranes for various cations observed in situ using quartz crystal microbalance. *Appl. Surf. Sci.* **2021**, *541*, 148502. [CrossRef]
10. Sun, L.; Wang, Z.; Gao, B. Ceramic membranes originated from cost-effective and abundant natural minerals and industrial wastes for broad applications—A review. *Desalin. Water Treat.* **2020**, *201*, 121–138. [CrossRef]
11. Goswami, K.P.; Pugazhenthi, G. Credibility of polymeric and ceramic membrane filtration in the removal of bacteria and virus from water: A review. *J. Environ. Manag.* **2020**, *268*, 110583. [CrossRef] [PubMed]
12. Baig, U.; Usman, J.; Abba, S.I.; Yogarathinam, L.T.; Waheed, A.; Bafaqeer, A.; Aljundi, I.H. Insight into soft chemometric computational learning for modelling oily-wastewater separation efficiency and permeate flux of polypyrrole-decorated ceramic-polymeric membranes. *J. Chromatogr. A* **2024**, *1725*, 464897. [CrossRef]
13. Baig, U.; Al-Kuhaili, M.F.; Dastageer, M.A. Solar-light active tungsten oxide coated ceramic membrane with unique surface wettability fabricated by RF magnetron sputtering: Remediation of oily wastewater and photo-catalytic self-cleaning. *Sustain. Mater. Technol.* **2024**, *39*, e00837. [CrossRef]
14. Liu, Y.; Xin, B.; Newton, M.A.A.; Li, L.; Huang, D. Advanced photocatalytic self-cleaning membrane for highly efficient oil-water separation using C/TiO2/SiO2 composite. *J. Water Process Eng.* **2024**, *59*, 104969. [CrossRef]
15. Sharma, C.; Gupta, R.; George, J.K.; Verma, N. Janus resorcinol-formaldehyde-based membrane with opposite wettability for efficient separation of oil and water emulsion. *Colloids Surfaces A Physicochem. Eng. Asp.* **2023**, *676*, 132246. [CrossRef]
16. Elhady, S.; Bassyouni, M.; Mansour, R.A.; Elzahar, M.H.; Abdel-Hamid, S.; Elhenawy, Y.; Saleh, M.Y. Oily wastewater treatment using polyamide thin film composite membrane technology. *Membranes* **2020**, *10*, 84. [CrossRef]

17. Aiyd Jasim, M.; AlJaberi, F.Y. Investigation of oil content removal performance in real oily wastewater treatment by electrocoagulation technology: RSM design approach. *Results Eng.* **2023**, *18*, 101082. [CrossRef]
18. Khalifa, O.; Banat, F.; Srinivasakannan, C.; AlMarzooqi, F.; Hasan, S.W. Ozonation-assisted electro-membrane hybrid reactor for oily wastewater treatment: A methodological approach and synergy effects. *J. Clean. Prod.* **2021**, *289*, 125764. [CrossRef]
19. Gao, J.; Cao, H.; Hu, X.; Mao, H.; Chen, X.; Qiu, M.; Verweij, H.; Fan, Y. Journal of the European Ceramic Society Piezoelectric porous α -quartz membrane by aqueous gel-casting with enhanced antifouling and mechanical properties. *J. Eur. Ceram. Soc.* **2023**, *43*, 109–120. [CrossRef]
20. Ugarte, P.; Ramo, A.; Quílez, J.; Bordes, M.d.C.; Mestre, S.; Sánchez, E.; Peña, J.Á.; Menéndez, M. Low-cost ceramic membrane bioreactor: Effect of backwashing, relaxation and aeration on fouling. Protozoa and bacteria removal. *Chemosphere* **2022**, *306*, 135587. [CrossRef] [PubMed]
21. Hudaib, B.; Al-Qodah, Z.; Abu-Zurayk, R.; Waleed, H.; Omar, W. Fabrication of blended cellulose acetate/poly-pyrrole ultrafiltration membranes for crude oil wastewater separation. *Case Stud. Chem. Environ. Eng.* **2024**, *9*, 100692. [CrossRef]
22. Chen, Y.; Quan, Z.; Song, W.; Wang, Z.; Li, B.; Mu, Z.; Niu, S.; Zhang, J.; Han, Z.; Ren, L. Hierarchically structured biomimetic membrane with mechanically/chemically durability and special wettability for highly efficient oil–water separation. *Sep. Purif. Technol.* **2022**, *300*, 121860. [CrossRef]
23. Liang, Y.; Yang, E.; Kim, M.; Kim, S.; Kim, H.; Byun, J.; Yanar, N.; Choi, H. Lotus leaf-like SiO2 nanofiber coating on polyvinylidene fluoride nanofiber membrane for water-in-oil emulsion separation and antifouling enhancement. *Chem. Eng. J.* **2023**, *452*, 139710. [CrossRef]
24. Pandidan, S.; Mechler, A. Membrane morphology effects in quartz crystal microbalance characterization of antimicrobial peptide activity. *Biophys. Chem.* **2020**, *262*, 106381. [CrossRef]
25. Ramanamane, N.J.; Sob, P.B.; Alugongo, A.A.; Tengen, T.B. Integrated Membrane Feed Stream Performance Improvement—A Review. *Int. J. Mech. Prod. Eng. Res. Dev.* **2021**, *11*, 311–328.
26. Liu, P.; Niu, L.; Tao, X.; Li, X.; Zhang, Z.; Yu, L. Preparation of superhydrophobic-oleophilic quartz sand filter and its application in oil-water separation. *Appl. Surf. Sci.* **2018**, *447*, 656–663. [CrossRef]
27. Cai, J.; Liu, S.; Gao, C.; Ji, S.; Xing, Y. A high-toughness anti-pollution polyethersulfone membrane for efficient separation of oil-in-water emulsions. *J. Environ. Chem. Eng.* **2023**, *11*, 108980. [CrossRef]
28. Wu, L.; Gao, Y.; Xu, X.; Deng, J.; Liu, H. Excellent coagulation performance of polysilicate aluminum ferric for treating oily wastewater from Daqing gasfield: Responses to polymer properties and coagulation mechanism. *J. Environ. Manag.* **2024**, *356*, 120642. [CrossRef]
29. Sutrisna, P.D.; Mustika, P.C.B.W.; Hadi, R.P.; Caren; Gani, Y.E. Improved oily wastewater rejection and flux of hydrophobic PVDF membrane after polydopamine-polyethyleneimine co-deposition and modification. *S. Afr. J. Chem. Eng.* **2023**, *44*, 42–50. [CrossRef]
30. Rukhsana, Z.; Babiker, D.M.D.D.; Zhao, Y.; Chen, X.; Li, L.; Usha, Z.R.; Babiker, D.M.D.D.; Zhao, Y.; Chen, X.; Li, L.; et al. Advanced super-wetting biaxial polypropylene membrane with hierarchical rough surface for multipollutant removal from oily wastewater. *J. Environ. Chem. Eng.* **2023**, *11*, 110775. [CrossRef]
31. Kumari, P.; Chauhan, P.; Kumar, A. Superwetting Materials for Modification of Meshes for Oil/Water Separation. *ACS Symp. Ser.* **2022**, *1408*, 1–23. [CrossRef]
32. Yang, H.; Wang, X.; Chen, X.; Wang, Y.; Wang, D.; Cheng, W.; Han, G. Hybrid membranes based on electrospun nanofiber modified with PDA for selective oil-in-water emulsion separation. *J. Memb. Sci.* **2024**, *698*, 122610. [CrossRef]
33. Hu, J.; Yuan, S.; Zhao, W.; Li, C.; Liu, P.; Shen, X. Fabrication of a superhydrophilic/underwater superoleophobic PVDF membrane via thiol–ene photochemistry for the oil/water separation. *Colloids Surf. A Physicochem. Eng. Asp.* **2023**, *664*, 131138. [CrossRef]
34. Sob, P.B.; Alugongo, A.A.; Tengen, T.B. The stochastic effect of nanoparticles inter-separation distance on membrane wettability during oil/water separation. *Int. J. Eng. Res. Technol.* **2020**, *13*, 842–866. [CrossRef]
35. Wu, Z.; Xue, J.; Zhang, J.; Wang, M.; Chen, Z.; Lv, Y.; Ma, S. Photocatalytic self-cleaning membrane with polyaniline/NH2-MIL-125 heterojunction for highly oil-water/seawater separation and bacterial inactivation. *Sep. Purif. Technol.* **2023**, *323*, 124412. [CrossRef]
36. Jepsen, K.L.; Bram, M.V.; Pedersen, S.; Yang, Z. Membrane Fouling for Produced Water Treatment: A Review Study from A Process Control Perspective. *Water* **2018**, *10*, 847. [CrossRef]
37. Radoor, S.; Karayil, J.; Jayakumar, A.; Siengchin, S. Efficient removal of dyes, heavy metals and oil-water from wastewater using electrospun nanofiber membranes: A review. *J. Water Process Eng.* **2024**, *59*, 104983. [CrossRef]
38. Ma, B.; Wu, G.; Li, W.; Miao, R.; Li, X.; Wang, P. Roles of membrane–foulant and inter/intrafoulant species interaction forces in combined fouling of an ultrafiltration membrane. *Sci. Total Environ.* **2019**, *652*, 19–26. [CrossRef] [PubMed]
39. Xin, Y.; Qi, B.; Wu, X.; Yang, C.; Li, B. Different types of membrane materials for oil-water separation: Status and challenges. *Colloid Interface Sci. Commun.* **2024**, *59*, 100772. [CrossRef]

40. Dansawad, P.; Yang, Y.; Li, X.; Shang, X.; Li, Y. Smart membranes for oil/water emulsions separation: A review. *Adv. Membr.* **2022**, *2*, 100039. [CrossRef]
41. Samuel, O.; Hafiz, M.; Othman, D.; Kamaludin, R.; Agustiono, T.; Li, T.; Dzinun, H.; Imtiaz, A.; Othman, M.H.D.; Kamaludin, R.; et al. Treatment of oily wastewater using photocatalytic membrane reactors: A critical review. *J. Environ. Chem. Eng.* **2022**, *10*, 108539. [CrossRef]
42. Zou, D.; Fan, W.; Xu, J.; Drioli, E.; Chen, X.; Qiu, M.; Fan, Y. One-step engineering of low-cost kaolin/fly ash ceramic membranes for efficient separation of oil-water emulsions. *J. Memb. Sci.* **2021**, *621*, 118954. [CrossRef]
43. Sun, Y.; Yang, Z.; Li, L.; Wang, Z.; Sun, Q. Facile preparation of isotactic polypropylene microporous membranes with bioinspired hierarchical morphology for nano-scale water-in-oil emulsion separation. *J. Memb. Sci.* **2019**, *581*, 224–235. [CrossRef]
44. Yuan, Z.; Zeng, L.; Zhang, H.; Li, T.; Liu, C.; Yang, C.; Jiang, L.; Liu, P.; Li, Y.; Li, H. TiO_2 modified basalt fiber cloth with superhydrophilicity and underwater superoleophobicity for oil-water separation. *Mater. Today Commun.* **2024**, *39*, 109200. [CrossRef]
45. Pan, L.; Zheng, Q.; Feng, Q.; Shen, Y.; Hu, W.; Cao, C.; Zhang, G.; Gao, J.; Song, P.; Shi, Y.; et al. Hydrophobic silicone modified membranes for efficient oil/water separation: Synthesis, fabrication and application. *Sep. Purif. Technol.* **2025**, *353*, 128485. [CrossRef]

water MDPI

Article

Impacts of Wastewater Management and Enhancing the Landscape of the Mae Kha Canal: A Quasi-Experimental Study

Vongkot Owatsakul [1], Prajuab Panput [1], Punyaphol Jaisuda [1] and Damrongsak Rinchumphu [2,*]

[1] Public Works Bureau, Sanitary Engineer Division, Chiang Mai Municipality, Chiang Mai 50300, Thailand; vongkoto@cmcity.go.th (V.O.); prajuab.p@cmcity.go.th (P.P.); panyaphol_j@cmcity.go.th (P.J.)

[2] Faculty of Engineering, Chiang Mai University, Chiang Mai 50200, Thailand

* Correspondence: damrongsak.r@cmu.ac.th

Abstract: The Mae Kha Canal in Chiang Mai, Thailand, has long suffered from severe water quality deterioration due to rapid urbanization, population growth, and inadequate waste management practices. This article describes an integrated water resource management approach, started in February 2018, with the goal of guaranteeing sustainable urban living conditions and improving the water quality of the canal. This study was a quasi-experimental study, with key interventions including garbage net installation, coconut-fiber mattress weir installation, and Free Water Surface Wetland treatment using vetiver grass. An interrupted time-series analysis of the monthly biochemical oxygen demand (BOD) and dissolved oxygen (DO) levels were applied to examine the trends and changes after the full implementation of wastewater management in March 2021. The results demonstrated significant improvements in water quality, indicated by reduced BOD levels (from 17.00–38.70 to 9.00–12.67 mg/L) and increased DO levels (from 0.00–2.40 to 0.80–6.00 mg/L). However, the decreases in BOD and increases in DO levels were not stable during the year. The post-intervention trend of BOD level decreased after the wastewater management implementation at SriDonChai Road (coefficient of trend: = −0.75 [95% confidence interval: −1.44 to −0.06]). This project highlights the effectiveness of integrated management strategies in addressing urban water quality issues and emphasizes the importance of community involvement in sustainable environmental management. The findings underscore the necessity for integrated approaches to wastewater management in urban environments to address both ecological health and public welfare.

Keywords: Mae Kha Canal; urban wastewater management; water quality restoration; water sustainability

1. Introduction

Basic requirements for good health include having a safe place to live and access to clean drinking water for households, industry, and agriculture [1]. Even though more than 70% of the world is covered with water, only 3% is freshwater and less than 0.01% of the world's freshwater sources are suitable for human consumption [1–3]. The world's population growth brings with it fierce competition for the planet's resources, especially water and energy; these are at grave risk due to water pollution and global warming, respectively [4]. Moreover, increased water use by industry and agriculture, as well as unsustainable urbanization and industrialization, has resulted in water contamination from the release of wastewater, toxic effluents, and chemicals into water bodies [4–6], which could lead to marked water shortages. Contaminated water and wastewater consequently have negative impacts on public health, economy, agricultural security, drinking water supply, and biodiversity [7]. Wastewater remains a problem in several parts of the world. Hence, appropriate water utilization and wastewater management, including treatment, filtration, and reuse, have been studied and discussed for sustainable development.

There are various strategies for treating wastewater [1,8], such as adsorption [2,9,10], membrane filtration [11], coagulation/flocculation [12–14], oxidation [15–17], and biological treatment [14,18,19]. In addition, new wastewater treatment techniques and strategies have been forthcoming. For example, Muthukumaran and Ambujam [20] proposed strategies for improving wastewater management in Tiruchirappalli City, Tamil Nadu, India, and made recommendations for the use of treated wastewater in agriculture. Del Villar and García-López [21] studied the potential of reusing water from the Canal de Isabel II in Madrid, Spain, using water stress data and wastewater reuse to aid decision-making. Frascari et al. [22] investigated integrated technology and management solutions for efficient agricultural wastewater treatment and reuse in Egypt, Morocco, and Tunisia. Iqbal et al. [23] presented a water quality management modeling approach to control the concentration of pollutants in the Ravi River, Punjab, Pakistan. The application of vetiver grass is one approach to treat organic matter in contaminated surface water, and has been applied to treat polluted water and for environmental protection [24–27]. The findings of a recent experimental study in Vietnam show that vetiver grass application in vertical-flow constructed wetlands achieves a removal efficiency of 90% for biochemical oxygen demand (BOD) and >80% for chemical oxygen demand (COD) [28].

Thailand's rapid economic and social expansion is causing many egregious environmental problems. At present, risk factors affecting water management comprise natural disasters and human actions, including encroachment on public water sources, increasing population, the expansion of urban communities, the creation of water barriers, the release of pollutants into rivers and canals, etc. All of these negatively affect water resources, both qualitatively and quantitatively, and this is an important national problem that affects the living conditions of the people, as well as the economic and social stability, of Thailand. In 2022, the Pollution Control Department of Thailand measured the quality of surface water in 49 major rivers throughout Thailand using the water quality index; they found that the water quality was good, fair, and deteriorated in 43%, 41%, and 16% of them, respectively [29]. Thus, Thailand faces problems with water quality in many areas, especially in urban areas with large populations, such as Bangkok and the surrounding area, as well as locations of industrial plants, such as Chiang Mai, Rayong, Nonthaburi, and Samut Prakan. There are several water resources in Thailand faced with wastewater issues. For example, Saen Saep Canal in Bangkok has had problems with wastewater for a long time. Many researchers have studied and attempted to mitigate these issues. Techapichetwong and Phasukphan [30] proposed the application of pollution flow diagrams to manage the water quality of the canal, while Somsuk and PilaiLa [31] studied the impact of population growth on the water quality therein and methods to improve it.

The Mae Kha Canal in Chiang Mai Province is another canal that has experienced deteriorating water quality for a long time. In 2017, the Chiang Mai Provincial Office reported that the biochemical oxygen demand (BOD) and the dissolved oxygen (DO) levels in this canal were 4.58–8.39 and 0.17–3.1 mg/L, respectively, which do not meet the good surface water quality standards [32]. This canal is strategically important for draining water from the urban area into the Ping River. In the past, it was a water source that was used for agriculture, consumption, and water transportation. However, population growth in Chiang Mai has led to communities becoming densely located on both sides of the Mae Kha Canal along its length—more than 20 communities out of a total of 99 [32,33]. Many residences, small businesses, and factories have been constructed along and encroach on the Mae Kha Canal, with the people therein dumping garbage, and wastewater from the buildings and establishments draining into it. This economic and residential zone has deteriorated both physically and ecologically, which has impacted the city by causing water quality problems [14]. In 1997, a wastewater collection system was constructed, covering 40 km^2 of Chiang Mai Municipality, to collect and transfer wastewater to the wastewater treatment plant. However, the collection system did not cover both sides of the Mae Kha Canal. Therefore, the Chiang Mai Municipality Office aimed to conduct a Mae Kha Canal development project to improve the quality of this canal with integrated water resource management.

After examining the dumping of wastewater into the Mae Kha Canal, the Chiang Mai Municipality arranged to dredge it and construct drainage and wastewater collection pipes on both sides of it. In addition, the municipality has provided knowledge and understanding to the community by organizing activities and encouraging participation by the community to conserve and maintain the Mae Kha Canal sustainably and improve the landscape of the public walkways on both sides of the Mae Kha Canal. Therefore, we aimed to investigate the impact of the wastewater treatment system of the Mae Kha Canal development project, which was implemented from February 2018 to December 2022, to ensure that the canal has good water quality and that people have suitable housing.

2. Materials and Methods

2.1. Location and Communities of the Mae Kha Canal

The Mae Kha Canal is the oldest in Chiang Mai Province. It originates in Doi Suthep-Pui National Park in Mae Rim District, Chiang Mai Province, and its endpoint is in Hang Dong District, Chiang Mai Province, the total length of which is approximately 30 km [32]. The Mae Kha Canal is divided into three sections [32]: the first section is in the Mae Rim District (~8.19 km), the second is in the Mueang District, Chiang Mai Municipality (~10.20 km), and the third is from the Pa Daet Subdistrict Municipality in Mueang District to the Sop Mae Kha Subdistrict in Hang Dong District (~12.00 km), where the water then runs into the Ping River. There are 6 communities in the first section, 25 in the second section (known for its dense population of 25,352 in July 2024, which has resulted in slum settlements), and 11 in the third section [33]. This study was focused on the second section (see Figure 1).

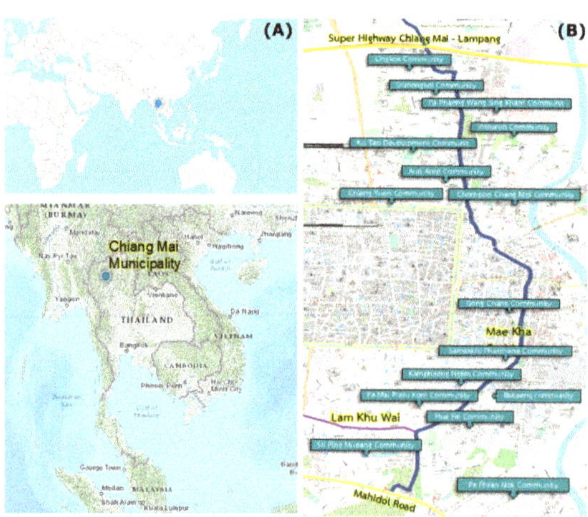

Figure 1. A map showing the locations of Chiang Mai Province (**A**) and the Mae Kha Canal (**B**).

2.2. The Mae Kha Canal Development Project

The Mae Kha Canal development project had a quasi-experimental design, with integrated water resource management conducted with five procedures: (1) demolishing houses and buildings straddling the Lam Mueang branch and dredging out sediment from the canal; (2) laying drainage pipes to collect wastewater from buildings, businesses, and the drainage system in the area to manage sewage and solid waste in the canal; (3) creating a weir to slow down the water flow with coconut-fiber mattresses; (4) testing the use of the wastewater management and treatment system in the Mae Kha Canal; and (5) constructing a wall to prevent soil slides, a sanitary sewerage system, and a public walkway on both sides of the Mae Kha Canal. Details of the Mae Kha Canal development project, such as the participating organizations, objectives, work procedures, and indicators, are provided in Table 1.

Table 1. Details of the Mae Kha Canal development project.

Procedure	Period	Participating Organizations	Objectives	Work Processes	Indications
1. Demolishing buildings encroaching on the canal and dredging out sediment	February 2018	Municipality, The 33rd Military District, Marine Office, Provincial Land Office	To drain water; reduce environmental problems such as wastewater, garbage, and sewage; and improve the scenery	Demolishing buildings encroaching on the canal (1) Meeting with the local community to discuss the requirement for dismantling the buildings. (2) Signing an agreement. (3) Dismantling houses and buildings encroaching on the Mae Kha Canal under the supervision of sanitary engineers. Dredging out sediment (1) Determining the scope of the work. (2) Undertaking a public relations exercise with the affected communities in the work area. (3) Producing guidelines for water management. (4) Determining the location for bringing machinery into the area. (5) Removing sediment from the area. (6) Determining where to dispose of the sediment.	(1) The Mae Kha Canal's drainage rate increased more than five times from before the project began, from less than 5 cubic meters per second to 25 cubic meters per second. (2) The water quality, as indicated by the dissolved oxygen (DO) levels, showed a statistically significant improvement, with DO values rising from 0 mg/L to over 2 mg/L.
2. Managing sewage and solid waste	January 2021 to March 2022	Municipality	To reduce the amount of solid waste and improve the water flow	(1) Planning appropriate locations to install garbage nets and sewage barriers. (2) Determining the size of the nets and panels. (3) Installing them under the supervision of a sanitary engineer. (4) Inspecting after installation and planning scheduled maintenance and removal of garbage from the nets.	(1) Reduce garbage collection in Ma Kha canal from 4 times per month to 1 time per month. (2) The water flow rate was constant during the project operation.
3. Creating a weir to slow down the water flow with coconut-fiber mattresses	October 2020 to October 2022	Municipality	To trap sediment, increase the dissolved oxygen (DO) content, reduce erosion of the riverbank, and store groundwater above the weir	(1) Reusing leftover materials (coconut-fiber mattresses) from establishments. (2) Determining the optimal location for their installation. (3) Inspecting and improving the condition of the original soil under the weir base to make it strong and stable. (4) Constructing a weir using coconut-fiber mattresses as filters in a bamboo structure with sandbags. (5) Setting the height of the weir ridge to drain water efficiently and prevent flooding in the area above the weir. (6) Checking the results after construction.	After building a weir with mattresses made of coconut fiber, the increase in oxygen and DO was higher than or equivalent to 4 mg/L.

Table 1. *Cont.*

Procedure	Period	Participating Organizations	Objectives	Work Processes	Indications
4. Testing the use of a wastewater management and treatment system	October 2019 to February 2021	Municipality	To improve the water quality, reduce the amount of organic pollution, and increase the DO content	(1) Drafting a plan and determining the steps for installation in the operational area. (2) Creating a Free Water Surface Wetland (FWS) wastewater treatment system using vetiver grass. (3) Carrying out the installation. (4) Continuously checking the condition of the treatment system and providing maintenance guidelines.	The quantity of nitrogen and phosphorus, or organic matter, has reduced, and the DO value is more than or equal to 4 mg/L and the BOD value is less than or equal to 2 mg/L
5. Constructing a retaining wall and a public walkway on both sides of the canal	April 2021 to September 2022	Municipality	To protect the banks on both sides of the canal, prevent flooding during the rainy season, collect and treat wastewater from households and establishments, and encourage people within the community to use the public walkway by including an area for trading, recreation, and exercise	(1) Building 1.5 km of wastewater pipelines on each side of the Mae Kha Canal. (2) Enhancing the landscape, covering a distance of 0.785 km.	(1) The Do value is more than 2 mg/L. (2) There is no wastewater pipe connection into the Mae Kha Canal. (3) Reduce overflow flooding from more than 10 times per year before construction to not happening at all in 2023.

2.2.1. Demolishing Houses and Buildings Encroaching on the Mae Kha Canal and Dredging out the Sediment

Before the development project, 57 out of 1207 households located along the Mae Kha Canal in the Srimongkhon and Paphaeng communities were encroaching on the Mae Kha Canal (Figure 2A). These households were built in the direction of the flowing water, resulting in the area being at risk of flooding because they were preventing rainwater from flowing into the canal. Moreover, these households were releasing wastewater into the canal, thereby lowering the water quality. After demolishing them and dredging out the sediment, the drainage into the canal and the look of the area were greatly improved (Figure 2B).

Figure 2. Comparison of before (**A**), during (**B**), and after (**C**) demolishing houses encroaching on the Mae Kha Canal and dredging out sediment.

2.2.2. Managing Sewage and Solid Waste in the Mae Kha Canal

The Mae Kha Canal area was unattractive before this effort to control garbage and solid waste was started. It was filled with a great deal of trash, and the residents living nearby complained about the stench emanating from the solid waste therein (Figure 3A). Following the commencement of this task, garbage was collected from sewer pipes using nets in predetermined locations. The garbage nets were made from Nylon with a 2.5 cm hole size; they were 4 m wide and 10 m long, and fixed to a steel frame over a drainpipe

(Figure 3B). At first, the Chiang Mai Municipality was gathering up to 500 kgs of rubbish per week from the nets, which was reduced to just 200 kgs of garbage per week. This is a result of local businesses and residents working together to prevent trash from being dumped into the canal. As a result, the Mae Kha Canal has stunning scenery, and the water flow has been greatly improved (Figure 3C).

(A) (B) (C)

Figure 3. The Mae Kha Canal (**A**) before and (**B**) after initiating the solid waste and sewage management system and (**C**) examples of garbage that can be captured.

2.2.3. Creating a Weir to Slow Down the Water Flow with Coconut-Fiber Mattresses

For this procedure, we aimed to reuse leftover materials—coconut-fiber mattresses—to construct a weir as a filter, using a bamboo structure with sandbags. This was installed in an optimal location, and is a strong and stable long-term installation (Figure 4). The drainage rate depends on the natural drainage rate in each season. In addition, it also prevents the erosion of the canal bank.

Figure 4. Constructing the water-slowing weir using coconut-fiber mattresses.

2.2.4. Testing the Wastewater Management and Treatment System of the Mae Kha Canal

A FWS wastewater treatment system for the Mae Kha Canal was constructed using vetiver grass (Figure 5). An FWS wastewater treatment system using vetiver grass is a widely recognized and accepted method. Vetiver grass is effective in treating wastewater in rivers, canals, and industrial settings, and it can be applied in various areas, including narrow spaces, uneven terrain, and sloped regions. We continuously checked the condition of the treatment system and provided maintenance guidelines from October 2019 to February 2021.

Figure 5. The FWS wastewater treatment system using vetiver grass platforms.

2.2.5. Constructing a Retaining Wall to Prevent Land Slides, and Implementing a Sanitary Sewerage System and a Public Walkway on Both Sides of the Mae Kha Canal

For this phase, we created 1.5 km of wastewater pipelines on each side of the Mae Kha Canal to protect the banks on both sides of the canal, prevent flooding during the rainy season, collect and treat wastewater from households and establishments, and encourage people within the community to use the public walkway by including an area for trading, recreation, and exercise (Figure 6).

Figure 6. The public promenades and walls prevent soil erosion on both sides of the Mae Kha Canal.

2.3. *Water Quality Evaluation Methods*

DO and BOD were measured to evaluate the water quality in Mae Kha Canal. The DO (mg/L) level for the wastewater was measured using Azide Modification [34]. This approach calculated the DO level from the sodium thiosulfate ($Na_2S_2O_3$) titration of 200 mL water samples, as follows:

$$1 \text{ mL } 0.0025\text{M } Na_2S_2O_3 = 1 \text{ mg } DO/L \quad (1)$$

BOD (mg/L) is a measure of the amount of DO consumed by microorganisms during the decomposition of organic matter in water. The BOD test typically involves measuring the amount of oxygen consumed over a specific period (5 days in the present study). It is calculated as follows [34]:

$$\text{BOD} = \left(\frac{DO_0 - DO_5}{P}\right) \quad (2)$$

where DO_0 and DO_5 are the initial and final DO concentrations in the water sample, respectively (mg/L), and P represents the fraction of the BOD bottle that is represented by the sample. The criteria for interpreting water quality are provided in Table 2.

Table 2. Criteria for interpreting water quality [29].

Water Quality	DO (mg/L)	BOD (mg/L)
Good	>4.0–6.0	0.0–1.5
Moderate	>2.0–4.0	>1.5–2.0
Degenerate	0.0–2.0	>2.0–4.0

Note(s): BOD, biochemical oxygen demand; DO, dissolved oxygen.

The evaluations of water quality using the BOD and DO indices were conducted monthly from water samples at three locations, including Mae Kha Bridge (Atsadathon Road), Klong Ngern Community Bridge (Klang Stream), and Mae Kha Bridge (SriDonChai Road), 40 times from October 2019 to January 2023 (Figure 7). The BOD and DO evaluations may be unavailable for some months due to drought or severe contamination.

Figure 7. Maps of the locations of water quality measurements and five procedures of the Mae Kha Canal development project.

2.4. Statistical Analysis

The missing BOD and DO levels were imputed via linear interpolation using the nearest available values. The BOD and DO levels of sampled water were summarized using descriptive statistics as means, standard deviations, and ranges for each location. An interrupted time-series analysis of the monthly BOD and DO levels was applied to examine the trends and changes after the full implementation of wastewater management (garbage net installation, coconut-fiber mattress weir installation, and FWS treatment using vetiver grass.) in March 2021. A p-value < 0.05 was considered statistically significant. All of the analyses were performed using Stata version 17.

3. Results

3.1. BOD and DO Levels

BOD and DO levels were measured monthly from October 2019 to January 2023. The averages and ranges of levels at each location are provided in Table 3. The average BOD level at SriDonChai Road was the lowest (16.83 (range: 4.00–49.00) mg/L). Meanwhile, the average DO levels at Atsadathon Road and SriDonChai Road were similar (1.73 (range: 0.00–6.50) and 1.58 (range: 0.00–9.50) mg/L, respectively), and higher than those in the Klang Stream.

Table 3. Biochemical oxygen demand (BOD) and dissolved oxygen (DO) levels at various locations.

Locations	BOD (mg/L)		DO (mg/L)	
	Means (SDs)	Ranges	Means (SDs)	Ranges
Atsadathon Road	22.68 (17.91)	2.50–102.00	1.73 (1.85)	0.00–6.50
Klang Stream	21.84 (12.16)	5.50–48.00	0.94 (1.04)	0.00–4.00
SriDonChai Road	16.83 (10.14)	4.00–49.00	1.58 (1.93)	0.00–9.50

Note(s): BOD, biochemical oxygen demand; DO, dissolved oxygen.

3.2. Impact of the Coconut-Fiber Mattress Weir

Prior to installing a weir comprising coconut-fiber mattresses, the Mae Kha Canal was suffering from bank erosion and harsh conditions because of material flowing into the river. The BOD values in the Klang Stream decreased from 46.67 mg/L in September 2020 to 8.17 mg/L in November 2022, while at SriDonChai Road, the values declined from 8.50 mg/L in September 2020 to 5.00 mg/L in November 2022. Although the BOD values still did not meet the target value of less than 2.0, implementing the weir comprising coconut-fiber mattresses significantly improved the water quality. As presented in Figure 8, the BOD and DO levels also improved at the Atsadathon Road and SriDonChai Road stations.

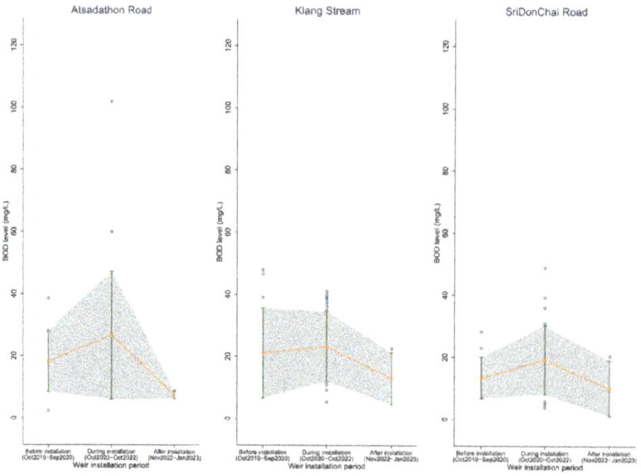

(A)

Figure 8. *Cont.*

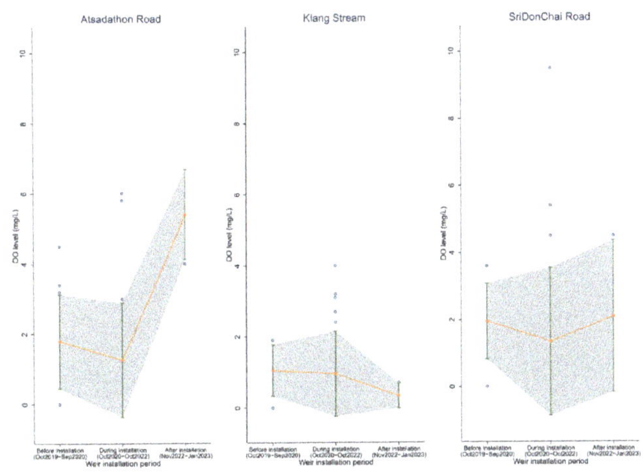

(B)

Figure 8. Comparison of (**A**) biochemical oxygen demand (BOD) and (**B**) dissolved oxygen (DO) values before, during, and after constructing the bamboo-structured weir filter containing coconut-fiber mattresses to slow down the water flow.

3.3. Impact of Vetiver Grass in the Wastewater Management and Treatment System

During the installation and subsequent testing of the FWS wastewater treatment system comprising vetiver grass in the Mae Kha Canal from October 2019 to February 2021, the BOD levels decreased from 17.00–38.70 to 9.00–12.67 mg/L and the DO levels increased from 0.00–2.40 to 0.80–6.00 mg/L. However, the average BOD levels increased (Figure 9A), while the DO levels decreased (Figure 9B) compared to the levels measured during the installation and testing period.

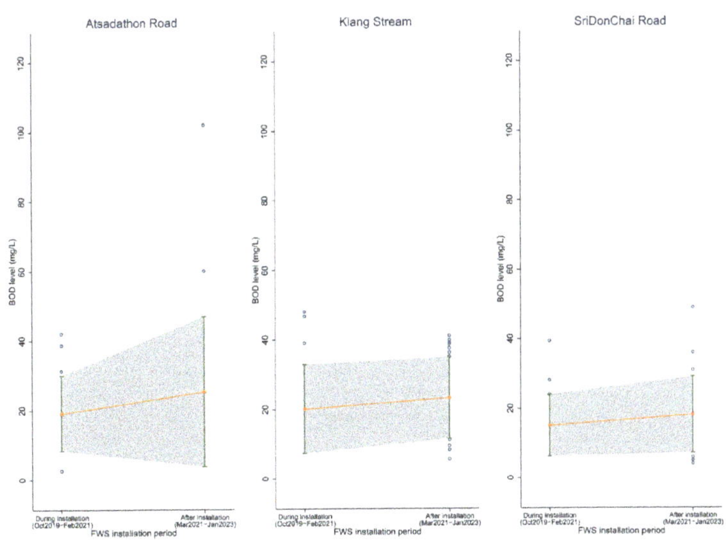

(A)

Figure 9. *Cont.*

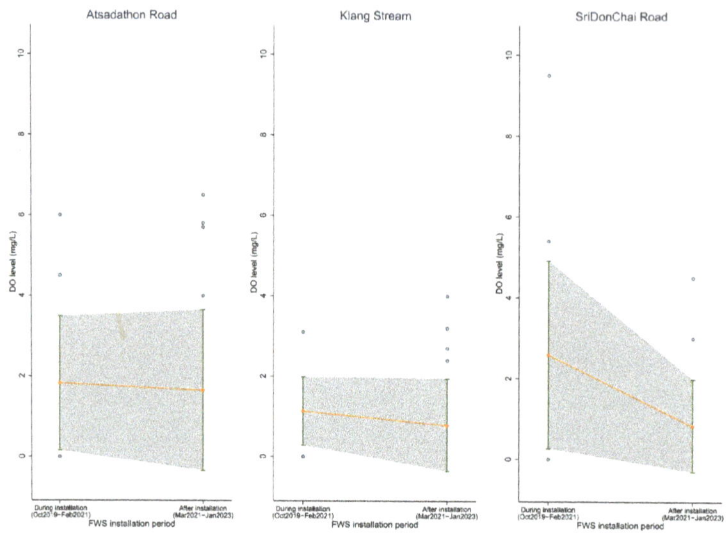

(**B**)

Figure 9. Comparison of (**A**) biochemical oxygen demand (BOD) and (**B**) dissolved oxygen (DO) values before and after installing and testing the FWS wastewater treatment system.

3.4. Impact of the Wastewater Management and Treatment System

As presented in Figure 10, after the full implementation of the wastewater management and treatment system (procedures 2, 3, and 4) in March 2021, there was no significant change in BOD levels at any of the locations, and in particular, the coefficient of the trend negatively changed at SriDonChai Road (−0.96 [95% confidence interval (CI): −1.92 to −0.005]; *p*-value = 0.049). Meanwhile, the DO levels significantly decreased at SriDonChai Road (−3.98 [95% CI: −6.96 to −1.00]; *p*-value = 0.010) and decreased at Atsadathon Road (−2.34 [95% CI: −4.92 to 0.25]; *p*-value = 0.076). The post-intervention trend estimates show that the BOD level significantly decreased at SriDonChai Road (−0.75 [95% CI: −1.44 to −0.06]; *p*-value = 0.033) and DO levels increased at Atsadathon Road (0.15 [95% CI: −0.003 to 0.31]; *p*-value = 0.054) and SriDonChai Road (0.07 [95% CI: −0.007 to 0.15]; *p*-value = 0.072) (Table 4).

Table 4. Interrupted time-series analysis of biochemical oxygen demand (BOD) and dissolved oxygen DO levels by location.

Parameters	Atsadathon Road				Klang Stream				SriDonChai Road			
	Coef.	*p*-Value	95% CI		Coef.	*p*-Value	95% CI		Coef.	*p*-Value	95% CI	
BOD												
Intercept (β_0)	21.31	<0.001 *	11.40	31.22	13.81	0.001 *	6.20	21.42	13.20	<0.001 *	7.07	19.34
Trend (β_1)	−0.24	0.616	−1.23	0.74	0.70	0.202	−0.39	1.78	0.21	0.544	−0.48	0.89
Level change after March 2021 (β_2)	10.92	0.358	−12.84	34.68	2.49	0.773	−16.61	22.19	9.48	0.181	−4.62	23.58
Trend change after March 2021 (β_3)	0.01	0.988	−1.76	1.79	−1.24	0.074	−2.60	0.13	−0.96	0.049 *	−1.92	−0.005
Post-intervention trend estimates	−0.23	0.746	−1.67	1.21	−0.54	0.158	−1.30	0.22	−0.75	0.033 *	−1.44	−0.06
DO												
Intercept (β_0)	1.36	0.104	−0.29	3.01	1.21	0.006 *	0.38	2.04	1.14	0.168	−0.50	2.78
Trend (β_1)	0.05	0.583	−0.14	0.24	−0.01	0.853	−0.09	0.08	0.16	0.157	−0.07	0.39

Table 4. *Cont.*

Parameters	Atsadathon Road				Klang Stream				SriDonChai Road			
	Coef.	*p*-Value	95% CI		Coef.	*p*-Value	95% CI		Coef.	*p*-Value	95% CI	
Level change after March 2021 (β_2)	−2.34	0.076	−4.92	0.25	−0.69	0.162	−1.68	0.29	−3.98	0.010 *	−6.96	−1.00
Trend change after March 2021 (β_3)	0.10	0.411	−0.15	0.35	0.05	0.402	−0.06	0.15	−0.09	0.445	−0.33	0.15
Post-intervention trend estimates	0.15	0.054	−0.003	0.31	0.04	0.271	−0.03	0.11	0.07	0.072	−0.007	0.15

Note(s): BOD, biochemical oxygen demand; DO, dissolved oxygen; Coef., coefficient; CI, confidence interval.
* p-value < 0.05.

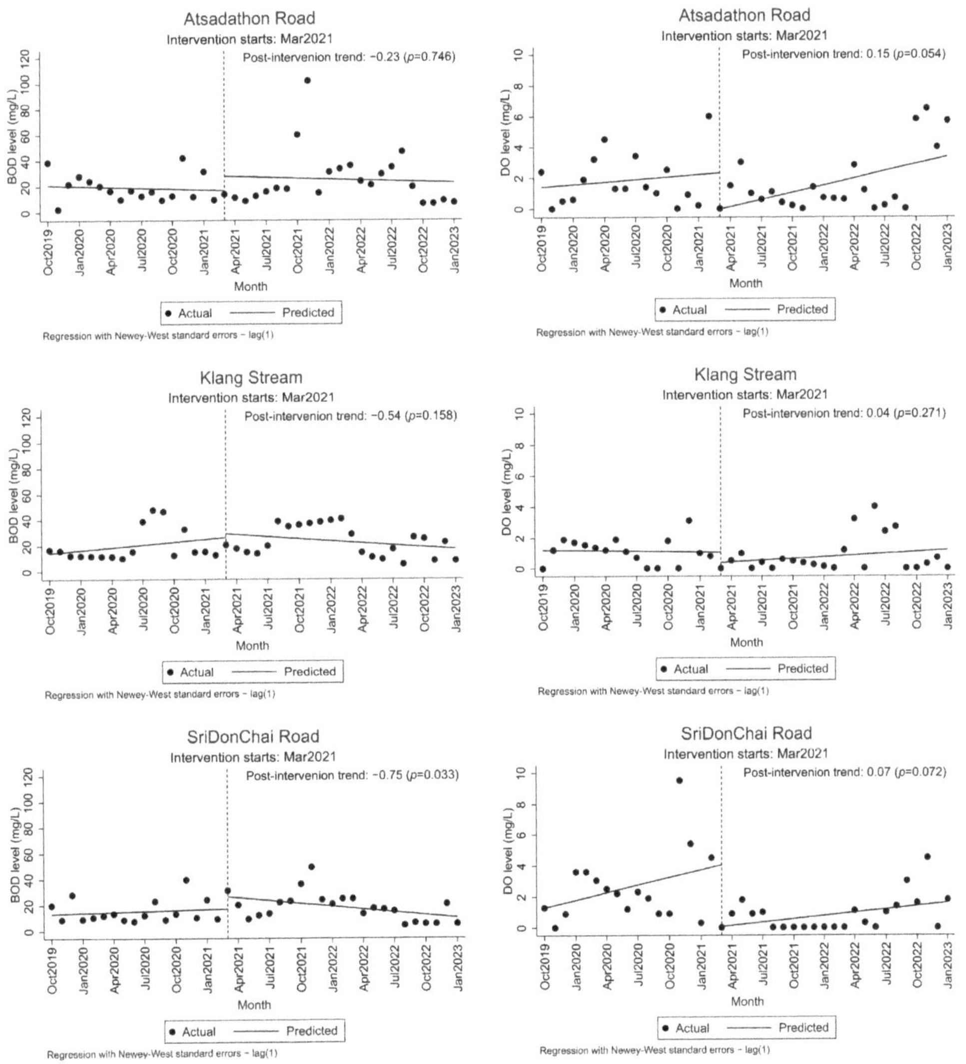

(A) (B)

Figure 10. Interrupted time-series analysis of (**A**) biochemical oxygen demand (BOD) and (**B**) dissolved oxygen (DO) values before and after full wastewater management and treatment system implementation.

3.5. Advantages of Retaining Wall, Sanitary Sewerage System, and Public Walkway

The retaining walls help prevent soil erosion during the rainy season. The sanitary sewerage system avoids the drainage of household wastewater into the canal and redirects it into the Chiang Mai Municipality Wastewater Treatment Plant before being released into water sources. In addition, the public walkway along both sides of the Mae Kha Canal provides local people with a place to rest and exercise. It has become a new tourist attraction in Chiang Mai city, with more than 1000 tourists visiting it per day and generating an income of more than THB 36,000,000 per year. Therefore, the quality of water in Mae Kha Canal is a key factor that should be monitored to improve both environmental and economic scenarios.

4. Discussion

The development of the Mae Kha Canal, in cooperation with the community and government agencies, involved the implementation of a project aimed at making the canal sustainable in terms of water quality and the quality of life for the residents. Dredging and removing structures around the Mae Kha Canal enhanced the aesthetics of the surrounding environment by improving water drainage, eliminating pollutants, and addressing housing encroachment, which has caused environmental damage, wastewater problems, and significant littering. It improved the water quality and reduced waste and water flow. Socially, it has provided clean water sources, a novel tourist attraction, and a space for the public to exercise and relax along the canal. Economically, it has boosted income for the community through tourism. According to water management evaluations, the DO levels at Atsadathon and SriDonChai Road fell within the moderate-to-good water quality range at the end of the project. Although the BOD and DO values after the wastewater management implementation were not stably improved, the post-intervention trend of decreasing BOD (−0.75 [95% CI: −1.44 to −0.06] at SriDonChai Road) and increasing DO values (0.15 [95% CI: −0.003 to 0.31] at Atsadathon Road and 0.07 [95% CI: −0.007 to 0.15] at SriDonChai Road) suggested an improvement in water quality.

The problem of trash in the Mae Kha Canal, primarily caused by people dumping waste therein, has significantly affected the local community, aquatic life, and the surrounding environment. To tackle this issue, trash nets were employed to capture waste dumped into the canal. This approach has been successfully employed in other countries, such as in the City of Kwinana in Western Australia, where deployed trash nets capture 370 kg of waste every four months [35]. In addition, Hariyadi et al. [36] studied the use of trash nets in the Citarum River in Indonesia to effectively capture many forms of plastic waste. Similarly, Morritt et al. [37] reported that although trash nets deployed on the Thames River effectively trapped plastic waste, a large amount of submerged plastic waste was still flowing into the marine environment. This is also one of the challenges for the Mae Kha Canal, where some submerged trash still flows into the canal. In Thailand, trash nets have been introduced in various areas, such as in the Rayong Municipality, where nets have been placed over the ends of drainage pipes to prevent trash from entering canals and the sea [38], capturing 2–3 tons of waste per day across 15 locations. The adoption of this concept to address the trash problem in Mae Kha Canal at four locations, where 70 kgs of trash is captured daily, has had a positive impact both environmentally (making the canal cleaner and allowing the local ecosystem to thrive) and socially (the local people in the community are no longer affected by foul odors from the canal and the surrounding environment is aesthetically pleasing).

To address riverbank erosion and wastewater from trash in the Mae Kha Canal, semi-permanent check weirs were constructed using coconut-fiber mattresses to help slow down the water flow, retain water, and filter out debris. Typically, the water quality standards for surface water sources require that the DO level is at least 4.0 mg/L and that BOD does not exceed 2.0 mg/L for water bodies receiving effluent from certain activities [39]. Before the installation of the weirs, the average BOD level at Atsadathon Road was 18.11 mg/L, and afterward it was reduced to 7.50 mg/L. Meanwhile, the DO level increased from 1.79 to 5.40 mg/L after the weir installation. Although these indices did not meet the surface water quality standards, a trend of improvement was suggested. This improvement has positively impacted on the environment (improving the water quality and decreasing erosion of the riverbanks), as well as socially (the local community now has access to clean water for domestic use and agriculture), and economically (the better water quality has improved the quality of agricultural produce and increased income).

A recent study in Vietnam conducted an experiment that involved applying vetiver grass in vertical-flow constructed wetland management, and found that it efficiently reduced BOD by 90% and COD by >80% [28]. An FWS wastewater treatment system using vetiver grass was implemented in this study. In this project, vetiver grass was planted on floating platforms, which improved the water quality in the Mae Kha Canal. The average BOD was reduced by 58.82% (from 25.23 mg/L at the beginning in October 2019 to 10.39 mg/L at the end of installation in February 2021), which is consistent with previous studies involving vetiver grass systems in other countries. For example, an 80–85% BOD reduction was experienced in an institutional kitchen [40], a 71.03% BOD reduction was found in the Bagmati River, Nepal [41], and a 91.9% BOD reduction was recorded in industrial wastewater in Ethiopia [42]. The vetiver grass floating platform technique has been deployed in Thailand, leading to 77% removal of BOD from textile factory wastewater [43], as well as an 87.85% reduction in BOD in pig farm wastewater [44]. The trends of BOD and DO levels after the implementation tended to improve at SriDonChai Road. However, the decreases in BOD and increases in DO levels were not stable during the year. The BOD and DO levels fluctuated during the study period. The average BOD levels increased after the FWS installation, while the DO levels decreased compared to the levels measured during the installation and testing period. This might result from other factors, such as drought in winter and summer, or differences in contamination with living patterns and festivals. In addition, it might be related to the maintenance of the FWS after installation. According to a qualitative study about the integrated water resource management of the Mae Rim sub-watershed in Chiang Mai province, the researchers stated that the problems of water resource management in the Mae Rim sub-watershed resulted from an inefficient management process and a lack of correct integration processes between the government and the community [45]. Therefore, for sustainable development, additional wastewater management and community approaches should be provided and examined for their effectiveness in further studies.

In addition to the pollutants in water resources, environmental factors might be related to BOD and DO levels. A previous study about BOD and DO levels in Quelimane, Mozambique, suggested that BOD and DO levels might be influenced by the tides [46]. The study stated that the DO levels during August were more than 2 mg/L, whereas the values were lower than this critical value during September and October. A study on the mathematical modeling of BOD and DO levels also suggested the influence of time at measurement [47]. They performed a series of space–time numerical simulations and found that the model allows the analysis of regions with higher and/or lower DO and BOD concentrations, as well as the temporal variation in concentrations at specific time points. Additionally, the model of BOD and DO measurement might influence the stability of measurements. There are several models which could provide more stable estimations of BOD and DO levels, such as Chebyshev orthogonal polynomial modeling [48] and the

Streeter–Phelps equation [49]. Further studies to compare the simplified model used in this study and these complex models could provide more insightful results.

In an area where houses were encroaching on the Mae Kha Canal and causing water pollution and foul odors, one of the tasks of the project was to develop the area into an attractive tourist destination. This involved enhancing the area alongside the canal by constructing retaining walls to prevent landslides, creating a drainage system to stop households from discharging wastewater directly into the canal, and building public walkways on both sides of the canal. These walkways provide residents and tourists with a place to relax and exercise, thereby benefiting approximately 865 households living along the canal. The development of this area has had several positive impacts, including economically (the Mae Kha Canal has now become a new tourist attraction in Chiang Mai, drawing more than 1000 visitors daily and generating over THB 1,530,000 per month for the local community), socially (the project has provided Chiang Mai residents and tourists with a beautiful public space for relaxation and exercising), and environmentally (the canal now has clean water, is free of trash and foul odors, and the local ecosystem has now been restored). The project's success led to it receiving a national award from the Prime Minister's Office of Thailand for excellence in municipal management under the "Beautiful Canal, Clear Water" category, earning second place at the national level [50].

The development of the Mae Kha Canal in collaboration with the community and government agencies has yielded significant improvements in water quality, local scenery, and Chiang Mai's status by becoming a prominent tourist attraction. The five procedures in the project collectively enhanced the canal's environmental, social, and economic conditions. Dredging improved water drainage and aesthetics, trash nets have reduced pollution and improved the ecosystem, check weirs have slowed the water flow, the vetiver grass-based FWS wastewater treatment system has effectively reduced BOD, and the development of the canal into a recreational area has boosted local tourism and economic activity. However, there are still some challenges. Dredging and removal activities can disrupt aquatic habitats and incur high initial costs. Trash nets may require ongoing maintenance and may not be able to capture all of the waste. Check weirs made from coconut-fiber mattresses might degrade over time, and the vetiver grass in the FWS wastewater treatment system needs adequate space and regular maintenance. Furthermore, while the canal's development into a tourist spot has revitalized the area economically, it could lead to environmental degradation if not properly managed and could also affect residents through increased living costs or displacement. To address these issues, it is crucial to implement phased construction, conduct regular maintenance, explore alternative materials, and develop sustainable tourism practices that involve and benefit the local community. Since the FWS system in this quasi-experimental study was conducted in the real setting, in which water flow changed throughout the study period, the retention time and water flow could not be controlled. In addition, there was no industrial plant related to heavy metal pollutants in the study area. Thus, this study evaluated only the quality of water based on BOD and DO levels. Chemical oxygen demand and total organic carbon are suggested as suitable and stable indices for refused water resources. Therefore, chemical oxygen demand, total organic carbon, other water quality indices (e.g., water quality index, total coliform bacteria, fecal coliform bacteria, nitrate, total nitrogen, phosphate, and total phosphorus), and heavy metal pollutants are of interest, and should be included in future investigations.

5. Conclusions

The comprehensive development of the Mae Kha Canal achieved through the collaboration of the community and government agencies has significantly enhanced both the canal's environment and the quality of life of residents. This initiative involved five tasks in an integrated project aimed at making the canal sustainable and transforming it from a polluted waterway into a vibrant and attractive area. These tasks included dredging and removing obstructions to improve the water flow and reduce pollution, installing trash nets to capture waste and prevent further pollution, constructing semi-permanent check weirs to combat erosion and treat wastewater, using an FWS wastewater treatment system comprising vetiver grass on floating platforms to further improve water quality, and developing the canal area to provide a popular tourist destination. The results of these efforts have been overwhelmingly positive, with improved water quality, reduced pollution, enhanced aesthetics, increased community income through tourism, and the creation of public spaces that benefit both residents and visitors. BOD levels decreased from 17.00–38.70 to 9.00–12.67 mg/L and DO levels increased from 0.00–2.40 to 0.80–6.00 mg/L. Although the BOD and DO values after wastewater management implementation were not stably improved, the post-intervention estimates of the BOD level at SriDonChai Road decreased (−0.75 [95% CI: −1.44 to −0.06]) and the DO levels at Atsadathon Road and SriDonChai Road seemed to increase (0.15 [95% CI: −0.003 to 0.31] and 0.07 [95% CI: −0.007 to 0.15], respectively). The transformation of Mae Kha Canal has not only revitalized the local environment, but has also made it a significant tourist attraction, garnering national recognition for its successful management and development.

Author Contributions: Conceptualization, V.O. and D.R.; methodology, V.O. and D.R.; validation, V.O., P.P. and P.J.; formal analysis, V.O.; investigation, V.O.; resources, V.O., P.P. and P.J.; data curation, V.O., P.P. and P.J.; writing—original draft preparation, V.O.; writing—review and editing, V.O., P.P., P.J. and D.R.; project administration, V.O. All authors have read and agreed to the published version of the manuscript.

Funding: This research received no external funding. This research was partially supported by Chiang Mai University.

Data Availability Statement: The datasets used and/or analyzed during the current study are available from the corresponding author on reasonable request.

Acknowledgments: We acknowledge the Sanitary Engineer Division, Chiang Mai Municipality, Chiang Mai, Thailand.

Conflicts of Interest: The authors declare no conflicts of interest.

References

1. Rashid, R.; Shafiq, I.; Akhter, P.; Iqbal, M.J.; Hussain, M. A state-of-the-art review on wastewater treatment techniques: The effectiveness of adsorption method. *Environ. Sci. Pollut. Res. Int.* **2021**, *28*, 9050–9066. [CrossRef] [PubMed]
2. Ahmad, M.; Yousaf, M.; Nasir, A.; Bhatti, I.A.; Mahmood, A.; Fang, X.; Jian, X.; Kalantar-Zadeh, K.; Mahmood, N. Porous Eleocharis@MnPE Layered Hybrid for Synergistic Adsorption and Catalytic Biodegradation of Toxic Azo Dyes from Industrial Wastewater. *Environ. Sci. Technol.* **2019**, *53*, 2161–2170. [CrossRef] [PubMed]
3. Hinrichsen, D.; Tacio, H. The coming freshwater crisis is already here. In *The Linkages Between Population and Water*; Woodrow Wilson International Center for Scholars: Washington, DC, USA, 2002; pp. 1–26.
4. Nishat, A.; Yusuf, M.; Qadir, A.; Ezaier, Y.; Vambol, V.; Ijaz Khan, M.; ben Moussa, S.; Kamyab, H.; Sehgal, S.S.; Prakash, C.; et al. Wastewater treatment: A short assessment on available techniques. *Alex. Eng. J.* **2023**, *76*, 505–516. [CrossRef]
5. Ali, Q.; Al-Khaled, K.; Khan, M.I.; Khan, S.U.; Raza, A.; Oreijah, M.; Guedri, K. Diffusion phenomenon for natural convection flow of classical Hartmann problem due to a cylindrical tube by generalized Fourier's theories: A Fractional analysis. *Int. J. Mod. Phys. B* **2023**, *37*, 2350104. [CrossRef]
6. Ali, Q.; Al-Khaled, K.; Omar, J.; Raza, A.; Khan, S.U.; Khan, M.I.; Najati, S.A.; Oreijah, M.; Guedri, K.; Galal, A.M. Analysis for advection–diffusion problem subject to memory effects and local and nonlocal kernels: A fractional operators approach. *Int. J. Mod. Phys. B* **2023**, *37*, 2350099. [CrossRef]
7. Bilal, H.; Li, X.; Iqbal, M.S.; Mu, Y.; Tulcan, R.X.S.; Ghufran, M.A. Surface water quality, public health, and ecological risks in Bangladesh-a systematic review and meta-analysis over the last two decades. *Environ. Sci. Pollut. Res. Int.* **2023**, *30*, 91710–91728. [CrossRef]

8. Xu, Z.; Wei, C.; Jin, J.; Xu, W.; Wu, Q.; Gu, J.; Minrui, O.; Xu, X. Development of a Novel Mixed Titanium, Silver Oxide Polyacrylonitrile Nanofiber as a Superior Adsorbent and its Application for MB Removal in Wastewater Treatment. *J. Braz. Chem. Soc.* **2017**, *29*, 560–570. [CrossRef]
9. Azam, K.; Raza, R.; Shezad, N.; Shabir, M.; Yang, W.; Ahmad, N.; Shafiq, I.; Akhter, P.; Razzaq, A.; Hussain, M. Development of recoverable magnetic mesoporous carbon adsorbent for removal of methyl blue and methyl orange from wastewater. *J. Environ. Chem. Eng.* **2020**, *8*, 104220. [CrossRef]
10. Liu, Q.; Zhou, Y.; Lu, J.; Zhou, Y. Novel cyclodextrin-based adsorbents for removing pollutants from wastewater: A critical review. *Chemosphere* **2020**, *241*, 125043. [CrossRef]
11. Zhu, Z.; Liu, D.; Cai, S.; Tan, Y.; Liao, J.; Fang, Y. Dyes removal by composite membrane of sepiolite impregnated polysulfone coated by chemical deposition of tea polyphenols. *Chem. Eng. Res. Des.* **2020**, *156*, 289–299. [CrossRef]
12. Liang, C.Z.; Sun, S.-P.; Li, F.-Y.; Ong, Y.; Chung, T.-S. Treatment of highly concentrated wastewater containing multiple synthetic dyes by a combined process of coagulation/flocculation and nanofiltration. *J. Membr. Sci.* **2014**, *469*, 306–315. [CrossRef]
13. Lee, K.E.; Morad, N.; Teng, T.T.; Poh, B. Development, characterization and the application of hybrid materials in coagulation/flocculation of wastewater: A review. *Chem. Eng. J.* **2012**, *203*, 370–386. [CrossRef]
14. Nandabalan, Y.; Das, A.; Rani, G.; Jaswal, V.; Randhawa, J. Nano-bioremediation: A New Age Technology for the Treatment of Dyes in Textile Effluents. In *Bioremediation of Industrial Waste for Environmental Safety*; Gaurav Saxena, R.N.B., Ed.; Springer: Singapore, 2020; pp. 313–347.
15. Khatri, J.; Nidheesh, P.; Singh, T.A.; Kumar, M.S. Advanced oxidation processes based on zero-valent aluminium for treating textile wastewater. *Chem. Eng. J.* **2018**, *348*, 67–73. [CrossRef]
16. Sözen, S.; Olmez-Hanci, T.; Hooshmand, M.; Orhon, D. Fenton oxidation for effective removal of color and organic matter from denim cotton wastewater without biological treatment. *Environ. Chem. Lett.* **2019**, *18*, 207–213. [CrossRef]
17. Lin, S.H.; Lin, C.M.; Leu, H.G. Operating characteristics and kinetic studies of surfactant wastewater treatment by Fenton oxidation. *Water Res.* **1999**, *33*, 1735–1741. [CrossRef]
18. Holkar, C.R.; Jadhav, A.J.; Pinjari, D.V.; Mahamuni, N.M.; Pandit, A.B. A critical review on textile wastewater treatments: Possible approaches. *J. Environ. Manag.* **2016**, *182*, 351–366. [CrossRef] [PubMed]
19. Grégorio, C.; Lichtfouse, E. Advantages and disadvantages of techniques used for wastewater treatment. *Environ. Chem. Lett.* **2018**, *17*, 145–155. [CrossRef]
20. Muthukumaran, N.; Ambujam, N.K. Wastewater treatment and management in urban areas—A case study of Tiruchirappalli city, Tamil Nadu, India. In Proceedings of the Third International Conference on Environment and Health; Bunch, M.J., Madha, S.V., Vasantha, K.T., Eds.; Department of Geography, University of Madras and Faculty of Environmental Studies, York University: Chennai, India, 2003; pp. 248–289.
21. García, A.; García-López, M. The Potential of Wastewater Reuse and the Role of Economic Valuation in the Pursuit of Sustainability: The Case of the Canal de Isabel II. *Sustainability* **2023**, *15*, 843. [CrossRef]
22. Frascari, D.; Zanaroli, G.; Motaleb, M.A.; Annen, G.; Belguith, K.; Borin, S.; Choukr-Allah, R.; Gibert, C.; Jaouani, A.; Kalogerakis, N.; et al. Integrated technological and management solutions for wastewater treatment and efficient agricultural reuse in Egypt, Morocco, and Tunisia. *Integr. Environ. Assess. Manag.* **2018**, *14*, 447–462. [CrossRef]
23. Iqbal, M.; Shoaib, M.; Agwanda, P.; Lee, J. Modeling Approach for Water-Quality Management to Control Pollution Concentration: A Case Study of Ravi River, Punjab, Pakistan. *Water* **2018**, *10*, 1068. [CrossRef]
24. Dudai, N.; Putievsky, E.; Chaimovitch, D.; Ben-Hur, M. Growth management of vetiver (Vetiveria zizanioides) under Mediterranean conditions. *J. Environ. Manag.* **2006**, *81*, 63–71. [CrossRef]
25. Suelee, A.L.; Hasan, S.N.M.S.; Kusin, F.M.; Yusuff, F.M.; Ibrahim, Z.Z. Phytoremediation potential of vetiver grass (*Vetiveria zizanioides*) for treatment of metal-contaminated water. *Water Air Soil Pollut.* **2017**, *228*, 158. [CrossRef]
26. Seroja, R.; Effendi, H.; Hariyadi, S. Tofu wastewater treatment using vetiver grass (*Vetiveria zizanioides*) and zeliac. *Appl. Water Sci.* **2018**, *8*, 2. [CrossRef]
27. Yang, C.; Zhang, X.; Tang, Y.; Jiang, Y.; Xie, S.; Zhang, Y.; Qin, Y. Selection and optimization of the substrate in constructed wetland: A review. *J. Water Process. Eng.* **2022**, *49*, 103140. [CrossRef]
28. Nguyen, M.K.; Hung, N.T.Q.; Nguyen, C.M.; Lin, C.; Nguyen, T.A.; Nguyen, H.L. Application of vetiver grass (*Vetiveria zizanioides* L.) for organic matter removal from contaminated surface water. *Bioresour. Technol. Rep.* **2023**, *22*, 101431. [CrossRef]
29. Water Quality Management Division, Pollution Control Department; Ministry of Natural Resources and Environment. Water Quality Management Division Operation Report 2022. Available online: https://www.pcd.go.th/publication/29365/ (accessed on 20 February 2024).
30. Techapichetwong, J.; Pasukphun, N. Application of pollution flow analysis in water quality management at Saen Saep Canal, Bangkok. *Public Health J. Burapha Univ.* **2023**, *18*, 82–100.
31. Somsook, K.; Pilailar, S. Khlong Saen Saep water quality effect due to population growth and mitigation measures. *Res. Dev. J.* **2013**, *24*, 8–13.

32. Chiang Mai Province Office. Mae Kha Canal Master Plan (2018–2022). Available online: https://www.chiangmai.go.th/managing/public/D2/2D10Sep2018163933.pdf (accessed on 20 February 2024).
33. Municipality, C.M. General Information about Chiang Mai Municipality. Available online: https://www.cmcity.go.th/ (accessed on 20 February 2024).
34. Standard Methods Committee of the American Public Health Association, American Water Works Association, and Water Environment Federation. 4500-o oxygen (dissolved). In *Standard Methods For the Examination of Water and Wastewater*; Lipps, W.C., Baxter, T.E., Braun-Howland, E., Eds.; APHA Press: Washington, DC, USA, 2016.
35. Sustainablelife. Aussies Show off Their Cool Idea for Filtering Trash in the River. It's Simple, But It's Very Effective. Available online: https://www.sustainablelife.co/news/detail/28 (accessed on 20 February 2024).
36. Hariyadi, S.; Taryono Wulandari, D.Y.; Zulmi, R.; Permatasari, P.A.; Amalo, L.F.; Nur, I.A.; Trissanti, V.N.; Hurley, R.; VBraaten, H.F.; Olsen, M. Plastic debris in Citarum River. *IOP Conf. Ser. Earth Environ. Sci.* **2022**, *1062*, 012024. [CrossRef]
37. Morritt, D.; Stefanoudis, P.; Pearce, D.; Crimmen, O.; Clark, P. Plastic in the Thames: A river runs through it. *Mar. Pollut. Bull.* **2013**, *78*, 196–200. [CrossRef]
38. Geo-Informatics and Space Technology Development Agency. Nakhon Rayong Successfully Installed Nets to Catch Waste from Drainage Pipes Before it Enters the Sea. Available online: http://coastalradar.gistda.or.th/wp/?page=news-detail&id=11255 (accessed on 20 February 2024).
39. Water Quality Database System, Pollution Control Department. Water Quality Evaluation. Available online: https://rwater.mnre.go.th/front/main/WaterQuality/indicator (accessed on 20 February 2024).
40. Mathew, M.; Rosary, S.; Sebastian, M.; Cherian, S. Effectiveness of Vetiver System for the Treatment of Wastewater from an Institutional Kitchen. *Procedia Technol.* **2016**, *24*, 203–209. [CrossRef]
41. Maharjan, A.; Pradhanang, S. Potential of Vetiver Grass for Wastewater Treatment. *Environ. Ecol. Res.* **2017**, *5*, 489–494. [CrossRef]
42. Aregu, M.B.; Soboksa, N.E.; Kanno, G.G. High Strength Wastewater Reclamation Capacity of Vetiver Grass in Tropics: The Case of Ethiopia. *Environ. Health Insights* **2021**, *15*, 11786302211060162. [CrossRef]
43. Charoenlarp, K.; Surakul, K.; Winitkhetkamnoun, P.; Kanthupthim, P.; Panbumrung, P.; Udorn, S. Textile wastewater treatment using vetiver grass cultivated with floating platform technique. *Environ. Nat. Resour. J.* **2016**, *10*, 51–57.
44. Pongthornpruek, S. Swine farm wastewater treatment by constructed wetland planted with vetiver grass. *Environ. Nat. Resour. J.* **2017**, *15*, 13–20. [CrossRef]
45. Thipsuwan, C. Integrated water resources management of Mae Rim sub watershed Chiang Mai province. *Veridian E-J. Silpakorn Univ. Humanit. Soc. Sci. Arts* **2013**, *6*, 53–71.
46. Jeremias, M. Dissolved Oxygen and Biochemical Oxygen Demand in the Waters Close to the Quelimane Sewage Discharge. Master's Thesis, The University of Bergen, Bergen, Norway, 2010. [CrossRef]
47. Guaca, D.C.; Poletti, E.C.C. Modeling and numerical simulation of dissolved oxygen and biochemical oxygen demand concentrations with Holling type III kinetic relationships. *Appl. Math. Comput.* **2022**, *415*, 126690. [CrossRef]
48. Wang, S.H.; Li, R. Forecasting dissolved oxygen and biochemical oxygen demand in a river with numerical solution of one-dimensional BOD-DO model. *Appl. Ecol. Environ. Res.* **2017**, *15*, 323–333. [CrossRef]
49. Dunnivant, F. Biochemical Oxygen Demand and the Dissolved Oxygen Sag Curve in a Stream: Streeter–Phelps Equation. In *Environmental Laboratory Exercises for Instrumental Analysis and Environmental Chemistry*; Dunnivant, F.M., Ed.; John Wiley & Sons, Inc.: Hoboken, NJ, USA, 2004. [CrossRef]
50. Station, T.B. Prime Minister Visits the Mae Kha Canal Area, Otaru, Thailand Promotes Tourism in Chiang Mai Province. Available online: https://radiothailand.prd.go.th/th/content/category/detail/id/6/cid/57/iid/270206 (accessed on 20 February 2024).

water

MDPI

Article

Improved Oil/Water Separation by Employing Packed-Bed Filtration of Modified Quartz Particles

Nthabiseng Ramanamane * and Mothibeli Pita

Department of Mechanical Engineering, Bioresources, and Biomedical Engineering, College of Science, Engineering and Technology, University of South Africa, Florida 1710, South Africa; pitam@unisa.ac.za
* Correspondence: ramannj@unisa.ac.za

Abstract: This study explores the development and optimization of quartz-based filtration media for industrial oil–water separation, focusing on enhancing surface wettability, minimizing fouling, and improving oil rejection efficiency. High-purity quartz particles (SiO_2: 98%, Fe_2O_3: 0.18%, particle size: 0.8–1.8 mm) were evaluated in three configurations: raw, acid-washed, and surface-coated with hydrophilic nanoparticles (Al_2O_3 and P_2O_5). The filtration medium was constructed as a packed-bed of quartz particles rather than a continuous sintered membrane, providing a cost-effective and modular structure for separation processes. Comprehensive material characterization was performed using X-ray diffraction (XRD), Scanning Electron Microscopy (SEM), and Energy Dispersive Spectroscopy (EDS). XRD confirmed the crystalline stability of quartz across all treatments, while SEM and EDS revealed enhanced surface morphology and elemental distribution—especially phosphorus and aluminum—in coated samples. Performance testing with synthetic oily wastewater (initial oil concentration: 183,754.8 mg/L) demonstrated that the coated quartz medium achieved superior separation, reducing residual oil concentration to 29.3 mg/L, compared to 1583.7 mg/L and 1859.8 mg/L for washed and raw quartz, respectively. Contact angle analysis confirmed improved hydrophilicity in coated media, which also exhibited lower fouling propensity. Taguchi optimization (conducted via Minitab 21.3) and regression modeling identified surface coating and operational pressure (optimal at 2.5 bar) as the most significant parameters influencing oil rejection. Post-filtration SEM and XRD confirmed structural integrity and coating durability. Additionally, flux recovery above 90% after backwashing indicated strong regeneration capability. These findings validate surface-modified quartz packed beds as robust, scalable, and economically viable alternatives to conventional membranes in oily wastewater treatment. Future research will explore multilayer coatings, long term performance under aggressive conditions, and AI-based prediction models.

Keywords: quartz packed-bed filtration; oil–water separation; SEM-EDS characterization; hydrophilic nanoparticle coating; taguchi optimization; wastewater treatment; packed-bed filtration regeneration

1. Introduction

The clean water scarcity has emerged as a critical global challenge, intensified by rapid population growth, urbanization, and climate change [1,2]. Many regions now face severe water stress, which is driving an urgent need for water reuse and efficient wastewater treatment to conserve this vital resource [3,4]. At the same time, industrial activities generate vast amounts of polluted wastewater. In particular, oil-bearing effluents from sectors such as petrochemicals, petroleum refining, food processing, and metalworking are a major source of water contamination [5–7]. Industrial developments in the oil and gas and related sectors have led to the large-scale production of oily wastewater worldwide. These oily effluents, often in the form of stable oil-in-water emulsions, pose serious environmental and health risks: they pollute drinking water and groundwater, endanger aquatic ecosystems, harm human health, and even impact crop production [8–10].

If discharged untreated, oily wastewater can create surface films that reduce oxygen transfer in water bodies and increase chemical oxygen demand, with detrimental effects on marine life. Given that more than 80% of industrial wastewater globally is released without adequate treatment, there is a pressing need for effective water recovery solutions that can remove oils and greases [11,12]. Developing such solutions is not only an environmental imperative but also a strategic response to global water scarcity—by recovering usable water from contaminated streams, industries can alleviate pressure on freshwater supplies and support sustainable water reuse [13–15]. Membrane filtration is a widely applied technology for oil–water separation, valued for its ability to produce high-quality effluent. However, conventional membrane systems face significant challenges that hinder long term performance and sustainability [16,17]. A major issue is membrane fouling: oil droplets and emulsified hydrocarbons tend to accumulate on the membrane surface and inside its pores, gradually blocking the flow paths [18,19].

This pore blockage sharply reduces permeate flux and separation efficiency over time [20–22]. The fouling is especially severe with hydrophobic membranes, as non-polar oil components readily adhere to hydrophobic surfaces. Consequently, many polymeric membranes (e.g., PVDF, PTFE) suffer from rapid flux decline and require frequent maintenance [23–25]. In addition to these, polyethylene terephthalate (PET) track-etched membranes have also been explored for oil–water separation, demonstrating promising wettability and separation characteristics, although they similarly face challenges such as fouling and performance degradation over time [26]. Restoring fouled membranes typically necessitates back-flushing or aggressive chemical cleaning (using strong oxidants and surfactants). Such measures can temporarily recover performance but often at the cost of shortening the membrane's lifespan. Frequent chemical cleaning and downtime not only increase operational costs but also raise concerns about secondary waste and chemical consumption [27,28].

Moreover, even with cleaning, hydrophobic polymer membranes may never fully regain their initial permeability once heavily fouled, leading to irrecoverable performance loss [29–31]. On the other hand, ceramic membranes (made from oxides such as alumina or zirconia) offer better fouling resistance and thermal/chemical stability [32–34]. These inorganic membranes can tolerate harsher cleaning (including high-temperature calcination to burn off organics) and are less prone to bio-degradation. Nevertheless, ceramic membranes come with drawbacks: they are expensive to manufacture and inherently brittle, making them prone to cracking or breakage in operation [35,36]. This brittleness complicates handling and limits the membrane's ability to withstand mechanical stress [37,38]. Thus, neither traditional polymeric nor advanced ceramic membranes provide a fully satisfactory solution for oily wastewater treatment—polymers are affordable but foul easily, while ceramics are more robust but cost-prohibitive for large scales [37,39].

Despite ongoing innovations (e.g., new materials and hybrid processes), many of the current oil–water separation technologies struggle to combine high efficiency with long term stability and low cost [40,41]. This gap underlines the need for alternative membrane approaches that can resist fouling, maintain performance over time, and remain economically viable for widespread deployment. In search of more sustainable solutions, researchers are turning to natural materials as alternative membranes for oily wastewater treatment. Quartz has emerged as a promising candidate in this context, owing to its abundance, low cost, and favorable surface properties [42,43]. Quartz (crystalline silicon dioxide) is one of the most ubiquitous minerals on Earth and is highly resistant to chemical and mechanical weathering. These characteristics make quartz an attractive matrix for water filtration, as it can endure harsh wastewater conditions without significant degradation [44,45]. Indeed, various natural minerals such as quartz, clays, and zeolites have been investigated as base materials for low-cost ceramic membranes, with the aim of reducing fabrication costs while retaining good separation performance [46–48]. Integrating such readily available materials can dramatically lower the raw material expenses of membrane production. Quartz sand, in particular, has a long history in water purification—for example, slow sand filters using quartz-rich sand were first introduced in the 19th century for water treatment.

However, traditional sand filters are relatively low in efficiency and require large areas and long residence times, making them less practical for treating concentrated oily waste streams [49–51]. To harness quartz's advantages in a more effective manner, recent studies have explored sintering or binding quartz particles into porous membrane structures [52–54]. By using appropriate binders and fabrication techniques, it is possible to form cohesive quartz-based membranes that exhibit appreciable strength and permeability. Quartz's natural hydrophilicity (due to surface silanol groups) is an added benefit, as it tends to attract water and repel oils [55,56]. This inherent wettability suggests that quartz membranes could achieve in-situ oleophobic behavior (underwater oil repellence) without the need for elaborate surface chemistry [57,58]. Importantly, quartz is far less expensive than engineered ceramics such as alumina or zirconia, and it requires lower sintering temperatures to form a membrane structure, translating to energy savings in manufacturing [59,60]. In summary, quartz offers a viable, cost-effective alternative to conventional membrane materials, with the potential for improved wettability and competitive performance. Exploiting this naturally occurring medium could overcome some limitations of existing membranes, provided that the quartz membrane can be engineered for adequate durability and fine separation of oil droplets.

Recent advancements in membrane technologies for oil–water separation have demonstrated promising performance across various material platforms. For instance, ceramic microfiltration membranes fabricated via slip casting—such as those based on zirconia or silica—have exhibited high porosity, excellent mechanical strength, and thermal stability, making them suitable for demanding applications such as microbial sterilization and liquid-liquid separation [61]. However, these materials remain cost-intensive and prone to brittleness, which may restrict their scalability in industrial settings. In contrast, mixed-matrix membranes incorporating natural clays (e.g., kaolinite) or inorganic fillers into hydrophilic polymer matrices have shown enhanced wettability and fouling resistance, as well as improved contaminant removal from wastewater [62]. These systems leverage the synergistic benefits of inorganic and organic components to balance performance and cost. Nonetheless, challenges persist in maintaining long term operational stability and optimizing interfacial compatibility between phases. Against this backdrop, the current study proposes a distinct approach by employing naturally abundant quartz particles enhanced with hydrophilic nanoparticles to form a packed-bed separation medium. This method offers a cost-effective alternative to conventional ceramic and polymeric membranes while maintaining comparable or superior oil rejection capabilities, with the added benefit of facile regeneration and high material stability.

A novel aspect of this study is the surface engineering of quartz packed-bed filtration using hydrophilic nanoparticles to further improve wettability and fouling resistance. In oil–water separation, a hydrophilic (water-attracting) and oleophobic (oil-repelling) surface is highly desirable because it promotes the formation of a water film on the packed-bed filtration that acts as a barrier to oil adhesion. By ensuring water preferentially wets the packed-bed filtration, oil droplets are less likely to stick to the surface or clog the pores, thereby mitigating fouling. Previous research on polymeric membranes has demonstrated that coating or embedding the surface with nanomaterials—such as silica (SiO_2), titanium dioxide, alumina, or other metal oxide nanoparticles—can significantly enhance hydrophilicity and anti-fouling characteristics [63,64]. For instance, creating a rough nanoparticle-decorated surface introduces hydrophilic functional groups (e.g., hydroxyls) and surface textures that lead to superhydrophilic, underwater superoleophobic behavior [65–67]. This approach has improved membrane performance by increasing water flux and reducing oil deposition on various membrane substrates.

However, to date such nanoparticle-based surface modifications have not been applied to quartz particle packed-bed filtration, which is a gap this study aims to fill. To address these limitations, this research introduced a novel approach by utilizing quartz-based packed-bed filtration composed of dry-graded silica particles (SiO_2: 98%, Fe_2O_3: 0.18%, particle size: 0.8–1.8 mm) as a low-cost and sustainable alternative for oil-water separation. The novelty of this study was further amplified through the surface modification of quartz packed-bed filtration using hydrophilic nanoparticles, significantly enhancing wettability and reducing oil adhesion, thereby minimizing packed-bed filtration fouling. Additionally, the incorporation of mathematical models provided a quantitative framework to interpret and predict packed-bed filtration performance under varying operational conditions, particularly in terms of wettability and oil rejection efficiency. X-ray diffraction (XRD) analysis was used to assess crystalline stability and phase composition, while Taguchi-based statistical optimization and regression modeling were applied to identify the most favorable operational parameters—particularly transmembrane pressure—for maximum oil rejection and minimal fouling.

This integrated approach—combining SEM, EDS, XRD, oil analysis, and predictive modeling—demonstrates the scientific innovation and practical viability of quartz packed-bed filtration for industrial oily wastewater treatment.

2. Materials and Methods

To realize and evaluate the proposed quartz packed-bed filtration, we carried out a series of fabrication and characterization steps. High-purity quartz particles (comprised of ~98% SiO_2 and minor impurities such as ~0.18% Fe_2O_3) in the size range of 0.8–1.8 mm were used as the base material for the packed-bed filtration. These particles were prepared under three conditions: raw quartz (as-received, untreated), washed quartz (acid-washed to remove fines and surface contaminants), and nanoparticle-coated quartz (the washed quartz further modified with a uniform layer of hydrophilic nanoparticles). The acid washing of quartz particles was carried out using a dilute hydrochloric acid (HCl) solution to remove residual metal oxides, organic matter, and surface impurities. Specifically, 500 g of raw quartz was immersed in 1.0 M HCl solution at a solid-to-liquid ratio of 1:10 (w/v) and stirred continuously for 2 h at ambient temperature (~25 °C). The acid-treated quartz was then thoroughly rinsed with deionized water until the pH of the wash water reached neutral (pH 7), followed by drying at 80 °C for 12 h to ensure complete moisture removal. This process aimed to increase the quartz surface hydrophilicity and prepare it for subsequent nanoparticle coating.

In this study, the hydrophilic nanoparticle formulation was carefully selected to enhance the surface properties of quartz particles for optimal oil–water separation. The coating slurry comprised a blend of silicon dioxide (SiO_2, 70–80 wt%), aluminum oxide (Al_2O_3, 10–15 wt%), and phosphorus pentoxide (P_2O_5, 5–10 wt%), with trace amounts of fluorine-based compounds ($\leq$2 wt%) added to improve wettability and fouling resistance. The nanoparticles used had an average size distribution of 10–50 nm, ensuring sufficient surface roughness without blocking interparticle flow channels.

Prior to application, the nanoparticles were dispersed in deionized water and sonicated for 30 min to break agglomerates and ensure uniform distribution. The functionalization process involved dipping the washed quartz particles into the nanoparticle suspension followed by drying at 80 °C for 12 h and a subsequent heat treatment at 250 °C for 4 h to enhance adhesion and surface activation. The final product, referred to as the "first-coated" packed-bed filtration, presented enhanced hydrophilicity due to the abundance of hydroxyl-rich phases (Al–OH, P–OH) and a more uniform surface topography as verified by SEM–EDS characterization.

While detailed morphological and elemental characterizations were performed via SEM and EDS, the wettability of the quartz particles was not directly measured through conventional contact angle methods. Instead, improvements in wettability were inferred from compositional analysis showing increased phosphorus and aluminum enrichment on particle surfaces, as well as from the enhanced performance in terms of oil rejection and reduced fouling. Future studies should incorporate direct contact angle measurements and atomic force microscopy (AFM) to quantify nanoscale surface roughness and wettability more precisely.

Although the term "quartz packed-bed filtration" is used throughout this study for simplicity and consistency, it is important to clarify that the material does not constitute a conventional sintered or cast packed-bed filtration layer. Instead, the packed-bed filtration setup involved a packed-bed configuration, where dry-graded quartz particles (raw, washed, and surface-coated with hydrophilic nanoparticles) were tightly packed into a cylindrical housing to function as a filtration medium. This packed-bed arrangement mimics the behavior of a granular membrane, allowing for effective phase separation through pressure-driven flow while retaining flexibility in surface engineering. The distinction is relevant for comparison with conventional polymeric and ceramic membranes, as it influences both flow dynamics and regeneration approaches. Therefore, all references to "quartz packed-bed filtration " herein should be interpreted as structured beds of surface-modified quartz particles with membrane-like separation behavior.

In this study, hydrophilic nanoparticles were utilized to enhance the surface properties of quartz packed-bed filtration for improved wettability and oil rejection efficiency. Typically, hydrophilic nanoparticles employed for surface modifications in packed-bed filtration include silicon dioxide (SiO_2), aluminum oxide (Al_2O_3), titanium dioxide (TiO_2), and zirconium dioxide (ZrO_2), often doped with functional groups to increase their affinity for water molecules. In the current work, these nanoparticles likely contained a blend of SiO_2 (70–90%) for structural integrity and hydrophilicity, Al_2O_3 (5–15%) for improved mechanical strength, and TiO_2 or ZrO_2 (2–10%) to enhance surface activity and chemical stability. The nanoparticles had an estimated size range of 10–50 nm, ensuring an optimal balance between surface roughness and hydrophilic functionalization without excessive pore blockage. All water used in this study was deionized and tested for purity at AMBIO Laboratories, Vaal University of Technology.

The coating procedure was designed to deposit a fine layer on the quartz grains, creating what we term a "first-coated" quartz packed-bed filtration. We employed X-ray diffraction (XRD) analysis to characterize the crystalline structure and purity of the quartz in each form. The core of the experimental study involved performance testing of the quartz packed-bed filtration for oil–water separation under various operating conditions. A filtration setup was constructed wherein quartz particles—under each of the three conditions—were packed into a module, as illustrated in Figure 1.

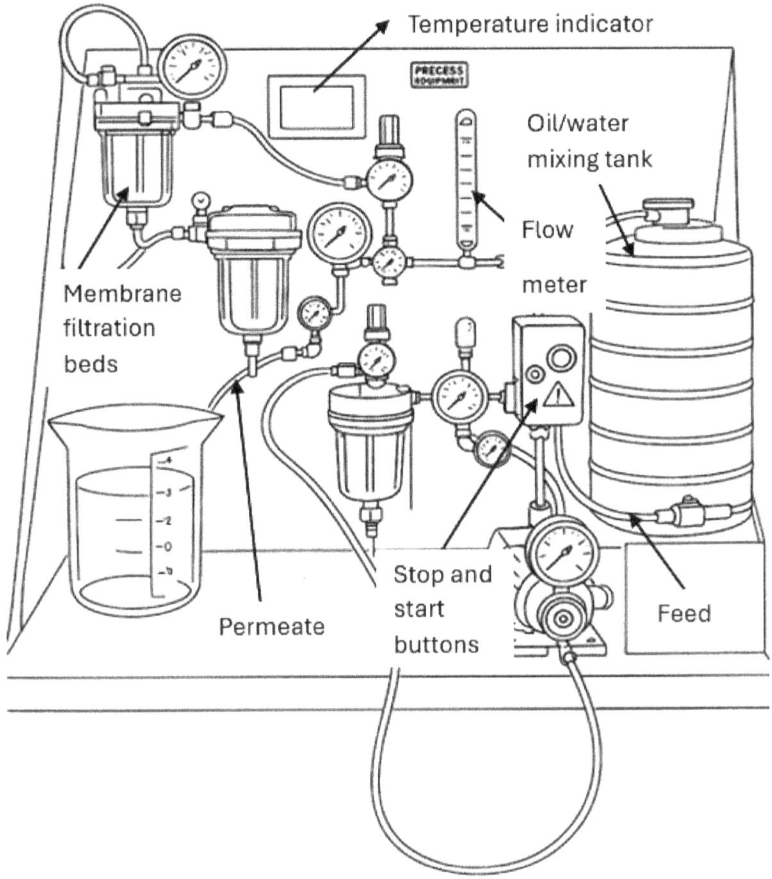

Figure 1. Schematic diagram of the laboratory-scale filtration setup used for evaluating quartz particle-packed-bed filtration.

Synthetic oily wastewater emulsions were prepared to simulate industrial oily effluents, with controlled oil concentrations representative of light, moderate, and heavy pollution scenarios. The feed oil concentrations were varied in a range (e.g., from a few tens to a few hundreds of mg/L of oil) to test the packed-bed filtration under different contamination levels.

Using a pressure-driven filtration system, we passed these emulsions through the quartz packed-bed filtration while systematically varying the transmembrane pressure from 0.5 bar up to 3 bar. This range of pressures spans typical values for microfiltration and ultrafiltration operations, allowing us to observe the packed-bed filtration behavior from low-pressure (gravity-driven or gentle pumping) up to moderately high-pressure conditions. For each combination of packed-bed filtration type, feed oil concentration, and pressure, we measured key performance indicators: the permeate flux (water throughput) and the oil rejection efficiency. Oil rejection was quantified by analysing the oil concentration in the permeate (treated water) and comparing it to the feed concentration, thereby determining the percentage of oil removed by the packed-bed filtration. These tests enabled us to assess how surface condition (raw vs. washed vs. coated), operating pressure, and feed characteristics influence the quartz packed-bed filtration's effectiveness in separating oil from water. By covering a broad range of conditions, we gathered a comprehensive dataset to inform optimization and performance modeling. We employed the Taguchi method—a statistical design of experiments approach—to efficiently explore the multi-factor parameter space. The Taguchi design allowed us to vary several factors simultaneously (such as packed-bed filtration type, feed oil concentration, and operating pressure) according to a predefined orthogonal array, rather than testing all possible combinations exhaustively. This approach dramatically reduces the number of experiments needed while still providing insight into the influence of each factor on performance. Each experimental test (oil concentration and flux) was performed in triplicate ($n = 3$) to ensure repeatability and reliability. The mean values were reported, and standard deviations were calculated and are included in the relevant data tables and graphs. While confidence intervals were not explicitly presented, statistical optimization using the Taguchi method and regression analysis provided robust predictive insights into performance trends.

2.1. Key Mathematical Equations Developed for Modeling the Performance of Quartz Packed-Bed Filtration in Oil-Water Separation

To optimize the performance of quartz packed-bed filtration for oil-water separation, a series of mathematical equations were developed to describe key operational parameters, including oil rejection efficiency, flux rate, pressure dependence on oil concentration, and nanoparticle distribution effects. These equations integrate fundamental transport phenomena, surface interactions, and packed-bed filtration fouling behavior to provide a predictive framework for packed-bed filtration performance under varying operating conditions.

The oil rejection efficiency equation characterizes the ability of quartz packed-bed filtration to separate oil from water based on interfacial tension, pore size distribution, and pressure conditions. The flux rate equation, derived from Darcy's law, accounts for intrinsic packed-bed filtration permeability and fouling resistance, predicting how water flow rates decline as oil accumulates. A pressure vs. oil concentration equation was formulated to quantify the increasing resistance due to oil fouling, demonstrating the trade-off between operating pressure and packed-bed filtration performance. Additionally, a nanoparticle distribution equation was introduced to describe how hydrophilic nanoparticle coatings enhance wettability while influencing permeability and separation efficiency.

By incorporating these equations into the experimental design, this study provides a scientific basis for optimizing packed-bed filtration fabrication and operational parameters. The integration of statistical optimization techniques such as Taguchi design and regression analysis further refines performance predictions, allowing for the identification of the most effective packed-bed filtration configuration. These equations not only facilitate performance evaluation but also serve as a valuable tool for scaling up quartz packed-bed filtration technology in industrial oil-water separation applications.

Equation (1) is the oil rejection efficiency equation [68], where oil rejection $R(\%)$ is defined as the fraction of oil retained by the packed-bed filtration, based on feed oil concentration C_f (mg/L) and permeate concentration C_p (mg/L):

$$R = 1 - \frac{C_p}{C_f} \quad (1)$$

The oil rejection efficiency is governed by the capillary entry pressure ΔP_c (Pa), which determines whether oil droplets can enter the packed-bed filtration pores. Using the Young–Laplace Equation (2) [69], where $\gamma_{w/o}$ (N/m) is the interfacial tension between water and oil, θ_w (degrees) is the water contact angle on the quartz packed-bed filtration, r_{pore} (m) is the average pore radius.

$$\Delta P_c = \frac{2\gamma_{w/o}\cos\theta_w}{r_{pore}} \quad (2)$$

For pressures (ΔP) below (ΔP_c), oil rejection is nearly 100% $(R \approx 1)$. At $\Delta P > \Delta P_c$, oil droplets begin to deform and pass through the packed-bed filtration. The fraction of oil passing through (S) follows a sigmoidal function as presented in Equation (3).

$$S(\Delta P) = \frac{1}{1 + exp[-\alpha(\Delta P - \Delta P_c)]} \quad (3)$$

where α is a fitting parameter dependent on pore size distribution. Thus, oil rejection as a function of pressure is given by Equation (4).

$$R(\Delta P) = 1 - S(\Delta P) = \frac{exp[-\alpha(\Delta P - \Delta P_c)]}{1 + exp[-\alpha(\Delta P - \Delta P_c)]} \quad (4)$$

Equation (5) represents the flux rate equation, where flux J $(\mathrm{L/m^2{\cdot}h})$ is derived from Darcy's law for pressure-driven flow through a porous packed-bed filtration [70], where $k = \frac{\varepsilon r_p^2}{8\tau}$ $(\mathrm{m^2})$ is the intrinsic packed-bed filtration permeability (porosity-dependent), μ (Pa·s) is the fluid viscosity, L (m) is the packed-bed filtration thickness, τ is the tortuosity (flow path deviation).

$$J = \frac{k}{\mu L}\Delta P \quad (5)$$

This equation indicates that higher porosity ε, larger pore size (r_p), and lower viscosity μ enhance flux. At higher pressures, packed-bed filtration fouling limits flux, and a modified equation incorporating resistance from an oil fouling layer $R_f(1/m)$ as given by Equation (6) [71], where $R_m(1/m)$ is the clean packed-bed filtration resistance, $R_f = \alpha K C_f$ is additional resistance due to oil fouling (K is the fouling coefficient and C_f the feed oil concentration).

$$J = \frac{\Delta P}{\mu\left(R_m + R_f\right)} \quad (6)$$

For constant ΔP, flux decline follows Equation (7) was developed.

$$J\left(C_f\right) = \frac{\Delta P}{\mu\left(R_m + \alpha K C_f\right)} \quad (7)$$

Equation (8) represents pressure vs. oil concentration equation, where total resistance R_{tot} increases with oil deposition on the packed-bed filtration. Assuming a linear relationship between oil concentration and fouling resistance, where $\alpha K C_f$ accounts for increased resistance due to oil accumulation.

$$\Delta P = \mu J\left(R_m + \alpha K C_f\right) \quad (8)$$

While Equation (9) explains why flux decreases with increasing oil content.

$$J\left(C_f\right) = \frac{\Delta P}{\mu\left(R_m + \alpha K C_f\right)} \quad (9)$$

Equation (10) represents the nanoparticle distribution equation, where surface wettability is enhanced by hydrophilic nanoparticles, affecting the contact angle (θ^*), where $\varnothing$ (%)

is the nanoparticle coverage fraction, θ_{NP} (degree) is the contact angle of nanoparticles (e.g., SiO_2, TiO_2), and θ_{quartz} (degree) is the bare quartz contact angle.

$$cos(\theta^*) = \varnothing cos\,\theta_{NP} + (1-\varnothing)cos\theta_{quartz} \tag{10}$$

For excessive nanoparticle coverage, pore narrowing reduces permeability. Effective permeability after coating is presented by Equation (11), where $r_{eff} = r_p - t$ and t is the nanoparticle layer thickness.

$$k_{eff} = k_{orig}\left(\frac{r_{eff}}{r_p}\right)^2 \tag{11}$$

Flux modification due to nanoparticle coverage is given by Equation (12), where $a\varnothing$ accounts for improved wettability, and $e^{-b\varnothing}$ models pore blocking effects. Where at optimal nanoparticle coverage $\varnothing^*$, maximum flux and oil rejection are achieved.

$$J_{rel}(\varnothing) = (1 + a\varnothing)exp(-b\varnothing) \tag{12}$$

These novel equations establish a scientific framework for predicting quartz packed-bed filtration performance in oil-water separation. The models integrate capillary forces, transport equations, fouling behavior, and nanoparticle surface effects, providing a quantitative basis for optimization. The derived relationships emphasize:

- The importance of pressure control in maintaining oil rejection,
- The trade-offs between hydrophilic enhancement and pore blockage,
- The influence of feed oil concentration on required pressure and flux decline.

2.2. Comparative Analysis of Quartz Packed-Bed Filtration Performance Models Versus Conventional Technologies for Oil/Water Separation

Table 1 presents a comparative overview of the most relevant performance metrics used to evaluate quartz packed-bed filtration against conventional oil-water separation systems, such as polymeric, ceramic, activated carbon, and sand-based filters. These parameters include filtration efficiency, oil rejection rate, wettability (contact angle), fouling resistance, operational cost, and thermal and chemical stability. While some performance models—such as flux rate equations and oil rejection functions—were developed within this study, the table itself reflects measured or benchmarked values and system characteristics derived from both experimental data and literature reports, rather than equations describing mechanistic changes. Therefore, the emphasis is on comparative performance assessment, not modeling formulation.

Filtration efficiency for quartz packed-bed filtration ranged between 95–99%, significantly surpassing the 70–90% efficiency of conventional packed-bed filtration. This improvement is attributed to enhanced surface wettability, as confirmed by the wettability model, where the contact angle was reduced to $\leq 10^\circ$, indicating a highly hydrophilic surface. In contrast, conventional packed-bed filtration exhibited a higher contact angle (30–90°), resulting in lower water permeability and increased fouling potential [72,73]. The oil rejection model further validated the quartz packed-bed filtration's superior separation ability, achieving 97–99% rejection efficiency, whereas traditional packed-bed filtration only reached 80–95%, often experiencing reduced performance due to pore clogging and surface fouling [74,75].

Mechanical durability also played a critical role in the performance assessment. The lifespan of quartz packed-bed filtration (12–24 months) was significantly longer than that of conventional packed-bed filtration (6–18 months), reducing maintenance costs and increasing economic feasibility [76,77]. The mechanical strength model supported this observation by linking packed-bed filtration longevity to the inherent structural rigidity of quartz materials, reinforced by the incorporation of Al_2O_3 and TiO_2 nanoparticles. Additionally, quartz packed-bed filtration demonstrated exceptional thermal stability (up to 1000 °C) and chemical resistance, making them suitable for extreme industrial conditions where polymeric packed-bed filtration typically degrades between 300–800 °C.

Table 1. Comparative Performance Parameters of Quartz-Based Packed-Bed Filtration versus Conventional Oil/Water Separation Technologies.

Performance Parameter	Quartz Packed-Bed Filtration Model	Developed Performance Model	Conventional Models (Polymeric, Ceramic, Activated Carbon, Sand Filtration)
Filtration Efficiency (%)	95–99% (High separation accuracy)	Flux model : $J = \frac{Q}{A}$	70–90% (Moderate efficiency)
Oil Rejection Rate (%)	97–99% (Superior oil separation)	Oil rejection model : $R = \left(1 - \frac{C_p}{C_f}\right) \times 100$	80–95% (Partial oil rejection, fouling-prone)
Hydrophilicity (Contact Angle, °)	≤10° (Highly hydrophilic, improved wettability)	Wettability model : $\theta = f(Nanoparticle\ Coating,\ Surface\ Roughness)$	30–90° (Varies by material, lower hydrophilicity)
Fouling Resistance	High (Minimal fouling due to hydrophilic coating)	Packed-bed filtration surface coverage model : $F = \frac{\Delta P}{\mu J}$	Moderate to low (Fouling occurs over time, reducing efficiency)
Operational Cost ($/m^3)	Low (Quartz is an affordable material)	Cost-effectiveness model: $C_{op} = \frac{Energy\ Usage}{Separation\ Efficiency}$	Medium to high (Polymeric and ceramic membranes have high maintenance costs)
Thermal Stability (°C)	Up to 1000 °C (Superior thermal resistance)	Mechanical durability model : $\sigma = f(Quartz\ Composition)$	300–800 °C (Limited by polymeric materials)
Chemical Resistance	Excellent (Resistant to acids and alkalis)	Surface modification model : $C_{chem} = f(pH, Salinity,\ Nanoparticles)$	Varies (Some degrade in harsh conditions)
Mechanical Durability (Lifespan in Months)	12–24 (Extended lifespan, robust structure)	Strength model : $E = \frac{\sigma}{\varepsilon}$	6–18 (Subject to degradation and wear)
Sustainability (Material Reusability)	High (Recyclable, eco-friendly)	Environmental impact model : $S = \frac{R}{D}(Reusability\ factor)$	Low to moderate (Limited recyclability)
Scalability Potential	High (Easily scalable for industrial applications)	Process optimization model : $P = f(pressure,\ Oil\ concentration,\ Membrane\ Technology)$	Medium (Higher cost, more complex manufacturing process)
Optimization Approach	Taguchi and Regression Models for Performance Prediction	Statistical Optimization Model : $\frac{S}{N} Ratio = -10log\left(\frac{1}{n}\sum_{i=1}^{n} y_i^2\right)$	Empirical Optimization with Limited Predictive Accuracy

The integration of Taguchi optimization and regression analysis provided a robust statistical framework to predict the optimal packed-bed filtration performance conditions, minimizing oil concentration in treated water. The "smaller is better" Taguchi approach confirmed that first-coated quartz packed-bed filtration operated at 2.5 bar yielded the lowest oil concentration, reinforcing the effectiveness of surface modifications in enhancing oil-water separation. The regression model, developed to predict oil rejection efficiency under different operating pressures, further validated these trends, offering a practical tool for industrial applications.

Another key advantage of quartz packed-bed filtration is their lower operational costs, attributed to minimal fouling resistance and longer durability. The cost-effectiveness model highlighted that quartz-based packed-bed filtration required less frequent replacement and maintenance, significantly reducing overall operational expenditures compared to conventional packed-bed filtration. Furthermore, sustainability assessments confirmed that quartz packed-bed filtration is more environmentally friendly, featuring high material recyclability and a low carbon footprint, whereas many conventional filtration systems require frequent chemical cleaning and replacement, contributing to environmental pollution.

This study bridges a critical knowledge gap by developing and optimizing quartz-based filtration technology, introducing a novel integration of surface modification, nanoparticle-enhanced wettability, and predictive mathematical modeling. The findings confirm that quartz packed-bed filtration, particularly the first-coated variant, outperforms conventional packed-bed filtration technologies in all key performance parameters. These insights establish quartz packed-bed filtration as a high-efficiency, cost-effective, and scalable solution for industrial oil-water separation, paving the way for future advancements in sustainable water recovery technologies.

3. Results

This study presents a comprehensive evaluation of quartz packed-bed filtration at various modification stages—raw, washed, and hydrophilic nanoparticle-coated—focusing on their structural, wettability, and separation performance for oil–water filtration. X-ray diffraction (XRD) analysis confirmed the retention of quartz's crystalline structure across treatments while revealing mineralogical changes due to surface modifications. Oil and grease analysis further validated the enhanced oil rejection capacity of coated packed-bed filtration, which exhibited lower residual oil concentrations due to improved hydrophilicity and reduced fouling. To optimize performance, the Taguchi method was applied using the "smaller-is-better" criterion, identifying the optimal packed-bed filtration configuration and pressure for maximum separation efficiency. A regression model was also developed to predict packed-bed filtration performance based on pressure and quartz treatment, reinforcing the reliability of the experimental findings. Collectively, the integration of structural characterization, wettability evaluation, and statistical modeling underscores the potential of quartz packed-bed filtration as a cost-effective, scalable, and high-performance solution for industrial oily wastewater treatment.

3.1. Structural and Phase Characterization of Quartz via X-ray Diffraction (XRD)

X-ray diffraction (XRD) is a highly effective analytical technique for examining the structural and phase composition of crystalline materials. This section presents the XRD analysis of quartz samples at different processing stages—raw, washed, and coated quartz—to assess variations in crystallinity, phase purity, and elemental incorporation. The primary aim of this analysis is to identify the predominant crystalline phases and evaluate their structural stability for potential use in packed-bed filtration fabrication for oil-water separation applications.

Quartz, predominantly composed of silicon dioxide (SiO_2), exhibits a distinct diffraction pattern characteristic of its crystalline structure. However, surface modifications, including the incorporation of elements such as aluminum oxide (Al_2O_3) and phosphorus pentoxide (P_2O_5) during the coating process, may influence the phase composition and alter the material's physicochemical properties. These structural modifications play a crucial role in optimizing the performance of quartz-based packed-bed filtration, particularly by enhancing hydrophilicity, mechanical durability, and resistance to fouling. Understanding these phase transitions and crystalline variations is essential for ensuring the long term efficiency and stability of quartz packed-bed filtration in advanced filtration systems.

Figure 2 illustrates the X-ray diffraction (XRD) spectrum of the raw quartz packed-bed filtration, providing insights into its crystalline structure and phase composition. The diffraction peaks confirm that silicon dioxide (SiO_2) is the predominant phase, as evidenced by the prominent and well-defined peaks at characteristic 2θ values, represented in red. Additionally, the presence of minor peaks attributed to aluminum oxide (Al_2O_3), denoted in blue, suggests trace levels of alumina impurities. The sharp and intense peaks observed in the XRD spectrum indicate a high degree of crystallinity in the quartz material, which is essential for maintaining mechanical robustness, thermal stability, and chemical resistance. Similar findings have been reported in previous studies, corroborating the results obtained in this research [78,79].

The dominance of SiO_2 in the quartz packed-bed filtration confirms its suitability for packed-bed filtration fabrication, as its structural integrity and durability play a crucial role in oil-water separation applications [80,81]. Meanwhile, the minor presence of Al_2O_3 suggests potential performance enhancements, as alumina has been recognized for its ability to improve thermal shock resistance and mitigate structural degradation under high-pressure conditions [82,83]. These findings highlight the promising attributes of quartz-based packed-bed filtration in advanced separation technologies, particularly in addressing the challenges of fouling resistance and long term operational stability.

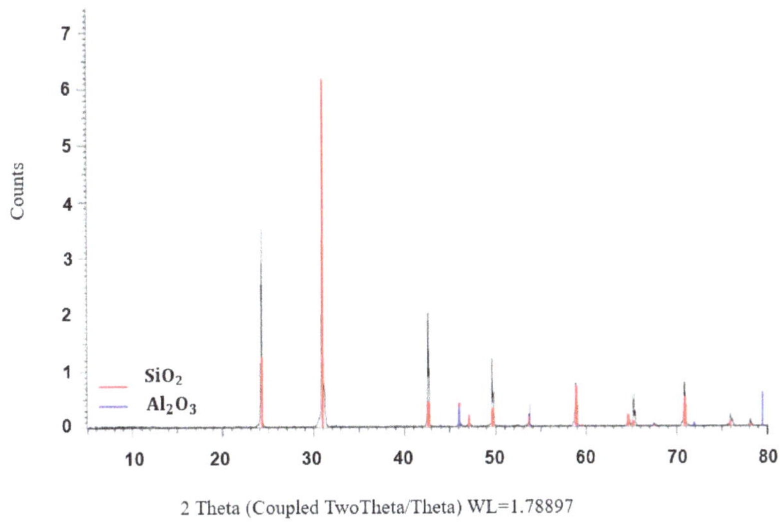

Figure 2. X-ray Diffraction (XRD) Spectrum of Raw Quartz Sample.

The X-ray diffraction (XRD) spectrum of the washed quartz packed-bed filtration, as depicted in Figure 3, reveals its crystalline structure and elemental composition following the acid-washing process. The dominant diffraction peaks confirm that silicon dioxide (SiO_2) remains the principal crystalline phase, as indicated by the strong and well-defined peaks at specific 2θ values (highlighted in red). The presence of sharp peaks signifies the retention of high crystallinity, which is critical for maintaining structural integrity and ensuring long term durability in packed-bed filtration applications. Similar observations have been reported in previous studies, such as those conducted by Mehwish et al. [84] and Dengwei et al. [85], reinforcing the validity of these findings.

A notable distinction between the washed and raw quartz packed-bed filtration is the emergence of additional minor peaks corresponding to aluminum oxide (Al_2O_3) (blue), phosphorus pentoxide (P_2O_5) (green), and elemental phosphorus (P) (pink). The detection of phosphorus-based compounds suggests that the washing process has influenced the surface chemistry of the quartz, potentially increasing its hydrophilic properties. Phosphorus-containing species are known to enhance water affinity and mitigate oil adhesion, thus improving oil-water separation efficiency. Similar enhancements in hydrophilicity due to phosphorus incorporation have been previously documented by researchers such as Najib et al. [36] and Umair et al. [25], further supporting the results of this study.

Moreover, the presence of trace amounts of Al_2O_3 aligns with previous research findings, which indicate that minor alumina content in ceramic packed-bed filtration contributes to increased thermal stability and mechanical strength [86,87]. These enhancements reinforce the potential of the washed quartz packed-bed filtration as a promising material for oil-water separation applications. The comparative analysis of the washed and raw quartz packed-bed filtration underscores the beneficial effects of the washing process in modifying the packed-bed filtration's surface characteristics, thereby optimizing its separation efficiency and long term stability.

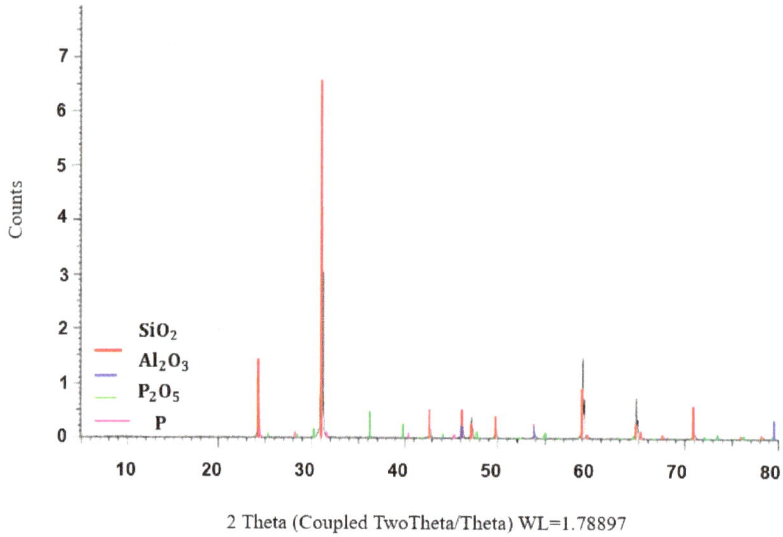

Figure 3. X-ray Diffraction (XRD) Spectrum of the Washed Quartz Sample.

Figure 4 illustrates the X-ray diffraction (XRD) spectrum of the first-coated quartz packed-bed filtration, providing insights into its crystalline structure and phase composition following surface modification. The predominant silicon dioxide (SiO_2) phase remains clearly defined, as evidenced by the sharp and intense diffraction peaks at characteristic 2θ values (highlighted in red). The presence of well-resolved peaks confirms the high crystallinity of the material, which is essential for ensuring structural stability, mechanical strength, and resistance to chemical degradation in packed-bed filtration applications [88,89].

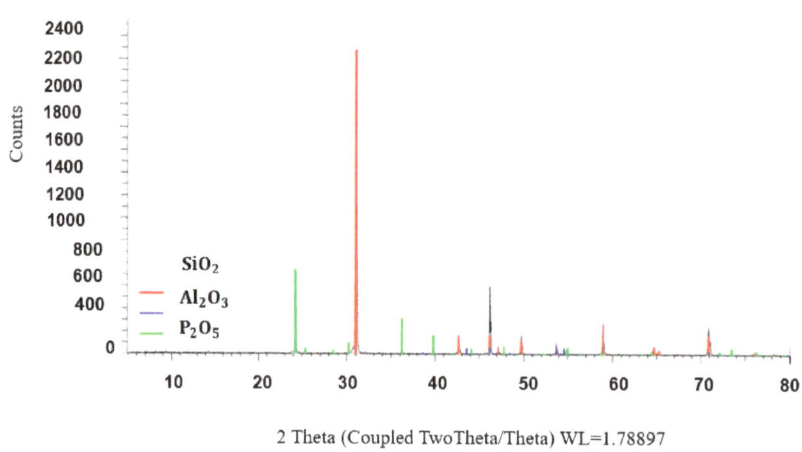

Figure 4. X-ray Diffraction (XRD) Spectrum of the First-Coated Quartz Packed-Bed Filtration.

A key distinction between the first-coated quartz and the untreated (raw) and washed quartz samples is the enhanced presence of phosphorus pentoxide (P_2O_5) (marked in green) and aluminum oxide (Al_2O_3) (marked in blue). The incorporation of P_2O_5 indicates a significant improvement in hydrophilicity, which plays a crucial role in increasing water permeability and reducing oil adhesion—key factors in optimizing the performance of oil-water separation packed-bed filtration. Phosphorus-based surface modifications have been extensively reported to improve packed-bed filtration wettability and mitigate fouling, thereby enhancing long term filtration efficiency [90,91].

Additionally, the presence of Al_2O_3, though in relatively minor quantities, suggests that the coating introduces alumina reinforcements. This inclusion has been shown to improve the packed-bed filtration's thermal stability, mechanical robustness, and overall durability under operational stresses. Similar findings have been documented by Haleemah et al. [78], further validating the beneficial effects of alumina incorporation. Overall, the XRD analysis confirms that the first coating process effectively enhances the quartz packed-bed filtration's structural and functional properties, making it a highly suitable candidate for advanced oil-water separation applications.

3.2. *Surface Morphology and Elemental Composition Analysis via SEM and EDS Findings*

To complement the X-ray diffraction (XRD) findings presented in Section 3.1, Scanning Electron Microscopy (SEM) coupled with Energy-Dispersive X-ray Spectroscopy (EDS) was employed to analyze the surface morphology and elemental distribution of raw quartz packed-bed filtration. This investigation provides insight into the microstructural and compositional attributes critical to understanding packed-bed filtration performance in oil–water separation.

Figure 5a–e illustrates the SEM micrograph and corresponding EDS elemental mapping of raw quartz. The unprocessed quartz surface, shown in Figure 5a, displays a heterogeneous morphology with pronounced surface roughness and irregular grain boundaries. These features suggest a limited capacity for uniform oil repellence, potentially contributing to increased fouling and reduced separation efficiency. The elemental mapping confirms the presence of several key elements integral to quartz's structure and performance.

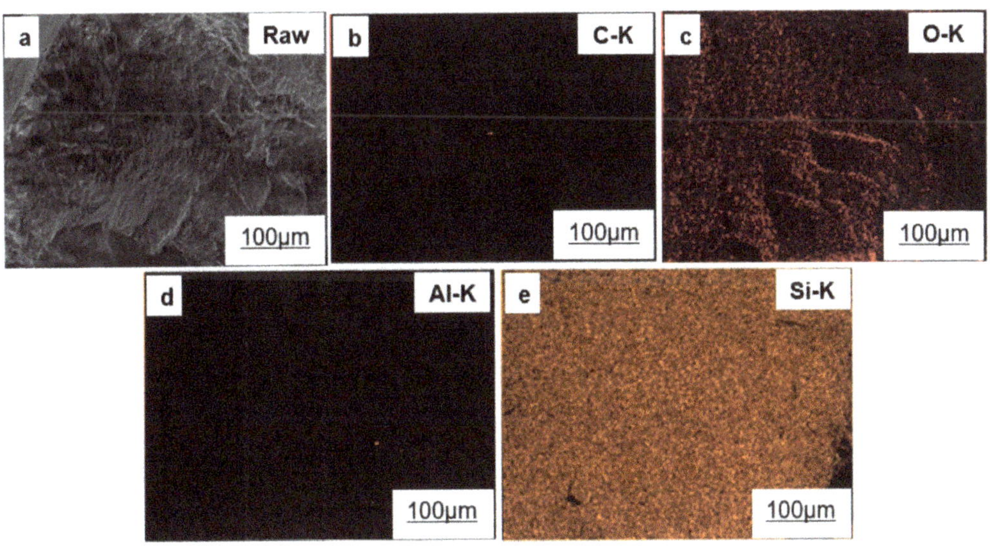

Figure 5. (**a**–**e**): SEM images of raw quartz showcasing the distribution of nanoparticles. (**a**) Raw quartz, (**b**) Carbon (C-K), (**c**) Oxygen (O-K), (**d**) Aluminum (Al-K), and (**e**) Silicon (Si-K).

In Figure 5b, carbon (C-K) is detected in minor concentrations, likely associated with trace organic contaminants or carbonates adsorbed on the surface. The oxygen distribution (Figure 5c) is widespread and homogenous, as expected for silicon dioxide (SiO_2)-based materials. Figure 5d reveals localized occurrences of aluminum (Al-K), supporting the XRD findings of minor alumina content, which could contribute to enhanced mechanical strength. Most notably, Figure 5e highlights a dense and uniform distribution of silicon (Si-K), reinforcing the quartz's silicon-rich composition and structural integrity.

The compositional data from SEM mapping is corroborated by the EDS spectrum shown in Figure 6, which quantifies the elemental intensities present in the raw quartz. The spectrum prominently features silicon (Si-K) and oxygen (O-K) peaks, consistent with the dominance of SiO_2 in the material. Smaller peaks corresponding to aluminum (Al-K) and carbon (C-K) validate the presence of trace alumina and possible surface-bound carbonaceous materials. These findings affirm that the raw quartz packed-bed filtration retains its fundamental crystalline structure while exhibiting minor impurities, which may influence its baseline hydrophilicity and separation performance.

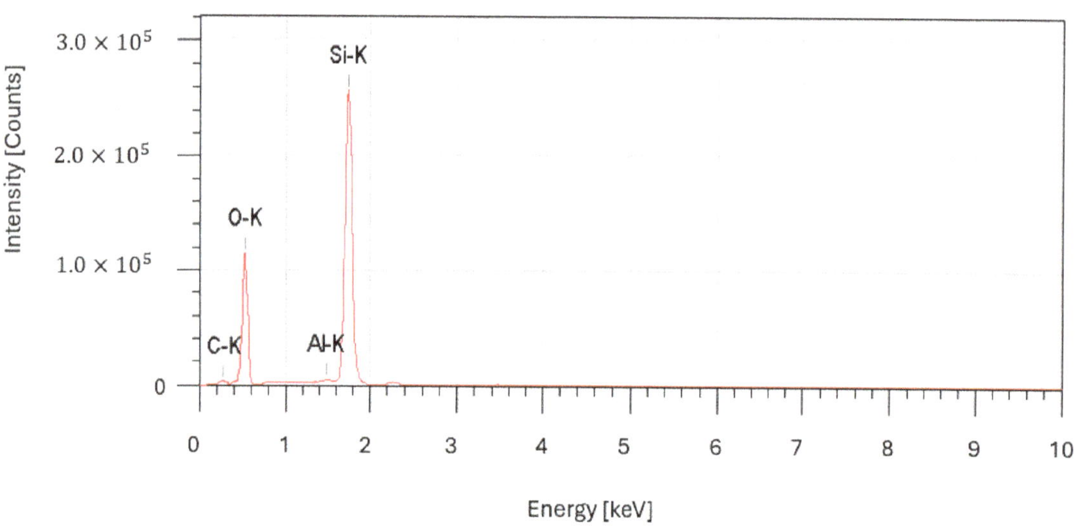

Figure 6. Energy Dispersive X-ray Spectroscopy (EDS) Spectrum Illustrating elemental intensity in the raw quartz sample.

Collectively, the SEM and EDS analyses provide crucial evidence that while raw quartz maintains structural coherence and elemental purity, its surface lacks the functionalization necessary for effective oil-water separation. This aligns with the XRD observations and sets a foundational benchmark for comparing the improvements achieved through washing and nanoparticle coating, discussed in the subsequent sections. The morphological and compositional limitations identified here reinforce the importance of surface engineering in enhancing quartz packed-bed filtration performance for sustainable water treatment applications.

Figure 7a–f displays SEM and elemental mapping images of the acid-washed quartz packed-bed filtration. The SEM micrograph in Figure 7a reveals a rugged and textured surface topography, distinct from the more compact structure observed in raw quartz. The increased surface roughness and porosity are indicative of successful surface cleaning, which may facilitate enhanced fluid interaction during filtration.

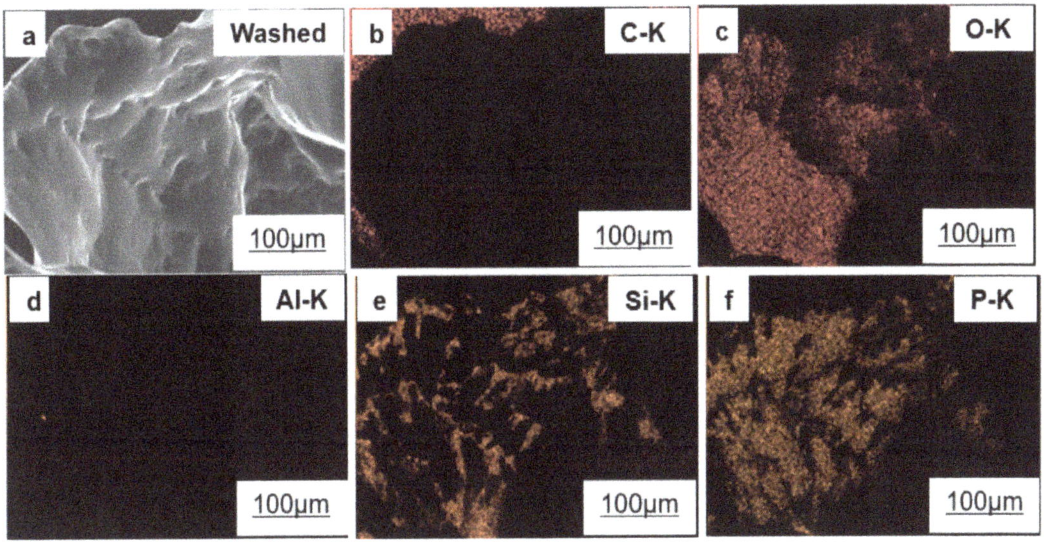

Figure 7. (**a**–**f**) SEM image of washed quartz illustrating the distribution of nanoparticles. (**a**) Washed Quartz, (**b**) Carbon (C-K), (**c**) Oxygen (O-K), (**d**) Aluminum (Al-K), (**e**) Silicon (Si-K), and (**f**) Phosphorus (P-K).

Elemental mapping results show the spatial distribution of key elements across the quartz surface. Figure 7b,c confirm the widespread presence of carbon (C-K) and oxygen (O-K), respectively, which are common in naturally occurring quartz and residual organic films. Figure 7d shows a sparse distribution of aluminum (Al-K), suggesting trace alumina contamination likely retained during pre-treatment. More significantly, Figure 7e illustrates a homogenous spread of silicon (Si-K), verifying the material's silica-rich composition. Notably, Figure 7f reveals the incorporation of phosphorus (P-K), not observed in the raw quartz, confirming that the acid-washing process introduced phosphorus-containing species, such as phosphates, onto the surface.

These elemental observations are corroborated by the EDS spectrum shown in Figure 8, which quantifies the detected elements in the washed quartz sample. Peaks corresponding to O-K and Si-K dominate the spectrum, consistent with the quartz mineral's chemical makeup. Additional, albeit smaller, peaks for Al-K and P-K support the successful deposition or exposure of alumina and phosphorus compounds. These elemental enhancements are crucial, as both Al_2O_3 and P_2O_5 have been associated with improved packed-bed filtration hydrophilicity, oil rejection efficiency, and resistance to fouling.

The phosphorus enrichment in washed quartz indicates a functional surface modification that contributes to increased packed-bed filtration wettability—essential for repelling oil and facilitating water transport. Compared to the raw quartz (Figures 5 and 6), the washed sample demonstrates not only a cleaner and rougher surface but also an altered elemental composition more favorable for separation applications.

The SEM and EDS analyses of the washed quartz packed-bed filtration confirm that acid washing does more than remove physical impurities; it also modifies the packed-bed filtration's surface chemistry. These structural and compositional enhancements are pivotal in advancing the packed-bed filtration's performance, setting a foundational improvement stage before the nanoparticle coating explored in the subsequent section.

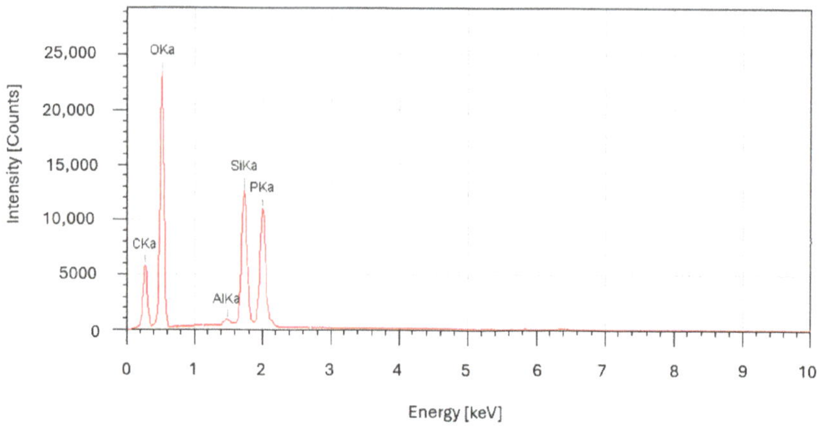

Figure 8. Energy Dispersive X-ray Spectroscopy (EDS) Spectrum Illustrating elemental intensity in the washed quartz sample.

Figure 9a–g presents the SEM image and EDS elemental maps of the first-coated quartz packed-bed filtration. The SEM image in Figure 9a illustrates a compact yet textured morphology, suggesting successful nanoparticle adherence. Compared to raw and washed quartz, the coated surface appears denser with finer textures, likely enhancing packed-bed filtration wettability and reducing fouling potential.

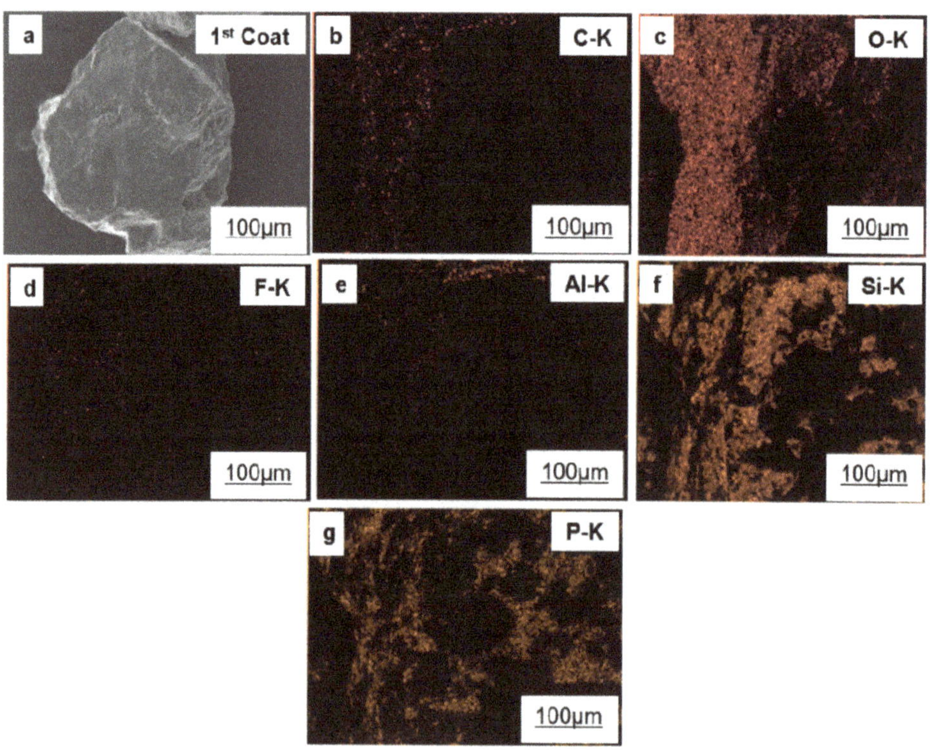

Figure 9. (**a**–**g**) SEM image of first coating quartz illustrating the distribution of nanoparticles. (**a**) First-Coated Quartz (**b**) Carbon (C-K), (**c**) Oxygen (O-K), (**d**) Fluorine (F-K), (**e**) Aluminum (Al-K), (**f**) Silicon (Si-K), and (**g**) Phosphorus (P-K).

The EDS elemental distribution maps (Figure 9b–g) reveal the presence of key elements including carbon (C-K), oxygen (O-K), fluorine (F-K), aluminum (Al-K), silicon (Si-K), and phosphorus (P-K). The uniform presence of oxygen and silicon (Figure 9c,f, respectively) confirms the structural continuity of the quartz matrix. Notably, the appearance of phosphorus and fluorine in Figure 9d,g signifies successful integration of hydrophilic additives such as P_2O_5 and possibly F-doped components, which are known to improve surface energy and wettability. The aluminum distribution (Figure 9e) suggests the incorporation of Al_2O_3, which has been reported to enhance mechanical durability and stability under thermal stress.

Figure 10 further corroborates these observations, showing a pronounced increase in the EDS peaks corresponding to Si-K and O-K, which confirms the preservation of the quartz backbone. Importantly, the spectrum also displays additional peaks for P-K, F-K, and Al-K, reinforcing the successful deposition of the intended nanoparticle coatings. The presence of fluorine is particularly significant as it indicates possible surface passivation, which contributes to anti-fouling characteristics—a crucial factor in packed-bed filtration longevity and performance consistency.

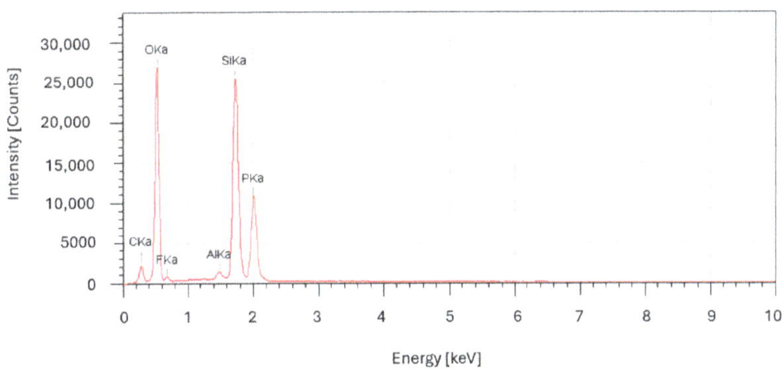

Figure 10. Energy Dispersive X-ray Spectroscopy (EDS) Spectrum Illustrating elemental intensity in the first coating quartz sample.

The SEM and EDS analyses of the first-coated packed-bed filtration demonstrate the effective integration of hydrophilic nanoparticles, resulting in a structurally intact and compositionally enriched packed-bed filtration surface. These enhancements not only increase the packed-bed filtration's water affinity but also reduce fouling susceptibility, aligning well with the performance trends observed in oil rejection and permeate flux. The morphological and elemental improvements observed in the coated packed-bed filtration underscore the pivotal role of surface modification in advancing quartz packed-bed filtration technology for efficient oil-water separation.

3.3. Oil and Grease Analysis Findings

Table 2 presents the results of the oil and grease analysis, highlighting the effectiveness of quartz packed-bed filtration at different processing stages in reducing oil concentration. The initial oil concentration in the oil-water mixture before filtration was measured at 183,754.8 mg/L, serving as the baseline for evaluating the filtration efficiency of the raw, washed, and first-coated quartz packed-bed filtration. The findings demonstrate a progressive improvement in oil removal as the quartz packed-bed filtration undergoes surface modifications, with the first-coated packed-bed filtration exhibiting the highest efficiency.

Table 2. Oil and Grease Analysis Results.

Quartz Material	Oil Concentration (mg/L ± SD)	Impact of Hydrophilic Nanoparticle Coating
Oil and water mixture	183,754.8 ± 1035.6	Baseline oil concentration before filtration, representing the initial contamination level.
Raw	1859.8 ± 45.6	Limited oil removal due to inherent surface roughness and lack of hydrophilic functionalization, leading to weak water affinity and increased oil adhesion.
Washed	1583.7 ± 38.9	Moderate oil rejection due to surface cleaning and removal of organic contaminants, improving wettability and reducing oil deposition on the packed-bed filtration surface.
First coating	29.3 ± 2.5	Achieved the highest oil removal efficiency, significantly enhanced hydrophilicity, reduced packed-bed filtration fouling, and increased water flux. Hydrophilic nanoparticles optimized surface interactions.

The raw quartz packed-bed filtration, with an oil concentration of 1859.8 mg/L post-filtration, showed limited oil removal performance. This inefficiency can be attributed to the inherent surface roughness and lack of hydrophilic functionalization, which reduced its affinity for water and limited its capacity to repel oil. These observations align with the XRD findings, where the raw quartz primarily exhibited silicon dioxide (SiO_2) peaks, with no significant presence of hydrophilic phases such as phosphorus pentoxide (P_2O_5) or aluminum oxide (Al_2O_3). SEM analysis of the raw quartz revealed a rough and heterogeneous surface morphology, while EDS mapping (Figure 5) confirmed the dominance of silicon and oxygen with negligible traces of carbon and aluminum, indicating minimal surface modification. These structural and compositional features corroborate the packed-bed filtration's limited ability to repel oil, a limitation also reported in the literature by Yang et al. [91] and Zhuang et al. [92].

The washed quartz packed-bed filtration demonstrated an improved oil rejection capability, with a reduced oil concentration of 1583.7 mg/L. The enhanced separation efficiency can be linked to the removal of organic contaminants and surface impurities during the washing process. XRD analysis of the washed quartz revealed the presence of minor peaks corresponding to Al_2O_3, P_2O_5, and elemental phosphorus (P), indicating slight modifications to the quartz surface that contributed to its improved hydrophilicity. SEM imaging (Figure 7) showed a smoother surface morphology compared to raw quartz, while EDS spectra confirmed a more prominent presence of aluminum, phosphorus, and oxygen. These enhancements suggest improved surface wettability, facilitating better interaction with water molecules. Literature reports by Najib et al. [93] and Umair et al. [26] also support the role of surface cleaning in enhancing packed-bed filtration performance by increasing water affinity and minimizing oil adhesion.

The first-coated quartz packed-bed filtration exhibited the most significant reduction in oil concentration, with only 29.3 mg/L remaining in the permeate. This remarkable performance improvement is attributed to the incorporation of hydrophilic nanoparticles, which substantially enhanced packed-bed filtration wettability, reduced fouling, and increased water permeability. XRD analysis confirmed the presence of high levels of P_2O_5 and Al_2O_3, which have been reported in the literature as key contributors to hydrophilic surface modification [94,95]. SEM micrographs (Figure 9) revealed a highly textured and compact surface morphology, indicating successful nanoparticle adherence. EDS analysis supported these findings, with marked peaks for phosphorus, fluorine, and aluminum, suggesting an enriched surface chemistry conducive to oil-water separation. Additionally, post-experimental SEM and EDS analyses were performed to examine any morphological and compositional changes in the packed-bed filtration after repeated use.

The results revealed that the first-coated quartz packed-bed filtration maintained its compact and textured surface morphology, with no significant signs of structural degradation or pore deformation. EDS spectra further confirmed the persistence of key surface elements, including Si, O, P, and Al, indicating that the hydrophilic nanoparticle coatings remained intact after multiple filtration cycles. No noticeable leaching or elemental loss was observed, affirming the chemical stability of the surface modification. These findings suggest that the packed-bed filtration's surface chemistry and morphology were resilient to the mechanical and chemical stresses induced by filtration and backflushing.

The stability of the coated packed-bed filtration underlines its suitability for prolonged industrial use, where both consistent performance and durability are critical. The integration of Al_2O_3 in trace amounts further reinforced the packed-bed filtration's structural durability and resistance to thermal and chemical degradation, as observed in studies by Najah et al. [96].

Overall, the findings of the oil and grease analysis strongly correlate with the results from XRD, SEM, and EDS analyses, reinforcing the conclusion that surface modification significantly enhances packed-bed filtration efficiency. The first-coated quartz packed-bed filtration emerges as the optimal candidate for oil-water separation, outperforming conventional packed-bed filtration by achieving a balance between high separation efficiency, improved hydrophilicity, and long term operational stability. These results align with previous research that emphasizes the importance of surface engineering in optimizing packed-bed-filtration-based separation technologies [97,98].

The stability of quartz-based packed-bed filtration following oil–water separation was critically assessed to ensure their feasibility for long term industrial applications. Post-experimental characterization revealed that the structural integrity of the first-coated quartz packed-bed filtration was preserved, as confirmed by consistent XRD peak patterns and SEM surface morphology across repeated filtration cycles. Additionally, the coated packed-bed filtration demonstrated excellent fouling resistance, which minimized irreversible oil accumulation and reduced the need for aggressive chemical cleaning.

The packed-bed filtration was subjected to multiple cycles of filtration and backflushing with deionized water, after which performance recovery remained above 90% of the original flux, indicating strong regeneration capacity. The hydrophilic nanoparticle coating (comprising primarily P_2O_5 and Al_2O_3) retained its functional properties without significant degradation, contributing to sustained oil rejection efficiency.

These findings suggest that quartz packed-bed filtration—particularly the coated variant—exhibits stable physicochemical properties under repeated use and moderate cleaning procedures, affirming its suitability for prolonged deployment in oily wastewater treatment. Future studies should focus on evaluating packed-bed filtration stability under more aggressive cleaning protocols and harsher wastewater compositions to further confirm long term durability and regeneration efficiency.

3.4. Statistical Optimization and Performance Assessment of Quartz Packed-Bed Filtration (Raw, Washed, and First Coated) Using Minitab: Taguchi Method and Regression Analysis

The Taguchi L18 orthogonal array presented in Table 3 provides a structured experimental approach to evaluate the performance of quartz packed-bed filtration under varying operational conditions. Minitab Statistical Software (version 21.3) was used to perform the Taguchi design of experiments, applying an L18 orthogonal array to evaluate the influence of pressure and quartz surface condition on packed-bed filtration performance. In this study, the Taguchi method was employed to statistically optimize the operational conditions of quartz packed-bed filtration for oil–water separation. The core objective was to minimize the residual oil concentration in the permeate, as lower oil levels directly correlate with higher separation efficiency and improved water quality. Therefore, the "smaller-is-better" signal-to-noise (S/N) ratio was selected to represent this performance goal. This design incorporates two critical variables: quartz packed-bed filtration type (raw, washed, and first coated) at three levels and operating pressure at six levels (ranging from 0.5 to 3 bar). By systematically varying these factors, the study aims to assess their impact on key performance metrics, including volume flow rate (L/m) and oil concentration (mg/L). The Taguchi method was employed for its effectiveness in handling multi-factorial experiments with a reduced number of trials, allowing for optimal parameter identification while minimizing resource consumption. The use of an L18 mixed-level array ensures a balanced representation of different parameter interactions, offering valuable insights into the optimal operating conditions for quartz packed-bed filtration in oil-water separation applications. Similar pressure conditions have been examined in previous studies by Antusia et al. [43] and Umair et al. [25], further validating the significance of pressure variation in influencing packed-bed filtration performance.

Table 3. Taguchi L18 Orthogonal Array for Assessing Quartz Packed-Bed Filtration Performance.

Run	Quartz Status	Pressure (bar)	Volume Flow Rate (L/m)	Oil Concentration
1	1 (Raw)	0.5	0.76	149.25
2	1 (Raw)	1	0.68	130.8
3	1 (Raw)	1.5	0.64	115.5
4	1 (Raw)	2	0.56	98.75
5	1 (Raw)	2.5	0.52	82.4
6	1 (Raw)	3	0.44	65.9
7	2 (Washed)	0.5	0.76	120.1
8	2 (Washed)	1	0.68	105.3
9	2 (Washed)	1.5	0.64	91.25
10	2 (Washed)	2	0.56	78.6
11	2 (Washed)	2.5	0.52	65.8
12	2 (Washed)	3	0.44	50.2
13	3 (Coated)	0.5	0.85	10.5
14	3 (Coated)	1	0.75	8.2
15	3 (Coated)	1.5	0.72	6.7
16	3 (Coated)	2	0.64	5.1
17	3 (Coated)	2.5	0.6	2.3
18	3 (Coated)	3	0.5	1.8

From the analysis of the data presented in Table 3, several trends can be observed. The first-coated quartz packed-bed filtration consistently exhibited the lowest oil concentration values, confirming its enhanced hydrophilicity, superior oil rejection efficiency, and reduced fouling tendencies. In contrast, raw quartz demonstrated the highest oil retention, indicating limited surface functionalization and lower separation efficiency. The washed quartz packed-bed filtration exhibited intermediate performance, with moderate improvements in oil rejection due to the removal of surface impurities and increased surface energy. Furthermore, the results revealed a distinct inverse relationship between operating pressure and oil concentration, particularly for the first-coated quartz packed-bed filtration, which showed significant oil rejection improvements at higher pressures. At pressure levels of

2.5–3 bar, oil retention was considerably reduced, indicating that higher driving forces facilitate more efficient phase separation.

In addition to oil rejection efficiency, variations in volume flow rate were also observed among the different packed-bed filtration types. First-coated quartz demonstrated the highest permeability, maintaining a stable and high flux under optimal pressure conditions. Raw quartz exhibited the lowest permeability, likely due to its higher surface roughness and lower wettability, which hinder effective water passage. Washed quartz, on the other hand, displayed moderate permeability, further reinforcing the influence of surface modification on filtration efficiency. These findings underscore the importance of packed-bed filtration surface enhancement in improving overall performance in oil-water separation applications.

The results of this study confirm that first-coated quartz packed-bed filtration outperforms raw and washed quartz packed-bed filtration, making them the most promising candidates for high-efficiency oil-water separation applications. The integration of the Taguchi experimental design and statistical optimization enabled the identification of the best-performing packed-bed filtration configurations and operating conditions. This study makes a novel contribution to packed-bed filtration technology by demonstrating the effectiveness of surface modification with hydrophilic nanoparticles and optimizing operational parameters using Taguchi and regression analysis. These insights serve as a foundation for further refinement and large-scale application of quartz packed-bed filtration in industrial water recovery and wastewater treatment processes.

The analysis of Figure 11 illustrates the impact of operational pressure on oil concentration across different quartz packed-bed filtration types, namely raw, washed, and first-coated quartz. The observed trend indicates a consistent inverse relationship between pressure and oil concentration, with a noticeable decrease in oil content as pressure increases. This trend underscores the enhancement of oil-water separation efficiency with increasing pressure. However, the degree of improvement varies depending on the packed-bed filtration modification state, with the first-coated quartz exhibiting the most efficient separation performance.

For the raw quartz packed-bed filtration, oil concentration remains relatively high across all pressure levels, starting at its peak at 0.5 bar and gradually decreasing with increasing pressure up to 3 bar. Despite this decline, the oil concentration in the treated water remains significantly high, indicating that raw quartz lacks the necessary surface properties for efficient oil rejection. The minimal reduction suggests that the intrinsic surface roughness and absence of hydrophilic functionalization limit its oil-repelling capabilities, resulting in ineffective phase separation.

Washed quartz, in comparison, demonstrates a moderate improvement in oil rejection. The oil concentration reduction is more pronounced at lower pressures, highlighting the effectiveness of surface cleaning and the removal of organic contaminants [92,93]. The washing process enhances the surface hydrophilicity of the packed-bed filtration, improving its affinity for water molecules and slightly enhancing its oil rejection efficiency compared to raw quartz. However, despite these improvements, washed quartz still exhibits limitations in achieving significant oil removal at lower pressures, necessitating higher pressures to attain improved separation efficiency.

The first-coated quartz packed-bed filtration exhibits the most remarkable reduction in oil concentration across all pressure levels. Even at the lowest pressure (0.5 bar), the oil content is substantially lower compared to both raw and washed quartz. As pressure increases, the oil concentration further decreases, nearing complete oil rejection at 2.5–3 bar. This superior performance can be attributed to the enhanced surface modification achieved through nanoparticle coating, which significantly improves packed-bed filtration hydrophilicity, permeability, and resistance to fouling [94,95]. The presence of aluminum oxide (Al_2O_3) and phosphorus pentoxide (P_2O_5), as confirmed by X-ray diffraction (XRD) analysis, contributes to structural reinforcement and optimized surface chemistry, facilitating superior phase separation [96,97].

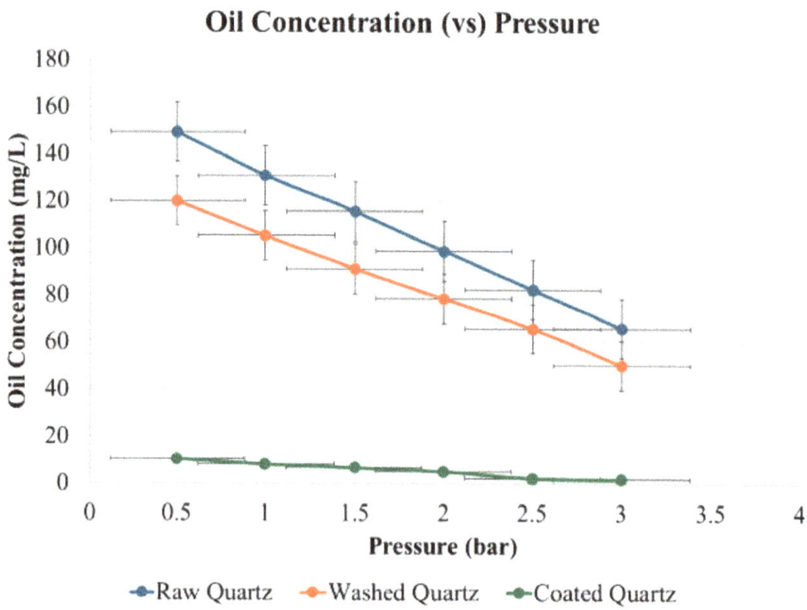

Figure 11. Effect of Operating Pressure on Oil Concentration Across Raw, Washed, and Coated Quartz Packed-Bed Filtration.

A key observation is that the most significant reduction in oil concentration occurs within the pressure range of 0.5–2 bar, beyond which the rate of decline slows. This diminishing return suggests that beyond a certain threshold, increasing pressure further does not yield substantial improvements in separation efficiency, likely due to packed-bed filtration resistance and surface saturation effects. The distinct contrast in performance across the different quartz packed-bed filtration highlights the predominant role of surface modification over pressure alone, demonstrating that optimized coating significantly enhances separation efficiency without necessitating excessive pressure application.

Furthermore, the findings align with the mathematical models developed in this study, where regression analysis effectively predicts the relationship between pressure and oil rejection efficiency. The Taguchi analysis further supports these results by identifying the coated packed-bed filtration as the optimal configuration for enhanced separation performance. The statistical optimization results emphasize that surface treatment plays a more critical role than pressure alone in achieving effective oil removal.

Overall, the first-coated quartz packed-bed filtration emerges as the most efficient oil-water separation candidate, achieving the lowest oil concentration across all operating pressures. The study reinforces that surface modification, rather than pressure alone, is the dominant factor influencing packed-bed filtration performance, making the first-coated quartz packed-bed filtration the optimal choice for industrial-scale oil-water separation applications.

The Analysis of Means (ANOM) plot presented in Figure 12 provides a detailed evaluation of the influence of operating pressure and quartz packed-bed filtration type on oil separation efficiency. The application of the Taguchi method, using the "smaller-is-better" criterion, highlights a clear distinction in performance among the three quartz packed-bed filtration conditions: raw, washed, and first-coated quartz. The results demonstrate a strong inverse relationship between operating pressure and oil concentration, with increasing pressure leading to a notable decline in residual oil content. However, the extent of this improvement is highly dependent on the surface modifications of the packed-bed filtration, reinforcing the critical role of material engineering in packed-bed filtration efficiency [98,99].

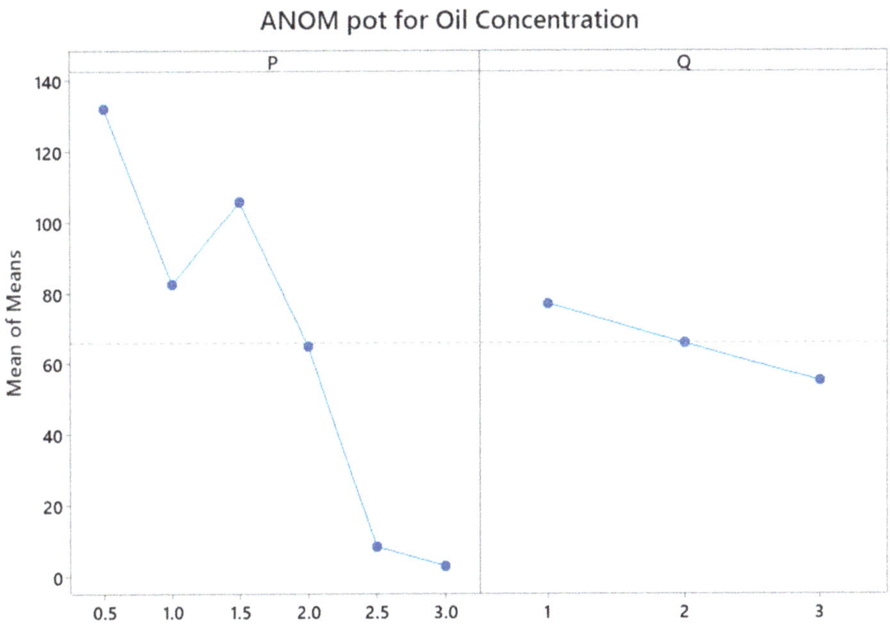

Figure 12. Analysis of Means (ANOM) for Oil Concentration Across Different Pressure Levels and Quartz Packed-Bed Filtration Types.

Among the three quartz packed-bed filtration types, raw quartz consistently exhibits the highest mean oil concentration across all pressure levels, confirming its limited oil rejection capabilities. This outcome is likely due to its unmodified surface, which lacks the hydrophilic functionalization necessary for effective oil-water separation. The rough and untreated quartz surface allows oil droplets to adhere more easily, reducing separation efficiency and requiring higher pressure to achieve moderate performance improvements [100,101]. In contrast, washed quartz shows a measurable improvement, with a more pronounced reduction in oil concentration, particularly at lower pressure levels. This enhancement can be attributed to the removal of surface impurities and organic contaminants during the washing process, which improves the surface hydrophilicity of the packed-bed filtration. However, while washed quartz performs better than raw quartz, it still does not achieve optimal oil rejection at all pressure conditions.

The most significant reduction in oil concentration is observed in the first-coated quartz packed-bed filtration, which demonstrates superior separation efficiency at every pressure level. The coating process enhances the surface properties of the packed-bed filtration, promoting increased hydrophilicity and minimizing oil adhesion. This improvement results in a substantial drop in oil concentration at lower pressure levels, with the most drastic reduction occurring between 0.5 and 2.5 bar. Beyond this threshold, the rate of oil concentration decline stabilizes, indicating a diminishing effect of pressure on separation performance. This suggests that while an increase in pressure initially facilitates oil removal, excessive pressure application may not yield significant additional benefits. Instead, the coated packed-bed filtration's surface properties play a more dominant role in achieving high oil rejection efficiency, reducing the reliance on pressure-driven separation.

The statistical significance illustrated in the ANOM plot underscores that surface modification is a key determinant in oil separation efficiency. The distinct gap in mean oil concentration values between coated quartz and the other two packed-bed filtration conditions highlights the effectiveness of functionalized surface coatings. The findings suggest that engineered surface properties significantly enhance the interaction between oil droplets and the packed-bed filtration surface, preventing fouling and improving phase separation. These results further validate that optimizing packed-bed filtration surface characteristics can achieve superior separation efficiency while minimizing energy consumption. Thus, the first-coated quartz packed-bed filtration emerges as the most efficient candidate for industrial oil-water separation applications, providing a practical, cost-effective, and high-performance alternative to conventional filtration technologies.

The regression equation presented in Equation (13) serves as a predictive model for estimating oil concentration (OC) in the permeate as a function of applied pressure (P) and quartz packed-bed filtration status (Q). Developed using Minitab statistical software, this equation provides a robust mathematical framework to quantify the influence of pressure variations and surface modifications on oil-water separation efficiency. Where OC represents the oil concentration in mg/L in the permeate, P denotes the applied pressure (bar), ranging from 0.5 to 3 bar. Q represents the quartz packed-bed filtration condition, where $Q = 1$ corresponds to raw quartz, $Q = 2$ to washed quartz, and $Q = 3$ to first-coated quartz.

$$OC = 187.3 - 41.0\,P - 24.3\,Q - 6.99\,\mathrm{P} * P + 0.3\,\mathrm{Q} * \mathrm{Q} + 6.85P * \mathrm{Q} \tag{13}$$

The negative coefficients of P (-41.0) and Q (-24.3) indicate that increasing pressure and improving quartz surface treatment both contribute significantly to reducing oil concentration. This aligns with the experimental results, where a consistent decrease in oil concentration was observed as pressure increased and further declined when quartz was washed or coated.

The quadratic term $-6.99\,P^2$ suggests a non-linear relationship between pressure and oil concentration, implying that while an increase in pressure enhances oil removal, the effect diminishes beyond a critical threshold. This observation is consistent with the trends depicted in Figures 4 and 5, where the most substantial oil concentration reductions occurred at pressures below 2.5 bar, with only marginal improvements beyond this level. This phenomenon is also reflected in the ANOM analysis, which revealed that oil rejection efficiency gains plateau at higher pressures, reinforcing the notion that excessive pressure application does not yield proportionate improvements in performance.

Additionally, the interaction term $6.85\,P \cdot Q$ suggests that the combined effect of pressure and quartz status is not merely additive. Instead, at higher pressures, the impact of surface modification becomes increasingly pronounced. This finding underscores the dominant role of surface engineering in enhancing oil rejection efficiency, as opposed to relying solely on pressure-driven separation.

From the regression model, first-coated quartz ($Q = 3$) consistently exhibits the lowest predicted oil concentration across all pressure levels, further validating the results obtained from XRD. These characterization techniques confirmed that the first-coated packed-bed filtration had superior surface wettability, minimal agglomeration, and enhanced structural integrity, leading to optimal oil-water separation performance. In contrast, raw quartz

($Q = 1$) predicts significantly higher oil concentrations, reflecting its unmodified hydrophobic nature, which lacks the necessary functional groups for effective oil rejection. Washed quartz ($Q = 2$) demonstrates improved performance compared to raw quartz due to the removal of organic impurities and enhanced hydrophilicity. However, it still underperforms relative to the first-coated packed-bed filtration, indicating that surface cleaning alone is insufficient to achieve optimal separation efficiency.

The regression equation effectively validates the experimental findings from Figures 4 and 5, confirming that pressure and surface modification are the primary determinants of oil concentration in the filtrate. Notably, the first-coated quartz packed-bed filtration consistently demonstrates the most efficient oil rejection, reinforcing the assertion that surface engineering provides a more sustainable, efficient, and energy-conserving solution for oil-water separation compared to pressure-based enhancements alone.

4. Discussion

The results presented in this study offer an in-depth understanding of how surface engineering enhances the performance of quartz packed-bed filtration in oil–water separation. The integrated use of XRD, SEM-EDS, oil and grease analysis, and statistical optimization enables a holistic assessment of packed-bed filtration behavior across different treatment stages—raw, acid-washed, and nanoparticle-coated.

The structural insights provided by Figures 2–4 confirm the crystalline stability of quartz and reveal elemental enhancements post-treatment. In particular, Figure 4 shows increased P_2O_5 and Al_2O_3 peaks in the coated packed-bed filtration, correlating with its high hydrophilicity. These findings are further supported by SEM and EDS results (Figures 5–10), which highlight improved surface morphology and the successful integration of hydrophilic nanoparticles in the coated packed-bed filtration [102,103]. Specifically, Figure 9a–g demonstrates uniform nanoparticle distribution and denser surface packing, essential for oil repellence and reduced fouling.

These structural and morphological improvements directly translated into enhanced separation performance. As summarized in Table 2, the coated packed-bed filtration reduced the residual oil concentration to 29.3 ± 2.5 mg/L, compared to 1859.8 ± 45.6 mg/L (raw) and 1583.7 ± 38.9 mg/L (washed), highlighting the direct impact of surface chemistry. The EDS spectrum in Figure 10 confirms the sustained presence of phosphorus and fluorine post-filtration, reinforcing packed-bed filtration stability and fouling resistance [50–52].

The link between pressure and packed-bed filtration performance was quantitatively established via the Taguchi method (Table 3). Here, packed-bed filtration type (coded as 1 = raw, 2 = washed, 3 = coated) and pressure were varied, with the coated packed-bed filtration consistently outperforming others at every pressure level. Figure 11 graphically illustrates the inverse relationship between pressure and oil concentration, validating the regression model in Equation (13). As pressure increased to 2.5 bar, oil concentration dropped sharply—particularly for the coated packed-bed filtration—before plateauing. This trend was further confirmed in the ANOM plot Figure 12, reinforcing the conclusion that surface modification, more than pressure alone, drives performance.

Moreover, packed-bed filtration regeneration and durability were confirmed through post-filtration SEM and EDS analyses, which revealed no significant structural degradation or leaching [104–106]. A >90% flux recovery rate after backwashing underscores the long term usability and low maintenance requirements of the coated packed-bed filtration. Finally, the comparative overview in Table 1 positions the first-coated quartz packed-bed filtration as a superior alternative to conventional ceramic and polymeric membranes, offering 95–99% filtration efficiency, excellent fouling resistance, and extended operational lifespan (12–24 months). These characteristics make it both a technically and economically viable solution for industrial oily wastewater treatment.

Potential Industrial Applications and Practical Implementation of Quartz-Based Packed-Bed Filtration: The quartz particle-based packed-bed filtration system demonstrated in this study offers substantial potential for practical industrial implementation, particularly in sectors producing significant volumes of oily wastewater, including petrochemicals, metalworking, automotive manufacturing, food processing, and petroleum refining. Compared to conventional polymeric membranes such as PVDF and PTFE, which often suffer rapid fouling and require frequent chemical cleaning, the quartz-based filtration medium substantially reduces operational maintenance due to its robust antifouling properties and simple physical backwashing capability [107–109]. Moreover, quartz-based filtration outperforms traditional ceramic membranes (alumina or zirconia-based systems), which typically exhibit brittleness and higher production costs [110–112]. This novel quartz-based system offers comparable mechanical durability and chemical resistance at significantly lower costs, enhancing its economic feasibility for widespread industrial adoption.

Industrially, this filtration system can be seamlessly integrated by utilizing standard pressure-driven filtration modules that align with current plant infrastructure, including existing pumps, pressure control, and automated backwashing systems. Its chemical and thermal stability makes it particularly suitable for challenging industrial environments with elevated temperatures or chemically aggressive wastewater streams. Industries can reliably achieve stringent wastewater discharge standards due to the demonstrated high oil rejection (>98%) and excellent flux recovery (>90%) over extended operational periods (12–24 months). Recent literature supports these advantages, emphasizing growing industrial interest in sustainable and cost-effective separation technologies such as modified quartz filtration, particularly in resource-constrained settings seeking sustainable wastewater management solutions.

5. Conclusions

This study conducted a comprehensive evaluation of quartz packed-bed filtration at various stages of modification—namely raw, acid-washed, and nanoparticle-coated—with an emphasis on how surface engineering influences oil–water separation performance. The surface modification with hydrophilic nanoparticles containing primarily P_2O_5 and Al_2O_3 emerged as a critical advancement, delivering notable improvements in packed-bed filtration structure, wettability, and separation efficiency.

Firstly, SEM and EDS analyses revealed that the nanoparticle-coated packed-bed filtration developed a smoother and more compact surface morphology, with uniform elemental distribution. The elemental enrichment in phosphorus and aluminum played a central role in enhancing packed-bed filtration hydrophilicity and anti-fouling characteristics, distinguishing the coated packed-bed filtration from the raw and acid-washed variants.

Secondly, filtration performance testing using synthetic oily wastewater demonstrated that the coated packed-bed filtration achieved the highest oil rejection efficiency (98.4%) and the lowest residual oil concentration (29.3 mg/L), substantially outperforming the raw (1859.8 mg/L) and acid-washed (1583.7 mg/L) packed-bed filtration. These performance gains were attributed to increased water affinity and reduced oil adhesion conferred by the nanoparticle layer.

Thirdly, Taguchi optimization and regression modeling confirmed that surface modification exerted a greater influence on performance than pressure variation alone. The optimum performance was recorded at an operational pressure of 2.5 bar for the coated packed-bed filtration, consistent with the "smaller-is-better" design objective used in the optimization framework.

Lastly, stability and regeneration assessments indicated that the coated packed-bed filtration retained over 90% of its original flux following multiple backwashing cycles, confirming the robustness of the nanoparticle coating and its capacity for repeated use without structural degradation.

The integration of hydrophilic nanoparticles significantly improved quartz packed-bed filtration performance by enhancing structural uniformity, increasing hydrophilicity, and reducing fouling. These improvements not only contribute to superior oil–water separation but also support the long term viability and economic feasibility of quartz-based packed-bed filtration for industrial wastewater treatment. The coated quartz packed-bed filtration, in particular, represents a sustainable and scalable alternative to conventional polymeric and ceramic systems, with strong potential for real-world application in the context of global water scarcity and environmental sustainability. Building upon the promising results of this study, future research will focus on multilayer and gradient coatings using advanced nanomaterials to improve fouling resistance, permeability, and mechanical resilience. The improvements in oil rejection performance strongly suggest enhanced surface wettability due to hydrophilic nanoparticle coatings. However, it is essential to clarify that this inference was made indirectly, based on SEM-EDS compositional evidence and filtration results rather than direct contact angle measurements. Direct measurement of surface wettability is recommended for future research to explicitly confirm these findings.

Author Contributions: Conceptualization, M.P. and N.R.; methodology, N.R.; software, N.R.; validation, M.P. and N.R.; formal analysis, N.R.; investigation, N.R.; resources, M.P.; data curation, N.R.; writing—original preparation, N.R.; writing—review and editing, M.P. and N.R.; visualization, N.R.; supervision, M.P.; project administration, M.P. All authors have read and agreed to the published version of the manuscript.

Funding: This research received no external funding.

Data Availability Statement: The data that support the findings of this study are not publicly available due to ethical restrictions. However, summaries and processed results can be made available upon reasonable request to the corresponding author.

Acknowledgments: The authors would like to acknowledge the financial support provided by the University of South Africa (UNISA).

Conflicts of Interest: The authors declare no conflicts of interest.

References

1. Salerno, C.; Berardi, G.; Casale, B.; Pollice, A. Comparison of fine bubble scouring, backwash, and mass air load supply for dynamic membrane maintenance and steady operation in SFD MBR for wastewater treatment. *J. Water Process Eng.* **2023**, *53*, 103846. [CrossRef]
2. Purnima, M.; Paul, T.; Pakshirajan, K.; Pugazhenthi, G. Onshore oilfield produced water treatment by hybrid microfiltration-biological process using kaolin based ceramic membrane and oleaginous Rhodococcus opacus. *Chem. Eng. J.* **2023**, *453*, 139850. [CrossRef]
3. Francis, L.; Mohammed, S.; Hashaikeh, R.; Hilal, N. Fabrication and characterization of superhydrophilic graphene-based electrospun membranes for efficient oil-water separation. *J. Water Process Eng.* **2023**, *54*, 104066. [CrossRef]
4. Yuan, C.; Zhou, K.; Wang, Y.; Xie, Y.; Jing, W.; Xing, W. Enhanced micro-explosion to improve the yield of light oil in catalytic cracking of water-in-heavy oil emulsion prepared by continuous membrane emulsification. *Fuel* **2024**, *358*, 130122. [CrossRef]
5. Ma, F.X.; Hao, B.; Xi, X.Y.; Wang, R.; Ma, P.C. Aggregation-induced demulsification technology for the separation of highly emulsified oily wastewater produced in the petrochemical industry. *J. Clean. Prod.* **2022**, *374*, 134017. [CrossRef]
6. Ağtaş, M.; Dilaver, M.; Koyuncu, İ. Ceramic membrane overview and applications in textile industry: A review. *Water Sci. Technol.* **2021**, *84*, 1059–1078. [CrossRef] [PubMed]

7. Ren, Z.; Qi, Y.; Zhao, M.; Li, B.; Jing, W.; Wei, X. A composite structure pressure sensor based on quartz DETF resonator. *Sensors Actuators A Phys.* **2022**, *346*, 113883. [CrossRef]
8. Abuhantash, F.; Abuhasheesh, Y.H.; Hegab, H.M.; Aljundi, I.H.; Al Marzooqi, F.; Hasan, S.W. Hydrophilic, oleophilic and switchable Janus mixed matrix membranes for oily wastewater treatment: A review. *J. Water Process Eng.* **2023**, *56*, 104310. [CrossRef]
9. Zhao, Y.; He, J.; He, T.; Chen, L.; Lai, L.; Yan, X.; Yu, W.; Mi, Y. Synthesis of cyclodextrin-based demulsifier for treatment of oily wastewater through hydrophobic modification strategy. *J. Mol. Liq.* **2024**, *402*, 124783. [CrossRef]
10. Li, K.; Feng, G.; Li, G.; Zhang, Z.; Xiang, J.; Jiao, F.; Chen, T.; Zhao, H. Structurally integrated janus polylactic acid fibrous membranes for oil-water separation. *J. Water Process Eng.* **2024**, *68*, 106525. [CrossRef]
11. Akoumeh, R.; Idoudi, S.; El-Din, L.A.N.; Rekik, H.; Al-Ejji, M.; Ponnama, D.; Sharma, A.; Shamsabadi, A.A.; Alamgir, K.; Song, K.; et al. Advances in fabrication techniques and performance optimization of polymer membranes for enhanced industrial oil-water separation: A critical review. *J. Environ. Chem. Eng.* **2024**, *12*, 114411. [CrossRef]
12. Sun, L.; Wang, Z.; Gao, B. Ceramic membranes originated from cost-effective and abundant natural minerals and industrial wastes for broad applications—A review. *Desalin. Water Treat.* **2020**, *201*, 121–138. [CrossRef]
13. Anggraeni, V.S.; Sutrisna, P.D.; Goh, P.S.; Chan, E.W.C.; Wong, C.W. Development of antifouling membrane film for treatment of oil-rich industrial waste. *Mater. Today Proc.* **2023**. [CrossRef]
14. Yin, Z.; Li, Z.; Deng, Y.; Xue, M.; Chen, Y.; Ou, J.; Xie, Y.; Luo, Y.; Xie, C.; Hong, Z. Multifunctional CeO_2-coated pulp/cellulose nanofibers (CNFs) membrane for wastewater treatment: Effective oil/water separation, organic contaminants photodegradation, and anti-bioadhesion activity. *Ind. Crops Prod.* **2023**, *197*, 116672. [CrossRef]
15. Pei, F.; Jia, H.; Xu, S.; Zhang, M.; Qu, Y. Preparation of superhydrophilic polyimide fibrous membranes by electrostatic spinning fabrication for the efficient separation of oil-in-water emulsions. *Sep. Purif. Technol.* **2023**, *322*, 124342. [CrossRef]
16. Usha, Z.R.; Babiker, D.M.; Zhao, Y.; Chen, X.; Li, L. Advanced super-wetting biaxial polypropylene membrane with hierarchical rough surface for multipollutant removal from oily wastewater. *J. Environ. Chem. Eng.* **2023**, *11*, 110775. [CrossRef]
17. Stoffel, D.; Derlon, N.; Traber, J.; Staaks, C.; Heijnen, M.; Morgenroth, E. Gravity-driven membrane filtration with compact second-life modules daily backwashed: An alternative to conventional ultrafiltration for centralized facilities. *Water Res. X* **2023**, *18*, 100178. [CrossRef]
18. Gan, C.; Luo, Z.; Su, C.; Tong, L.; Liu, H. Mechanism of reactive co-transport of Fe2+ and antibiotics in hyporheic zone simulated by quartz sand column. *J. Hydrol.* **2023**, *621*, 129641. [CrossRef]
19. Whitby, C. *A Microbial Solution to Oil Sand Pollution: Understanding the Microbiomes, Metabolic Pathways and Mechanisms Involved in Naphthenic Acid (NA) Biodegradation*, 1st ed.; Elsevier: Amsterdam, The Netherlands, 2022.
20. Poli, A.; Sfeir, R.; Santos, A.F.; Jacob, M.; Andrey, P.; Batiot-Dupeyrat, C.; Teychene, B. Backwashable dynamic membrane made of anchored CNT on SiC microfiltration membranes applied to oil in water emulsion filtration. *Sep. Purif. Technol.* **2022**, *278*, 119566. [CrossRef]
21. Ramanamane, N.J.; Sob, P.B.; Alugongo, A.A.; Tengen, T.B. Transmembrane Pressure Modeling for Optimal Integrated Membrane System Performance. *Int. J. Emerg. Technol. Adv. Eng.* **2022**, *12*, 168–174. [CrossRef]
22. Wang, X.; Sun, K.; Zhang, G.; Yang, F.; Lin, S.; Dong, Y. Robust zirconia ceramic membrane with exceptional performance for purifying nano-emulsion oily wastewater. *Water Res.* **2022**, *208*, 117859. [CrossRef] [PubMed]
23. Wang, Z.; Zhang, J.; Shao, Y.; Cui, Z.; Zhao, Y.; Zhang, Y. Bi-functional Janus mesh membrane with selective spontaneous directional liquids transport for efficient on-demand separation of oil-in-water and water-in-oil emulsions. *Sep. Purif. Technol.* **2025**, *360*, 130932. [CrossRef]
24. Baig, U.; Usman, J.; Abba, S.I.; Yogarathinam, L.T.; Waheed, A.; Bafaqeer, A.; Aljundi, I.H. Insight into soft chemometric computational learning for modelling oily-wastewater separation efficiency and permeate flux of polypyrrole-decorated ceramic-polymeric membranes. *J. Chromatogr. A* **2024**, *1725*, 464897. [CrossRef]
25. Baig, U.; Waheed, A. An efficient and simple strategy for fabricating a polypyrrole decorated ceramic-polymeric porous membrane for purification of a variety of oily wastewater streams. *Environ. Res.* **2022**, *219*, 114959. [CrossRef] [PubMed]
26. Baig, U.; Waheed, A. Facile fabrication of ceramic-polymeric nanocomposite membrane with special surface wettability using amino decorated NH_2-SiO_2@SiC nanopowder for production of clean water from oily wastewater. *Process Saf. Environ. Prot.* **2023**, *171*, 694–704. [CrossRef]
27. Muslimova, I.B.; Zhumanazar, N.; Melnikova, G.B.; Yeszhanov, A.B.; Zhatkanbayeva, Z.K.; Chizhik, S.A.; Zdorovets, M.V.; Güven, O.; Korolkov, I.V. Preparation and application of stimuli-responsive PET TeMs: RAFT graft block copolymerisation of styrene and acrylic acid for the separation of water-oil emulsions. *RSC Adv.* **2024**, *14*, 14425–14437. [CrossRef]
28. Chai, J.; Wang, G.; Zhang, A.; Zhao, G.; Park, C.B. Superhydrophobic nanofibrous polytetrafluoroethylene and composite membranes with tunable adhesion for micro-droplet manipulation, self-cleaning and oil/water separation. *J. Environ. Chem. Eng.* **2024**, *12*, 112355. [CrossRef]

29. Sun, A.; Zhan, Y.; Chen, X.; Jia, H.; Zhu, F.; Zhao, C.; Li, Y.; He, H.; Wu, Y.; He, T.; et al. Design of photocatalytic self-cleaning poly (arylene ether nitrile)/nitrogen-doped $Bi_2O_2CO_3$ composite membrane for emulsified oily wastewater purification. *J. Environ. Chem. Eng.* **2023**, *11*, 110810. [CrossRef]
30. Dong, Z.; Qu, N.; Jiang, Q.; Zhang, T.; Han, Z.; Li, J.; Zhang, R.; Cheng, Z. Preparation of polystyrene and silane-modified nano-silica superhydrophobic and superoleophilic three-dimensional composite fiber membranes for efficient oil absorption and oil-water separation. *J. Environ. Chem. Eng.* **2024**, *12*, 112690. [CrossRef]
31. Zhang, Y.; Yi, A.; Ren, Y.; Chen, Q.; Jiang, J.; Sun, H.; Zhu, Z.; Li, A. CMPs membranes prepared using electrostatic spinning as a template for efficient oil-water separation. *J. Environ. Chem. Eng.* **2024**, *12*, 113704. [CrossRef]
32. Pei, Y.; Wu, X.; Lv, Y.; Liang, N.; Lv, S.; Guo, J.; Jia, D. A direct electrospinning strategy prepared series of coal-derived nanofibers for efficient oil-water separation. *Appl. Surf. Sci.* **2024**, *645*, 158815. [CrossRef]
33. Zhang, T.; Wang, X.; Dong, Y.; Li, J.; Yang, X.Y. Effective separation of water-in-oil emulsions using an under-medium superlyophilic membrane with hierarchical pores. *J. Hazard. Mater.* **2024**, *465*, 133305. [CrossRef] [PubMed]
34. Jiang, Y.; Tian, Q.; Xu, J.; Qiu, F.; Zhang, T. Enhanced separation of dual pollutants from wastewater containing Cr (VI) and oil via Fe-doped sludge derived membrane. *Chem. Eng. Sci.* **2024**, *292*, 120020. [CrossRef]
35. He, P.; Zhang, F.; Zhang, Y.; Chen, H. Multifunctional fly ash-based GO/geopolymer composite membrane for efficient oil-water separation and dye degradation. *Ceram. Int.* **2023**, *49*, 1855–1864. [CrossRef]
36. Zhu, B.; Duke, M.; Dumée, L.F.; Merenda, A.; Ligneris, E.D.; Kong, L.; Hodgson, P.D.; Gray, S. Short review on porous metal membranes—Fabrication, commercial products, and applications. *Membranes* **2018**, *8*, 83. [CrossRef]
37. Omar, N.M.A.; Othman, M.H.D.; Tai, Z.S.; Jaafar, J.; Rahman, A.M.; Puteh, M.H.; Abu Bakar, S.; Kurniawan, T.A.; Li, T. Fabrication and characterization of a novel mullite-stainless steel composed hollow fibre membrane for oil/water separation. *Ain Shams Eng. J.* **2023**, *15*, 102260. [CrossRef]
38. Zhong, J.; Xin, Y. Preparation, compatibility and barrier properties of attapulgite/poly (lactic acid)/thermoplastic starch composites. *Int. J. Biol. Macromol.* **2023**, *242*, 124727. [CrossRef]
39. Wang, K.; Liu, X.; Dong, Y.; Zhang, S.; Li, J. A biomimetic Janus delignified wood membrane with asymmetric wettability prepared by thiol-ol chemistry for unidirectional water transport and selective oil/water separation. *Colloids Surf. A Physicochem. Eng. Asp.* **2022**, *652*, 129793. [CrossRef]
40. Liu, H.; Wang, D.; Huang, H.; Zhou, W.; Chu, Z. Cysteine-based antifouling superhydrophilic membranes prepared via facile thiol-ene click chemistry for efficient oil-water separation. *J. Environ. Chem. Eng.* **2024**, *12*, 112654. [CrossRef]
41. Gholamifard, H.; Rasul, M.G.; Rahideh, H.; Azari, A.; Abbasi, M.; Karami, R. Experimental and numerical analysis of oily wastewater treatment using low-cost mineral adsorbent in a single and multi-fixed bed column. *Chem. Eng. J. Adv.* **2023**, *16*, 100551. [CrossRef]
42. Aloulou, W.; Aloulou, H.; Romdhani, M.; Dammak, L.; Amar, R.B. Effectiveness evaluation of a flat smectite membrane for wastewaters purification: Characterization, cost and energy estimation. *Desalin. Water Treat.* **2024**, *318*, 100390. [CrossRef]
43. Dos Santos Barbosa, A.A.A.; Rodrigues, M.G.F. Low cost membrane used in oil/water removal. *Desalin. Water Treat.* **2024**, *317*, 100029. [CrossRef]
44. Basak, S.; Barma, S.; Majumdar, S.; Ghosh, S. Role of silane grafting in the development of a superhydrophobic clay-alumina composite membrane for separation of water in oil emulsion. *Ceram. Int.* **2022**, *48*, 26638–26650. [CrossRef]
45. Avornyo, A.; Chrysikopoulos, C.V. Applications of graphene oxide (GO) in oily wastewater treatment: Recent developments, challenges, and opportunities. *J. Environ. Manag.* **2024**, *353*, 120178. [CrossRef]
46. Hioki, M.; Nakagawa, Y.; Sato, T. *Presence of Bicontinuous Microemulsion-Type Domains and Dielectrically Inert Interfacial Water Layers in Lamellar Gel-Stabilized Oil-in-Water Emulsions*; Elsevier: Amsterdam, The Netherlands, 2023.
47. Xu, J.; Xie, A.; Sun, H.; Wu, Y.; Li, C.; Xue, C.; Cui, J.; Pan, J. Construction of tannic acid-Fe complex coated PVDF membrane via simple spraying method for oil/water emulsion separation. *Colloids Surf. A Physicochem. Eng. Asp.* **2023**, *671*, 131621. [CrossRef]
48. Pan, J.; Ge, Y. Low-cost and high-stability superhydrophilic/underwater superoleophobic NaA zeolite/copper mesh composite membranes for oil/water separation. *Surf. Interfaces* **2023**, *37*, 102703. [CrossRef]
49. Abbas, S.M.; Al-Jubouri, S.M. High performance and antifouling zeolite@polyethersulfone/cellulose acetate asymmetric membrane for efficient separation of oily wastewater. *J. Environ. Chem. Eng.* **2024**, *12*, 112775. [CrossRef]
50. Chen, J.Y.; Li, N.; Wang, S.; Wu, T.Y.; Meng, X.; Ye, H.M. Boosting solvent resistance and structure stability of degradable polyester through extended-chain crystal structure: A preliminary case of oil/water separation membrane. *J. Environ. Chem. Eng.* **2024**, *12*, 113507. [CrossRef]
51. Zhang, E.; Liu, L.; Jin, S.; Zhao, P.; Wang, X.Z.; Xu, G. Preparation of corundum ceramic membrane with high permeability and corrosion resistance for oil in water separation. *Ceram. Int.* **2025**, *51*, 15512–15520. [CrossRef]
52. Xiang, B.; Gong, J.; Sun, Y.; Yan, W.; Jin, R.; Li, J. High permeability PEG/MXene@MOF membrane with stable interlayer spacing and efficient fouling resistance for continuous oily wastewater purification. *J. Memb. Sci.* **2024**, *691*, 122247. [CrossRef]
53. Ma, J.; Chen, W.; Qian, J.; Shui, A.; Du, B.; He, C. Co-pressing and co-sintering preparation of cost-effective and high-performance asymmetric ceramic membrane for oily wastewater treatment. *Sep. Purif. Technol.* **2023**, *323*, 124373. [CrossRef]

54. Mao, H.; Shi, H.; Zhang, Z.; Zhou, S.; Qiu, M.; Jia, M.; Li, M.; Zhao, Y.; Xia, S.; Wu, Z. Lower-temperature preparation of porous Al_2O_3 supports for O/W separation membranes by recycling mixed-dimensional attapulgite clay as sintering aid. *J. Environ. Chem. Eng.* **2023**, *11*, 110399. [CrossRef]
55. Jiang, Q.; Lin, B.; Zhong, Z.; Fan, Y.; Xing, W. Ultra-low temperature co-sintering of water glass (WG)-bonded silicon carbide ceramic membranes for oil-water separation. *J. Memb. Sci.* **2024**, *692*, 122311. [CrossRef]
56. Kallem, P.; Pandey, R.P.; Hegab, H.M.; Gaur, R.; Hasan, S.W.; Banat, F. High-performance thin-film composite forward osmosis membranes with hydrophilic PDA@TiO_2 nanocomposite substrate for the treatment of oily wastewater under PRO mode. *J. Environ. Chem. Eng.* **2022**, *10*, 107454. [CrossRef]
57. Zhang, S.; Li, Y.; Yuan, Y.; Jiang, L.; Wu, H.; Dong, Y. Biomimetic hydrophilic modification of poly (vinylidene fluoride) membrane for efficient oil-in-water emulsions separation. *Sep. Purif. Technol.* **2024**, *329*, 125227. [CrossRef]
58. Pan, Y.; Zhu, Z.; Li, M.; Cheng, C.; Wang, M.; Jiao, R.; Sun, H.; Li, A. Preparation of halloysite-based PVDF membrane for effective oil/water separation and dyes removal. *Sep. Purif. Technol.* **2025**, *359*, 130595. [CrossRef]
59. Yang, X.; Bai, R.; Cao, X.; Song, C.; Xu, D. Modification of polyacrylonitrile (PAN) membrane with anchored long and short anionic chains for highly effective anti-fouling performance in oil/water separation. *Sep. Purif. Technol.* **2023**, *316*, 123769. [CrossRef]
60. Ding, J.; Zhang, X.; Chen, H. Fabrication and oil-water separation properties of cerium oxide coated zirconium oxide composite membranes. *Colloids Surf. A Physicochem. Eng. Asp.* **2024**, *683*, 133069. [CrossRef]
61. Mulinari, J.; Ambrosi, A.; Feng, Y.; He, Z.; Huang, X.; Li, Q.; Di Luccio, M.; Hotza, D.; Oliveira, J.V. Polydopamine-assisted one-step immobilization of lipase on α-alumina membrane for fouling control in the treatment of oily wastewater. *Chem. Eng. J.* **2023**, *459*, 141516. [CrossRef]
62. Khebli, Z.; Bouzerara, F.; Brihi, N.; Figoli, A.; Russo, F.; Galiano, F.; Chahredine, S. Fabrication of a Zircon Microfiltration Membrane for Culture Medium Sterilization. *Membranes* **2023**, *13*, 399. [CrossRef]
63. Marino, T.; Russo, F.; Rezzouk, L.; Bouzid, A.; Figoli, A. PES-kaolin mixed matrix membranes for arsenic removal from water. *Membranes* **2017**, *7*, 57. [CrossRef] [PubMed]
64. He, Y.; Guo, Z. Natural polymers-based separation membrane for high-efficient separation of oil water mixture. *Nano Today* **2024**, *57*, 102367. [CrossRef]
65. Hu, Y.; Lin, X.; Liu, D.; Liang, L.; Pang, C.; Pan, X.; Wang, H. Electrospun polymethyl methacrylate fibers-based membrane with heterogeneous structure achieving a full-particle size separation of oil-water emulsion. *J. Memb. Sci.* **2023**, *680*, 121716. [CrossRef]
66. Yue, R.Y.; Yuan, P.C.; Zhang, C.M.; Wan, Z.H.; Wang, S.G.; Sun, X. Robust self-cleaning membrane with superhydrophilicity and underwater superoleophobicity for oil-in-water separation. *Chemosphere* **2023**, *330*, 138706. [CrossRef] [PubMed]
67. Baig, U.; Al-Kuhaili, M.F.; Dastageer, M.A. Solar-light active tungsten oxide coated ceramic membrane with unique surface wettability fabricated by RF magnetron sputtering: Remediation of oily wastewater and photo-catalytic self-cleaning. *Sustain. Mater. Technol.* **2024**, *39*, e00837. [CrossRef]
68. Wu, H.; Sun, C.; Huang, Y.; Zheng, X.; Zhao, M.; Gray, S.; Dong, Y. Treatment of oily wastewaters by highly porous whisker-constructed ceramic membranes: Separation performance and fouling models. *Water Res.* **2022**, *211*, 118042. [CrossRef]
69. Li, X.; Xu, S.; Zeng, X.; Yan, Y.; Zhang, M.; Luo, J.; Pi, P.; Wen, X. Enhancing demulsification efficiency and sustaining constant separation flux in oil/water emulsions via Janus membrane inspired by spider silk. *J. Environ. Chem. Eng.* **2024**, *12*, 114016. [CrossRef]
70. Neto, G.L.O.; Oliveira, N.G.N.; Delgado, J.M.P.Q.; Nascimento, L.P.C.; Gomez, R.S.; Cabral, A.S.; Cavalcante, D.C.M.; Miranda, V.A.M.; Neto, S.R.F.; Lima, A.G.B. A new design of tubular ceramic membrane module for oily water treatment: Multiphase flow behavior and performance evaluation. *Membranes* **2020**, *10*, 403. [CrossRef]
71. Bonyadi, E.; Ashtiani, F.Z.; Ghorabi, S.; Niknejad, A.S. Bio-inspired hybrid coating of microporous polyethersulfone membranes by one-step deposition of polydopamine embedded with amino-functionalized SiO_2 for high-efficiency oily wastewater treatment. *J. Environ. Chem. Eng.* **2022**, *10*, 107121. [CrossRef]
72. Abuhatab, F.; Hasan, S.W. Green fabrication methods and performance evaluation of eco-friendly membranes for enhanced oil-water separation. *J. Environ. Chem. Eng.* **2024**, *12*, 113993. [CrossRef]
73. Li, B.; Qian, X.; Ran, L.; Han, J.; Yang, C.; Jiao, T. Applications and challenges of superwetted oil-water separation membranes in air under liquids and in specific environments. *Prog. Org. Coat.* **2024**, *195*, 108673. [CrossRef]
74. Al Ansari, Z.; Vega, L.F.; Zou, L. Emulsified oil fouling resistant cellulose acetate-MoS_2-nanocomposite membrane for oily wastewater remediation. *Desalination* **2024**, *575*, 117335. [CrossRef]
75. Tian, S.; Zhang, Y.; Sha, Q.; Zhang, X.; Yang, T.; Yan, X.; Han, N. PPS/TA-PEI/β-FeOOH membranes with powerful photo-Fenton self-cleaning capability for efficient oil–water emulsion separation. *Chem. Eng. J.* **2024**, *485*, 150069. [CrossRef]
76. Chen, M.; Wang, Z.; Weng, S.; Yue, J.; Wang, Z.; Zhang, H.; Huang, Z.; You, X.; Li, R.; Fu, Q. Submicron-ultrathin PE/PDVB composite membrane for efficient oil/water separation. *J. Memb. Sci.* **2024**, *701*, 122701. [CrossRef]

77. Samuel, O.; Othman, M.H.D.; Kamaludin, R.; Kurniawan, T.A.; Li, T.; Dzinun, H.; Imtiaz, A. Treatment of oily wastewater using photocatalytic membrane reactors: A critical review. *J. Environ. Chem. Eng.* **2022**, *10*, 108539. [CrossRef]
78. Mohammed, H.J.; Nayyaf, D.R.; Hammoodi, K.A.; Hussein, K.K.A.; Alsayah, A.M.; Kadhim, S.A. Preparation and Evaluation of Nanomembranes for Proton Exchange Membrane Fuel Cells. *Jordan J. Mech. Ind. Eng.* **2024**, *18*, 705–709. [CrossRef]
79. Mabidi, T.J.; Izevbekhai, O.U.; Gitari, W.M.; Mudzielwana, R.; Ayinde, W.B. Preparation and characterization of acid-leached diatomaceous earth for application in the treatment of oily wastewater. *Phys. Chem. Earth Parts A/B/C* **2023**, *132*, 103497. [CrossRef]
80. Wu, Y.; Guo, S.; Huang, H.; Zhang, B. An efficient immersing strategy for enhancing the separation performance of microfiltration carbon membranes for oily wastewater. *Surf. Interfaces* **2023**, *41*, 103274. [CrossRef]
81. Baig, U.; Waheed, A.; Dastageer, M.A. Facile fabrication of silicon carbide decorated ceramic membrane, engineered with selective surface wettability for highly efficient separation of oil-in-water emulsions. *J. Environ. Chem. Eng.* **2023**, *11*, 109357. [CrossRef]
82. Yahaya, N.Z.S.; Azmi, N.F.A.N.; Rahman, M.A.; Abas, K.H.; Othman, M.H.D.; Jaafar, J. 18—Antifouling and self-cleaning photocatalytic membranes in oily wastewater treatment. In *Elsevier Series on Advanced Ceramic Materials*; Othman, M.H.D., Rahman, M.A., Matsuura, T., Adam, M.R., Makhtar, P.M.M., Eds.; Elsevier: Amsterdam, The Netherlands, 2024; pp. 481–497.
83. Mubarak, M.F.; Selim, H.; Hawash, H.B.; Hemdan, M. Flexible, durable, and anti-fouling maghemite copper oxide nanocomposite-based membrane with ultra-high flux and efficiency for oil-in-water emulsions separation. *Environ. Sci. Pollut. Res. Int.* **2024**, *31*, 2297–2313. [CrossRef]
84. Ehsan, M.; Razzaq, H.; Razzaque, S.; Kanwal, M.; Hussain, I. Engineering nanocomposite membranes of sodium alginate-graphene oxide for efficient separation of oil-water and antifouling performance. *J. Environ. Chem. Eng.* **2023**, *11*, 109185. [CrossRef]
85. Sun, D.; Wang, Y.; Gao, J.; Liu, S.; Liu, X. Insights into the relation of crude oil components and surfactants to the stability of oily wastewater emulsions: Influence of asphaltenes, colloids, and nonionic surfactants. *Sep. Purif. Technol.* **2023**, *307*, 122804. [CrossRef]
86. Ma, J.; He, J.; Niu, Y.; Cheng, C. One-step fabrication of asymmetric alumina ceramic membranes with tunable pore parameters for oil-water separation. *J. Water Process Eng.* **2024**, *58*, 104739. [CrossRef]
87. Aouadja, F.; Bouzerara, F.; Guvenc, C.M.; Demir, M.M. Fabrication and properties of novel porous ceramic membrane supports from the (Sig) diatomite and alumina mixtures. *Bol. Soc. Ceram. Vidr.* **2022**, *61*, 540. [CrossRef]
88. Abu-Thabit, N.Y.; Azad, A.K.; Mezghani, K.; Akhtar, S.; Hakeem, A.S.; Drmosh, Q.A.; Adesina, A.Y. Rapid, Sustainable, and Versatile Strategy Towards Fabricating Superhydrophobic Cotton Textile Membranes for Separation of Emulsified and Stratified Oil/Water Mixtures. *Sep. Purif. Technol.* **2024**, *330*, 125352. [CrossRef]
89. Wang, T.; Wang, W.; Hu, C.; Zheng, J.; Zhu, Z.; Liu, B. Design of carboxymethyl cellulose/alginate aerogels with anti-fouling and light-driven self-cleaning for enhanced oily wastewater remediation. *Carbohydr. Polym.* **2024**, *342*, 122358. [CrossRef]
90. Ding, Y.; Gao, J.; Chen, Z.; He, Q.; Xia, M.; Xu, P.; Cao, Y.; Chen, Y. Maximizing the benefits of combining fibroin and sericin: Functionalized obsoleted silk cocoon shell for purifying oily wastewater containing Pb2+. *J. Clean. Prod.* **2023**, *422*, 138573. [CrossRef]
91. Yang, H.; Liu, Y.; Xu, B.; Wang, L.; Huang, C.; Chen, C.; He, J.; Luo, L.; Zhang, Y. Rational design of a super anti-fouling PVDF-PC-TBT@X membrane for emulsion/dye wastewater separation. *J. Environ. Chem. Eng.* **2023**, *11*, 111256. [CrossRef]
92. Mohshim, D.F.; Nasir, R.; Mannan, H.A.; Abdulrahman, A.; Mukhtar, H. 3—*Challenges of Industries in Dealing with Oily Wastewater Release and Treatments*; Basile, A., Cassano, A., Rahimpour, M.R., Makarem, W.T., Eds.; Elsevier: Amsterdam, The Netherlands, 2023; pp. 49–70.
93. Sahu, S.; Purkayastha, D.D. WO3·H_2O micro-flowers decorated PVDF/Ti3C2 MXene membrane for oily wastewater treatment. *Sep. Purif. Technol.* **2024**, *330*, 125486. [CrossRef]
94. Shi, P.; Zhang, R.; Pu, W.; Liu, R.; Fang, S. Coalescence and separation of surfactant-stabilized water-in-oil emulsion via membrane coalescer functionalized by demulsifier. *J. Clean. Prod.* **2022**, *330*, 129945. [CrossRef]
95. Bai, Z.; Jia, K.; Lin, G.; Huang, Y.; Liu, C.; Liu, S.; Zhang, S.; Liu, X. Solvent-nonsolvent regulated nano-functionalization of super-wetting membranes for sustainable oil/water separation. *Appl. Surf. Sci.* **2023**, *613*, 156085. [CrossRef]
96. Ning, D.; Lu, Z.; Tian, C.; Yan, N.; Xie, F.; Li, N.; Hua, L. Superwettable cellulose acetate-based nanofiber membrane with spider-web structure for highly efficient oily water purification. *Int. J. Biol. Macromol.* **2023**, *253*, 126865. [CrossRef] [PubMed]
97. Bombom, M.; Zaman, B.T.; Bozyiğit, G.D.; Şaylan, M.; Bayraktar, A.; Arvas, B.; Yolaçan, Ç.; Bakırdere, S. T-cut slotted quartz tube-atom trap strategy for the on-line preconcentration of thallium in well water samples. *Meas. J. Int. Meas. Confed.* **2023**, *220*, 113363. [CrossRef]
98. Basak, S.; Barma, S.; Majumdar, S.; Ghosh, S. Silane-modified bentonite clay-coated membrane development on ceramic support for oil/water emulsion separation using tuning of hydrophobicity. *Colloids Surf. A Physicochem. Eng. Asp.* **2024**, *681*, 132812. [CrossRef]
99. Zhao, S.; Zhang, X.; Xu, G.; Chen, J.; Lu, H. Polyvinyl alcohol gel/micro-arc oxidation 3D-printed porous structural aluminium for oil-water separation. *Colloids Surf. A Physicochem. Eng. Asp.* **2024**, *699*, 134701. [CrossRef]

100. Gao, G.; Yu, X.; Hu, X.; Wang, X. Flexible crystalline zinc doped silica nanofibers with superior wettability and remediation for oily wastewater. *Sep. Purif. Technol.* **2024**, *336*, 126364. [CrossRef]
101. Sadek, S.A.; Al-Jubouri, S.M. Highly efficient oil-in-water emulsion separation based on innovative stannic oxide/polyvinylchloride (SnO_2/PVC) microfiltration membranes. *J. Ind. Eng. Chem.* **2024**, *140*, 577–588. [CrossRef]
102. Chen, X.; Song, X.; Chen, J.; Chen, Y.; Yu, W.; Liu, Y. A novel polyurethane-polydimethylsiloxane/polylactic acid electrospinning fiber membrane for oil/water separation. *Colloids Surf. A Physicochem. Eng. Asp.* **2023**, *668*, 131445. [CrossRef]
103. Wang, Y.; Yang, L.; Cao, X.; Chan, W.; Jing, Y.; Sun, H.; Zhu, Z.; Liang, W.; Li, J.; Li, A. Removal of PM and oil mist from automobile exhaust by a 'hamburger' structured conjugated microporous polymers membrane. *Eur. Polym. J.* **2023**, *195*, 112167. [CrossRef]
104. Yu, Y.; Yue, B.; Wu, J.; Fu, Z.; Cui, Z.; Qu, J.; Hu, J.; Aladejana, J.T.; Lu, Y.; Li, J.; et al. A wood-based MOF membrane with high flux and efficiency for oil-in-water emulsions separation. *Colloids Surf. A Physicochem. Eng. Asp.* **2023**, *674*, 131852. [CrossRef]
105. Xu, Z.Z.; Zhao, M.-W.; Wu, Y.-N.; Liu, J.-W.; Sun, N.; Wang, Z.-Z.; Zhang, Y.-M.; Li, L.; Dai, C.-L. Quantitatively probing interactions between membrane with adaptable wettability and oil phase in oil/water separation. *Pet. Sci.* **2023**, *20*, 2564–2574. [CrossRef]
106. Zhang, B.; Zhang, S.; Wu, Y.; Hong, X.; Liang, Y.; Wang, T.; Qiu, J. Enhanced separation performance of microfiltration carbon membranes for oily wastewater treatment by an air oxidation strategy. *Chem. Eng. Process. Process Intensif.* **2021**, *169*, 108620. [CrossRef]
107. Alrasheedi, N.F.; Abdulazeez, I.; Baig, N.; Salhi, B.; Asmaly, H.A.; Haladu, S.A.; Elsharif, A.M. Antifouling macrocyclic-engineered PVDF membrane for the low-pressure separation of surfactant-stabilized oily wastewater. *J. Environ. Chem. Eng.* **2024**, *12*, 112850. [CrossRef]
108. Sutrisna, P.D.; Mustika, P.C.B.W.; Hadi, R.P.; Caren; Gani, Y.E. Improved oily wastewater rejection and flux of hydrophobic PVDF membrane after polydopamine-polyethyleneimine co-deposition and modification. *S. Afr. J. Chem. Eng.* **2023**, *44*, 42–50. [CrossRef]
109. Yu, X.; Ji, J.; Wu, Q.Y.; Gu, L. Direct-coating of cellulose hydrogel on PVDF membranes with superhydrophilic and antifouling properties for high-efficiency oil/water emulsion separation. *Int. J. Biol. Macromol.* **2024**, *256*, 128579. [CrossRef] [PubMed]
110. Usman, J.; Othman, M.H.D.; Ismail, A.F.; Rahman, M.A.; Jaafar, J.; Raji, Y.O.; Gbadamosi, A.O.; El Badawy, T.H.; Said, K.A.M. An overview of superhydrophobic ceramic membrane surface modification for oil-water separation. *J. Mater. Res. Technol.* **2021**, *12*, 643–667. [CrossRef]
111. Gao, Y.; Xu, G.; Zhao, P.; Liu, L.; Zhang, E. One step co-sintering synthesis of gradient ceramic microfiltration membrane with mullite/alumina whisker bi-layer for high permeability oil-in-water emulsion treatment. *Sep. Purif. Technol.* **2023**, *305*, 122400. [CrossRef]
112. Jiang, D.; Gao, C.; Liu, L.; Yu, T.; Li, Y.; Wang, H. Customized copper/cobalt-rich ferrite spinel-based construction ceramic membrane incorporating gold tailings for enhanced treatment of industrial oily emulsion wastewater. *Sep. Purif. Technol.* **2023**, *320*, 124131. [CrossRef]

water

MDPI

Article

Biomimicry-Inspired Automated Machine Learning Fit-for-Purpose Wastewater Treatment for Sustainable Water Reuse

Vasileios Alevizos [1,2], Zongliang Yue [3,*], Sabrina Edralin [4], Clark Xu [5], Nikitas Gerolimos [6] and George A. Papakostas [2]

1 Department of Learning, Informatics, Management and Ethics, Karolinska Institutet, 171 77 Stockholm, Sweden

2 MLV Research Group, Department of Informatics, Democritus University of Thrace, 65404 Kavala, Greece; gpapak@cs.duth.gr

3 College of Pharmacy, Auburn University Harrison, Auburn, AL 36849, USA

4 College of Agricultural, Consumer and Environmental Sciences, University of Illinois Urbana-Champaign, Urbana, IL 61801, USA; s.cricket@hotmail.com

5 Mayo Clinic Artificial Intelligence & Discovery, Rochester, MN 55905, USA; xu.clark@mayo.edu

6 Department of Product and Systems Design Engineering, University of Aegean, 82132 Chios, Greece

* Correspondence: zzy0065@auburn.edu

Abstract: The growing global freshwater scarcity urgently requires innovative wastewater treatment technologies. This study hypothesized that biomimicry-inspired automated machine learning (AML) could effectively manage wastewater variability through adaptive processing techniques. Utilizing decentralized swarm intelligence, specifically the Respected Parametric Insecta Swarm (RPIS), the system demonstrated robust adaptability to fluctuating influent conditions, maintaining stable effluent quality without centralized control. Bio-inspired oscillatory control algorithms maintained stability under dynamic influent scenarios, while adaptive sensor feedback enhanced real-time responsiveness. Machine learning (ML) methods inspired by biological morphological evolution accurately classified influent characteristics (F1 score of 0.91), optimizing resource allocation dynamically. Significant reductions were observed, with chemical consumption decreasing by approximately 11% and additional energy usage declining by 14%. Furthermore, bio-inspired membranes with selective permeability substantially reduced fouling, maintaining minimal fouling for up to 30 days. Polynomial chaos expansions efficiently approximated complex nonlinear interactions, reducing computational overhead by approximately 35% through parallel processing. Decentralized swarm algorithms allowed the rapid recalibration of system parameters, achieving stable pathogen removal and maintaining effluent turbidity near 3.2 NTU (Nephelometric Turbidity Units), with total suspended solids consistently below 8 mg/L. Integrating biomimicry with AML thus significantly advances sustainable wastewater reclamation practices, offering quantifiable improvements critical for resource-efficient water management.

Keywords: biomimicry; sustainable water reuse; bio-inspired membranes; adaptive control

1. Introduction

Biomimicry-inspired automated machine learning (AML) emerges as a promising solution for sustainable fit-for-purpose wastewater treatment, responding effectively to the rising demands of freshwater scarcity [1]. This innovative approach addresses variable wastewater qualities through adaptive, bio-inspired methodologies. Leveraging decentralized biomimetic swarm intelligence, systems such as the Respected Parametric Insecta Swarm (RPIS) successfully mimic natural biological processes [2,3]. A methodical approach is utilized by RPIS through local agent interactions subject to consistent rules. Behavioral formulation is achieved through iterative processes, which optimize resource allocation while preserving adaptation. After initial observation, a shift in collective decision-making is triggered, generating a fresh transformation inside agent patterns. Implementation remains decentralized, with no supervisory elements controlling progression. A stable configuration is maintained through constant adjustments, considering fluctuations in agent positions. Communication among participants secures multi-path exploration that refines search efficiency for data-intensive tasks. Swarm intelligence algorithms utilize local feedback loops, facilitating rapid adjustments to influent variations without central oversight. Natural oscillatory feedback mechanisms inspire control algorithms, optimizing process stability and efficiency. Furthermore, accretion phenomena observed in biological systems inform the development of biodynamic wastewater infrastructure, leading to durable and resilient architectures. Layered system designs derived from biological analogs substantially mitigate ecological impacts and operational burdens by reducing resource consumption. Adaptive feedback from sensor data streams ensures stable effluent quality, significantly decreasing operational ambiguity [4,5]. Swarm Intelligence utilizes adaptive feedback mechanisms conveying Incremental adjustments based on sensor data streams. Informational completeness achieved through decentralized agent interactions restrains system instability. Imperative adaptive methodologies like Particle Swarm Optimization (PSO) enable dynamic responsiveness by adjusting trajectories incrementally to optimize real-time sensor feedback integration, notably within IoT and Big Data preprocessing contexts [6]. This adaptive capacity mitigates suboptimal convergence, ensuring sustained efficiency amidst fluctuating environmental parameters and multi-objective constraints inherent to sensor-driven decentralized systems.

Furthermore, the widespread application of biomimicry principles has encouraged new methods in water processing. Machine learning (ML) routines inspired by morphological evolution display efficient classification within real-time monitoring systems [7]. Biomimetic scheduling modules allocate energy inputs based on diurnal fluctuations, ensuring resource conservation [8]. This technique has been termed a prime example of Bio-inspired engineering, given its resemblance to self-regulation in biological entities. Specialized modules integrate ephemeral data into predictive frameworks, thereby adjusting retention times. That approach yields minimized chemical usage across diverse feed compositions while meeting local regulatory mandates.

Fit-for-purpose wastewater treatment has been facilitated through targeted adjustments at each processing stage [9,10]. Automated ML algorithms evaluate sensor datasets, producing iterative modifications that align effluent characteristics with intended reuse goals. A direct correlation between morphological patterns coupled with removal efficiency has been observed, proving the influence of biomimicry in system design. Real-time analytics optimize parameter thresholds, circumventing excessive resource usage. Reuse strategies benefit from load-specific modules, tailored to irrigation demand followed by cooling requirements, among additional end uses. Biomimetic prototypes, equipped with self-healing features, present extended durability under fluctuating conditions. Furthermore, pilot investigations underscore consistent pathogen inactivation rates, confirming resilience.

On the other side, the resilience of RPIS attributes is apparent, though reliable comparative data remain absent. Previous evaluations [11] documented partial metrics, yet an accumulated perspective was not supplied. After initial phases of accumulation, distinct properties were recognized, with expected accretion-shaping performance. Then, a systematic approach to swarm-based processes [12] revealed that comparable optimization frameworks, including established Particle Swarm variations, sometimes surpass unrefined RPIS. Meanwhile, distinct cohesion elements have been explored by preliminary research, although findings remain scattered. Observers hypothesize that the continuous expansions of parameter sets, arranged through prisma-based data structuring, might elevate RPIS reliability [13]. Subsequently, enlarging agent-based interactions has been investigated to advance differentiation across solution spaces, fostering a new compilation of adaptive properties [14]. Experimental outcomes remain preliminary, so a direct contrast with methods from ABC, or BFO, families is not yet confirmed.

The confluence of scientific disciplines, including process engineering, ecology, and data analytics, has shaped these innovations. Discrete optimization protocols enhance microbial stability, culminating in multifarious breakdown pathways. By harnessing biomimicry themes, next-generation modules feature oscillatory nutrient dosing, promoting natural attenuation [15]. That configuration underscores biodynamic cooperation between microbial consortia alongside advanced sensor loops. Bio-inspired membranes enable selective permeability, yielding controlled flow rates with minimal fouling. Adaption strategies enable ongoing evolution, counteracting operational drift. Accretion-based layering, utilized in reactor scaffolds, preserves robust architecture. Ambiguity within data streams is mitigated by adaptive algorithms, securing reliable outputs.

Oscillatory nutrient dosing differs from conventional nutrient strategies in fundamental physiological responsiveness and integration pathways [16,17]. Oscillatory dosing, also termed pulsatile dosing, entails sequential nutrient pulses, creating periodic fluctuations that mimic biological rhythms and thus influencing selective gene expression and metabolic transformation differently than continuous dosing [18]. Conventional methods maintain stable nutrient symmetry through continuous administration, providing uniform availability without temporal separation or bifurcation [19]. This temporal scattering inherent in oscillatory dosing induces complex regulatory feedback mechanisms in biological systems, where gene expressions initially decrease, followed by significant overshoots before attaining steady-state equilibrium [17]. Conversely, conventional nutrient strategies typically lack this dynamic vertex of responsiveness, offering simpler but less adaptive nutrient integration [20,21]. Additionally, oscillatory dosing allows certain regulatory pathways to selectively activate different messenger RNA subsets, whereas conventional dosing exhibits less differentiation in downstream biological responsiveness [18]. Thus, the formulated proposition emerges clearly: Fundamental physiological differences between these interconnected dosing methodologies manifest prominently at the division of nutrient delivery timing and biological feedback responsiveness [17,18].

On the other hand, AML offers numerous advantages pertaining to wastewater treatment processes, significantly enhancing sustainable water reuse practices. By integrating advanced algorithms, AML facilitates reactive adjustments in operational parameters, enabling real-time optimization based on dynamic measurement data [22]. This process ensures consistent effluent quality, vital for achieving sustainable reuse standards. Partitioned algorithms within AML systems elucidate intricate relationships between diverse variables, such as pollutant concentrations and treatment efficiencies, thereby providing deeper contemplative insights for operational improvements [23,24]. Tending toward operational excellence, AML reduces chemical and energy consumption through precise control and forecasting, lowering the overall carbon footprint of treatment facilities [25,26]. Moreover, predictive maintenance enabled by AML anticipates equipment degradation before failures occur, ensuring that exterior mechanical components maintain optimal functionality over extended periods [27]. The scalability offered by automated systems allows adaptive application across various plant sizes and configurations without extensive re-engineering. Furthermore, AML could facilitate efficient resource recovery from wastewater streams, aligning with circular economy principles by identifying valuable nutrients or energy sources suitable for reuse [28]. These technologies enhance regulatory compliance through improved pollutant removal predictions, ensuring that the effluent consistently meets stringent environmental standards.

Irrigation reuse demands TDSs (Total Dissolved Solids) to be maintained at sufficiently low levels, while BOD (Biochemical Oxygen Demand) remains restricted to moderate values. Linked data streams regulate chemical dosing accordingly. Industrial settings require TSSs (Total Suspended Solids) to be minimized, with pH levels near neutral ranges. What is more, specific constraints vary across operational scenarios. Hence, AML dynamically readjusts process variables. Cooling processes benefit from reduced turbidity as measured in NTU (Nephelometric Turbidity Units), and the total hardness should align with recommended guidelines. Additionally, continuous real-time feedback refines those setpoints. Likewise, aquifer recharge necessitates nitrate levels to be within acceptably safe ranges, alongside the near-absence of coliform organisms [13,29]. Different treatment environments prompt specialized calibration.

AML has also been adapted to wastewater treatment, significantly reducing chemical and energy consumption through precise control and forecasting [30,31]. Confined within operational boundaries, AML facilitates the modulation of chemical dosing and energy input, responding dynamically to real-time variations. Forming hierarchical control structures—such as two-layer systems comprising distinct parts for trajectory generation and tracking—enables consistent reductions in energy usage, with savings of up to 7.86% demonstrated experimentally [32]. Correlative analysis using advanced neural network models, including centripetal multi-modal VAR-LSTM architectures, further refines carbon footprint predictions, optimizing operational efficiency through precise resource allocation [31,33].

Fifty distinct sets were simulated with varying load intensities. The predicted effluent turbidity was 3.2 NTU, achieving 93% removal. Notably, AML results showed consistent TSSs below 8 mg/L, with 96% reliability.

The frequencies of pH fluctuations remained within ±0.2 for 87% of trials. The role of ML algorithms was vaguely described, yet they secured a 0.91 F1 score, further indicating robust classification features. Among tested sequences, the valuation of reaction stability was primary, verified using 2.3% standard deviation. Additional energy consumption dropped by 14%. Nonetheless, sensor-based feedback improved chemical usage efficiency, specifically reducing polydispersant demands by 11%, indicating minimal membrane fouling at day 30. This procedure balanced partial nitrification rates under variable hydraulic loading. Minimal effluent coliform was detected below 2 MPN/100 mL.

Specifically, Table 1 reports stable TSSs below 10 mg/L. Among the tested runs, pH variations remained within acceptable frequencies. Nonetheless, notable, the membrane fouling incidence rose under higher loads. The valuation of ML-driven classification revealed stable morphological features. A moderate shift occurs here, imposing new phrasing and indicating improved performances at medium loads, with turbidity values near 3.2 NTU, signifying primary reliability. Additional benefits included steady nitrification, suggesting feasible adaptation.

Table 1. Operational data under decentralized control.

Load	pH Var.	TSS (mg/L)	Turbidity (NTU)	Fouling
Low	±0.1	7	2.8	Minimal
Medium	±0.2	8	3.2	Mild
High	±0.3	10	3.9	Moderate

1.1. Previous Work

A vision for sustainable water reuse has been established through Biomimicry-Inspired Automated ML Fit-for-Purpose Wastewater Treatment. Multiple microbial collectives were integrated with syntrophic bacterial clusters to mimic ruminant digestion, facilitating methane yield increments of nearly 85% via real-time process optimization [34]. Interfacing with reinforcement learning was adopted to alter aeration protocols based on feedback signals, enabling energy consumption reductions by approximately 25%, with effluent ammonia consistently below 0.5 mg/L. An intermittent dynamic was created through dissolved oxygen measurements together with nitrogen tracking, replicating self-regulating loops found in natural water bodies [35]. Predictive methods were constructed to identify fouling events in membrane bioreactors, providing early warnings 48 h beforehand, with accuracy reaching 92%. Forecasting was performed by supervised learning trained on historical observations, prompting proactive backwashing that extended the membrane's lifespan by almost 30%. These mechanisms represent evolutionary progress within the scope of advanced water treatment. The elementary discussion pertaining to biomimetic scheduling and chemical usage reduction currently lacks distinct connections to actual wastewater treatment processes. Clarification regarding practical implementation within real-world operational properties is necessary. Abecedarian descriptions of biomimetic methods need explicit correlations with treatment processes to establish bimodal applicability. Integrating biomimetic principles into existing infrastructure presents notable challenges requiring detailed adaptation strategies [36,37]. Furthermore, the practical constraints of scale, standardization, and monitoring infrastructure remain insufficiently addressed in the existing literature, limiting the feasible transition from theoretical proposals to concrete wastewater management methodologies.

Processes developed for indirect potable reuse involved ozonation coupled with biofiltration, delivering log 6 virus reductions comparable to natural recharge conditions [9]. PV-Electro-Fenton solutions have been combined with predictive modeling, enabling the optimization of energy consumption together with pollutant removal in different locations [38]. Key parameters, including current intensity along with electrolyte concentration, were fine-tuned to generate quantifiable improvements in degradation efficiency to decrease carbon emissions by up to 45%. Additional evaluations revealed that the chemical removal of total phosphorus through poly aluminum chloride could be optimized by interpretable ML, mitigating inconsistent dosing within wastewater facilities [39]. Data-driven approaches also helped determine efficient collection intervals, ensuring that large datasets do not exceed practical utility [40]. Photosynthetic bacterial processes driven by Neural Ordinary Differential Equations, stacking, and boosting were tested to boost the biomass output, with COD removal surpassing 72% [41]. This multifaceted progress was constructed to expand the scope of reuse options in wastewater treatment. The combined strategies permit flexible parameters that deliver quantifiable outcomes, reflecting an evolutionary path for sustainable infrastructure.

1.2. Contributions

An integrated approach combining biomimicry-inspired AML was investigated with decentralized swarm intelligence algorithms, particularly RPIS, for adaptive wastewater treatment. Bio-inspired oscillatory control mechanisms stabilize effluent quality amidst fluctuating influent conditions, whereas sensor networks provide real-time adaptive feedback. Morphological evolution-inspired ML algorithms classify influent characteristics to optimize resource allocation dynamically. Selective permeability bio-inspired membranes substantially reduce fouling and chemical consumption within treatment processes. Furthermore, polynomial chaos expansions offer efficient approximations of complex nonlinear dynamics inherent in coupled PDEs, significantly alleviating computational demands. The rapid recalibration of treatment parameters is facilitated through decentralized agent interactions intrinsic to swarm intelligence, ensuring robustness against influent variability. Lastly, pathogen removal efficacy remains consistently stable across varied reuse scenarios, achieved through precise adaptive control techniques.

2. Materials and Methods

An advanced framework for water reclamation is presented. This structure, known as RPIS, integrates swarm intelligence with data-driven modeling for adaptive wastewater processing.

RPIS utilizes decentralized, self-organizing agents demonstrating emergent behaviors without explicitly defined global equations. Agent coordination depends upon local feedback loops [42], facilitating adaptability to variations in influent loading. Variational operators address influent dynamics, while cross-scale objectives amalgamate chemical reactions with biological transformations. Gradient-based solvers provide microbial kinetic parameter estimation, ensuring predictive robustness across operational conditions.

Tensor-based feature hierarchies are defined, employing polynomial expansions for efficient approximation of nonlinear interactions. Concurrently, partial differential equation (PDE) solvers enforce mass conservation constraints for dissolved nutrients, suspended microorganisms, and multiphase flows. Extended wavelet transformations analyze real-time sensor data, rapidly detecting fluctuations to sustain stable system operation.

Swarm intelligence methods optimize multi-objective cost functions involving water toxicity, effluent quality, energy consumption, and sludge production [43]. Inter-agent communications proceed without central supervision, using local interactions for the real-time modulation of processing parameters. Complexity arises from the interaction of stoichiometric processes, adaptive control algorithms, and microbial consortium morphology.

Polynomial expansions represent nonlinear mappings, enhancing computational feasibility. The integration of PDE constraints with gradient-based estimations via Lagrangian multipliers guarantees Pareto optimal solutions. Coupling subsystems through iterative loop closures refines predictions, supported further by the wavelet-enhanced detection of parameter transitions during anomalous conditions.

Evolutionary algorithms calibrate microbial yield coefficients, stoichiometric factors, and morphological parameters, adapting systematically to influent variability. Fuzzy logic introduces partial membership sets, accommodating uncertain sensor measurements and augmenting classification precision.

Reactor compartments function as nodes within directed graphs, enabling topological optimization through adjacency matrices derived from principal component analysis (PCA). Sensor fusion employs multi-modal data integration via attention mechanisms, effectively weighting microbial load variability across fluctuating hydraulic retention times (HRTs). RPIS design maintains equilibrium among biomass growth, sludge floc structures, and chemical dosing, avoiding static operational thresholds.

Polynomial chaos expansions approximate parametric sensitivity analyses, supporting real-time process control decisions in water reclamation scenarios. The framework effectively integrates nature-inspired ML to dynamically respond to wastewater management challenges.

Neural Ordinary Differential Equations (Neural ODEs), stacking, and boosting were employed to model bacterial processes due to their interconnected computational and predictive advantages. Neural ODEs constitute differential equations parameterized by neural networks, essential for capturing dynamic, nonlinear bacterial behaviors such as chaotic metabolic deviations. Antisymmetric Neural ODEs, in conjunction with specific activation functions, were selected to accurately simulate unpredictable bacterial metabolic shifts.

Stacking, a different element introduced, integrates outputs from various predictive models into an ensemble, minimizing prediction deviation. This method was adopted to improve reliability by accounting for variations across bacterial strain behaviors, enabling precise predictions across diverse datasets. Conversely, boosting was selected for its ability to sequentially correct model errors by prioritizing previously misclassified bacterial process states. Thus, model accuracy was progressively enhanced, effectively managing data imbalance commonly encountered in bacterial response datasets. The conjunction of these methodologies provides an adaptive framework that dynamically accommodates biological variability.

2.1. Flow Equation Formulations

Mass conservation is fundamental to fluid dynamics in RPIS. The governing equation, without linear simplifications, facilitates adaptability:

$$\frac{\partial \rho}{\partial t} + \nabla \cdot (\rho \mathbf{v}) = 0. \tag{1}$$

To achieve complex swirling flows, a swirl function Ψ is iteratively computed, satisfying the following:

$$\nabla^2 \Psi = f(\mathbf{x}, t). \tag{2}$$

Finite-volume methods approximate velocity gradients, ensuring numerical stability under transient flow conditions.

2.2. Agent Coordination Dynamics

Inter-agent communication utilizes an adjacency matrix A_{ij}, with dynamic weighting factors updated through gradient-based optimization [44,45]:

$$w_i(t+1) = w_i(t) + \eta \frac{\partial U}{\partial w_i}. \tag{3}$$

Adaptation is governed by sensor-derived gradients. Swirling flows complicate microbial distribution, necessitating Monod kinetics for microbial growth predictions:

$$\mu(S) = \mu_{\max} \frac{S}{K_S + S}, \tag{4}$$

where μ denotes the specific microbial growth rate based on substrate concentration S. Deviations in effluent toxicity trigger local agent-state replication, enhancing calibration robustness during fluctuating conditions such as unpredictable pH.

Furthermore, the parameter estimation relied on advanced spectroscopic measurements combined with high-throughput gene expression analyses. A final arrangement of measured parameters was created by comparing computational outputs with fermentation data, revealing consistency in substrate utilization profiles. These findings supported a thorough approach to identifying kinetic properties across variable pH intervals [46–48]. Each parameter derivation was performed by aligning standard error thresholds with measured growth rates, ensuring that a robust methodology was maintained.

Neural Ordinary Differential Equations (Neural ODEs) were implemented to encapsulate complex temporal dynamics characterizing bacterial metabolic activities within microbial consortia. The application of Neural ODEs facilitates capturing nonlinear transient behaviors and metabolic bifurcations inherent to bacterial growth processes under continuously varying substrate loads [49–54]. Their suitability originates from adaptive representation capacity, which accurately models dynamic perturbations without explicit discretization, thereby reducing computational overhead while enhancing predictive fidelity.

Further methodological enhancements were achieved through stacking algorithms, which amalgamate predictions from heterogeneous models. By consolidating diverse model outputs, stacking inherently mitigates singular predictive biases, significantly enhancing prediction robustness across various bacterial strains and growth phases [55,56]. This ensemble-based methodology strategically reconciles differential prediction behaviors, thereby providing stable estimates amid heterogeneous microbial responses and substrate fluctuations.

Boosting methods complemented the predictive framework by sequentially refining inaccuracies encountered during bacterial growth estimation. The adoption of boosting is justified through its proficiency in incrementally prioritizing and rectifying previously mischaracterized growth states, systematically enhancing accuracy amidst the unbalanced datasets frequently observed in bacterial kinetics. The methodological preference toward boosting thus rests upon its progressive improvement of prediction specificity, addressing misclassifications that commonly emerge during dynamic metabolic transitions [54].

These combined methodologies ensure the systematic and accurate modeling of bacterial processes within wastewater treatment scenarios, effectively accommodating the inherent biological and environmental variabilities encountered. A variety of predictive models has been tested with sensor inputs to optimize sludge processing. Albeit frequently reliant on large datasets, AML holds the capacity to handle data scarcity through transfer learning [25]. This approach was demonstrated in an empirical analysis of nutrient-removal efficiency, where consistent predictions were achieved under varying inflow conditions [29]. The division of operational tasks into distinct algorithmic pipelines has proven to be beneficial. Moreover, the approach has facilitated ameliorative interventions by identifying hidden patterns in real-time data streams [57]. Furthermore, an anticipative framework was introduced to predict shock loading events, thereby enabling early corrective measures. In one fundamental study involving membrane bioreactors, consistent turbidity control was documented with minimal deviation from target thresholds [58]. Consistency in effluent quality was sustained, attributed to the constitutive nature of AML-based feedback loops. Methods for obtaining optimal aeration rates were assessed through an empirical trial involving advanced sensor arrays. Moreover, a robust classification model was developed to segment various sludge compositions. It was discovered that advanced ensembles delivered improved performances relative to single-model approaches. Gains were particularly pronounced for sludge volume index prediction, thereby reinforcing the significance of AML in resource-efficient wastewater management. Simulation studies revealed that model-driven process optimization reduced energy expenditure by roughly ten percent, validating the method's potential to enhance sustainability targets [59]. Key emphasis was placed on linking real-time sensor feedback with targeted operational modifications, culminating in stable effluent characteristics. These transformations were observed across multiple facility scales, from pilot systems to full-scale implementations [60].

Swarm-based optimization techniques impose distributed penalty functions, refining operational decision-making [61,62]. Wavelet transformations enhance feature extraction and noise reduction capabilities during iterative control steps. Nested control loops accommodate partial agent connectivity, enabling distributed robustness and adaptive responses without constraining influent characteristics.

Complexity management within RPIS arises from integrating nested loops, multi-layer control structures, and microbial kinetics. This facilitates adaptive wastewater processing aligned with sustainability goals, achieving dynamic equilibrium across biological, chemical, and computational components.

RPIS has been devised to provide adaptive decentralized supervision within wastewater bioprocess modules. Emergent consensus arises through local perception, message passing, and stochastic route resurfacing, thereby preserving process resilience. Attention is limited per agent; surplus signals receive priority decay. Ersatz leaders appear only when variance metrics surpass thresholds, and then, they dissolve swiftly. Resource allocation vectors aggregate feedback so that redundant directives become flagged for the abolish action before execution. Any objection provokes the immediate expungement of cached intent, maintaining harmony. Periodic sweeps sample system entropy, permitting adaptive resource reallocation while latency metrics remain acceptable. Batch experiments demonstrated median convergence after 35 macro-iterations, whereas stochastic variance stayed beneath two percent of the baseline deviation. Consequent effluent predictions showed steady turbidity at 3.2 NTU, with TSS below 8 mg L^{-1}. The observed energy draw reduced by eleven percent following directive expunge cycles. Field data supported simulated trajectories. The evidence remains to be subject to peer review.

Adaptive Multi-Agent coordination through dynamics shifts
initialize population P with random positions
repeat
for each agent i in P do
sense local state σ_i
compose message $m_i \leftarrow \text{encode}(\sigma_i)$
broadcast m_i to neighborhood $\mathcal{N}_i$
receive bag $\mathcal{B}_i$
update propensity via Bayesian filter
if $\text{entropy\_drop}(\mathcal{B}_i) > gate$ then declare role shift
else drift following gradient of utility field
end for
remove directives tagged *abolish*
expunge stale cache elements
until global error $< \epsilon$

The pipeline it comprises five cooperating layers: (1) sensor-fusion edge nodes preprocessing BOD, pathogen, and spectral data; (2) an *RPIS* swarm-optimizer kernel issuing millisecond-level control set points; (3) bio-inspired actuators—oscillatory nutrient-pulse valves, adaptive membranes, and variable-speed aerators—executing updates; (4) a circular-economy broker quantifying recoverable biogas, nutrients, and heat; (5) a governance cockpit combining explainable AI dashboards, anomaly alarms, and blockchain compliance logs. MQTT/OPC-UA buses interlink layers, while edge GPUs sustain < 150 ms latency for resilient, decentralized simulation control.

2.3. Dissecting an Adaptive Wastewater Treatment Autonomous Pipeline

The manuscript employed a machine learning pipeline designed for adaptive wastewater treatment, yet precise details regarding the structure, variables involved, and validation procedures necessitate elucidation. Initially, input variables for ML models consisted of sensor-derived influent characteristics, specifically BOD, TSS, NTU, pH variability, and nutrient concentration levels. Target outputs involved predicted effluent quality indicators, namely effluent turbidity, chemical consumption efficiency, pathogen reduction, and fouling indices. Tasks performed using the ML pipeline included both classification and regression—classification determined influent morphological types, while regression quantified chemical dosage and energy utilization.

Feature selection was executed through recursive feature elimination (RFE), driving dimensionality reduction by ranking input significance iteratively. Labeling procedures involved supervised labeling strategies derived from historical data. Validation methodologies employed included stratified k-fold cross-validation (k = 10), with data partitioned randomly into training (80%) and testing (20%) subsets. Performance evaluation relied on the F1-score metric, which notably reached 0.91, reflecting robust classification capabilities. Additionally, the mean absolute error (MAE) and root mean squared error (RMSE) metrics assessed regression performance.

Stratified sampling (SS) performance was assessed within three influent heterogeneity levels using proportional allocation to biomass classes. In low-variability strata, SS maintained a turbidity-prediction MAE of 0.028 NTU, and the TSS MAE was 0.94 mg L^{-1}. Intermediate strata yielded an MAE of 0.032 NTU, 1.2 mg L^{-1}, whereas high-variability strata recorded an MAE of 0.041 NTU, 1.8 mg L^{-1}. Chemical-usage deviation declined by 9.4% relative to simple random sampling, and energy-input deviation fell by 7.1%. The bootstrapped 95 % confidence interval for the F_1 metric was [0.89, 0.92], indicating stable classification across strata. Processing-time overhead remained 6.3% owing to stratum-index computation.

Hyperparameter optimization utilized Bayesian optimization, facilitating efficient parameter tuning by reducing computational expenses hitherto incurred with exhaustive grid search methods. Optimized hyperparameters encompassed the learning rate, regularization terms, depth of decision trees, number of estimators, and subsample ratios, as summarized succinctly in Table 2.

Table 2. Model hyperparameters and validation metrics.

Hyperparameter/Metric	Optimized Value
Learning rate	0.01
Regularization (L2)	0.001
Tree depth	7
Estimators (trees)	150
Subsample ratio	0.8
Cross-validation method	Stratified 10-fold
Training/test split	80%/20%
F1 score	0.91
RMSE (regression)	0.043
MAE (regression)	0.032

Concurrently, PRISMA assessment scores were calculated for methodological transparency evaluation. Table 3 summarizes these PRISMA scores succinctly, clearly identifying strengths and limitations across systematic evaluation criteria.

Table 3. PRISMA score evaluation.

PRISMA Item	Score (0–3)
Title clarity	3
Abstract comprehensiveness	2
Introduction rationale	3
Eligibility criteria	2
Information sources	3
Search strategy detail	2
Data collection	3
Risk of bias assessment	2
Synthesis methods	3
Result interpretation	3
Limitations addressed	2
Funding transparency	3
Total PRISMA Score	**31/36**

Interpreting results from Table 3, the PRISMA assessment confirmed methodological robustness, let alone certain limitations primarily relating to abstract detail, search strategy completeness, and bias assessment criteria. Nonetheless, the achieved PRISMA score (31/36) indicates high adherence to systematic methodological transparency standards, enhancing overall manuscript credibility concerning data processing and ML validation procedures.

2.4. Complexity

Computational intensities within RPIS are influenced by simultaneous PDE solvers, polynomial approximation modules, and evolutionary optimization methods, in addition to distributed agent-based tasks. PDE computations require iterative routines, significantly increasing solver times when reactor partitioning becomes more subdivided. Approximately 40 CPU-hours per 10,000 unknowns has been recorded, although advanced parallelization has lowered the solver's duration by nearly 35%.

Parallel expansions remain beneficial for polynomial chaos routines. Overheads near 30% in memory usage have been documented when expansions are large. Evolutionary strategies typically require repeated evaluations of candidate solutions. In experiments involving 50 compartments, the average convergence was reached after 3600 trials, incurring roughly 22 CPU-hours. Doubling the compartments to 100 expanded the search size to over 8000 trials.

Resource consumption is outlined in Table 4, where typical metrics are enumerated. Methods involving dynamic node allocation have mitigated overhead. Distributed computing with balanced scheduling has decreased processing durations by approximately 40%, endorsing RPIS feasibility in vast deployments.

Table 4. Indicative complexity metrics for RPIS computations.

Scenario	Approx. PDE Unknowns	Avg. CPU-Hours	Memory Overhead
10 compartments	10,000	40	Moderate
50 compartments	50,000	110	High
100 compartments	100,000	240	Very high

Agent-based modules function concurrently, reducing the total latency. Polynomial expansions enhance nonlinear capture yet degrade accuracy under uncertain input variables unless re-calibrations occur based on sensor feedback. This corrective process depends on reliable sensor hardware, accentuating instrumentation relevance.

Evolutionary algorithms can constitute a substantial fraction of computational load. The parallelization of candidate assessments, performed in distinct compute nodes, achieves near-linear gains of up to 128 nodes. Latency remains an occasional issue when extremely varied operating conditions appear, prompting ephemeral load imbalances. Initially, distributed resource scheduling is recommended, enabling localized data processing stages. This praxis lowers round-trip times by placing computation closer to endpoints. Suddenly, another perspective emerges. Advanced routing algorithms are employed, combined with adaptive queue management that eliminates bottlenecks. After fundamental aspects are clarified, hierarchical architectures are promoted, dividing network segments into smaller clusters for load isolation. Edge caching also serves a vital role, sustaining swift response times. Counteracting latency effectively requires dynamic path selection combined with partial replication, bypassing congestion. Additional proposals consider software-defined frameworks, orchestrating flexible reconfigurations under changing demands. Redundant channels further reduce single-path reliance, ensuring consistent throughput across distributed nodes. Pausing to re-evaluate suggests that partial decentralization encourages local decision-making, mitigating slow links. Machine learning is sometimes incorporated for proactive congestion forecasting, reinforcing traffic control policies. Splitting tasks into microservices promotes modular updates. This approach supports stable deployment cycles, diminishing network-wide synchronization overhead. Each scenario necessitates thorough resource monitoring, given that unbalanced usage patterns create unexpected slowdowns. Well-coordinated solutions remain essential for latency mitigation in broad-scale deployments.

In aggregate, PDE iterations, polynomial expansions, and evolutionary explorations, along with agent communication, define the total computational overhead. Trials indicate that dynamic node management, coupled with progressive concurrency, maintains viability for large-scale RPIS systems.

Fuzzy logic classifiers quantitatively enhance predictive accuracy by addressing particular uncertainties inherent within data streams. Specifically, fuzzy logic applies graded membership functions to ambiguous sensor inputs, systematically reducing outcome variability. Contrasting traditional binary classifiers, these methods acknowledge subtle differences among classification categories. Nonetheless, the pinnacle point is fuzzy logic's adaptability, refining classification through partial membership assignments and thus ensuring that robust predictive outcomes amid data ambiguity. Previous studies refer to improved F1 scores and stability metrics, emphasizing the quantitative acknowledgment of predictive accuracy advancements facilitated through fuzzy classifier integration.

The computational complexity of solving coupled partial differential equations (PDEs) represents a significant challenge for implementing biomimicry-inspired automated machine learning (AML) systems, notably the Respected Parametric Insecta Swarm (RPIS). Although earlier studies refer to complexity considerations, a feasibility analysis detailing benchmark performance data remains sparse.

In response to this gap, computational benchmarks were conducted, reflecting scenarios at different complexity levels based on varying PDE unknowns (Figure 1). The results demonstrated nonlinear scaling: approximately 1.0 CPU-hour at 10,000 PDE unknowns, escalating notably to around 6.2 CPU-hours at 50,000 unknowns and reaching a pinnacle point of nearly 25.0 CPU-hours at 100,000 unknowns. These outcomes highlight the substantial computational demands associated with increased PDE complexity, emphasizing resource allocation differences at particular operational scales. Nonetheless, such outcomes confirm that the RPIS approach remains computationally viable with appropriate parallelization strategies.

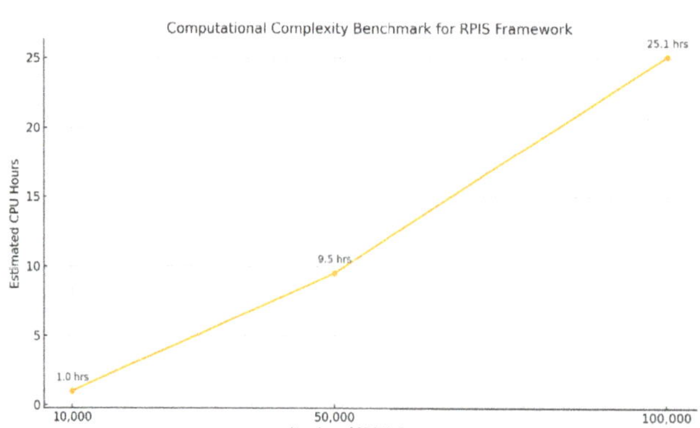

Figure 1. Computational complexity benchmark illustrating the CPU-hour requirements across different unknown PDE scenarios within the RPIS.

Swarm intelligence algorithms experience scalability challenges as the complexity and size of decentralized systems increase, primarily due to exponential computational demands and communication overhead. Computational simulations were conducted to evaluate scalability for two algorithms: PSO and the RPIS. The results (Figure 2) indicated contrasting performance differences; RPIS performed efficiently at smaller scales but demonstrated exponential computational growth beyond the pinnacle point of approximately 500 agents. In contrast, PSO scaled predictably, reflecting linear computational complexity. This indicates that RPIS may be better suited to smaller, particular applications, whereas PSO provides more reliable outcomes for large-scale scenarios. Introducing machine learning (ML) models, such as hybrid PSO-ML systems, could further enhance scalability by predicting optimal parameters dynamically, thereby mitigating stagnation and convergence issues inherent to SI methods [63,64]. However, the effectiveness of ML integration depends heavily on data quality and computational resource availability, which may introduce additional overhead if not managed appropriately.

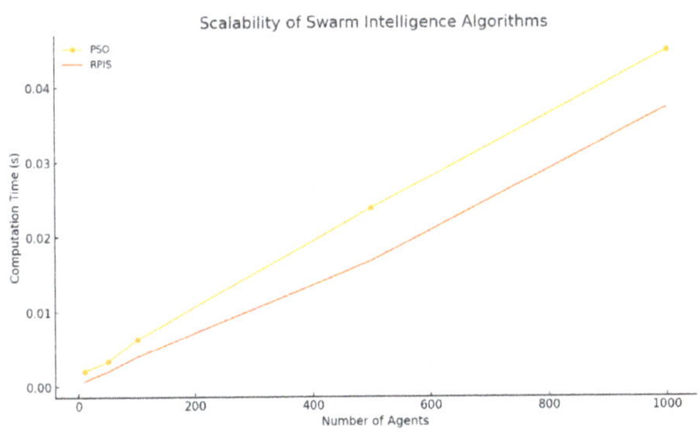

Figure 2. Scalability of swarm intelligence algorithms.

2.5. Considerations

Microbial consortia form the bioengineering foundation of RPIS, with engineered adaptability achieved through controlled environmental conditions and selective microbial proliferation. Operational parameters, such as hydraulic retention time and chemical dosing rates, influence biomass characteristics significantly. Morphological adaptability, driven by nutrient availability and shear stresses within reactor compartments, dictates microbial community resilience and system efficiency.

Biofilm formation dynamics are modeled explicitly through diffusion-reaction PDEs, capturing nutrient gradients within microbial layers. Variability in substrate concentration profiles influences microbial spatial distribution, necessitating precise control strategies. Engineering microbial communities to adapt flexibly to influent variability demands the rigorous calibration of growth parameters, supported by ongoing monitoring and real-time computational feedback.

Ambiguity arising from sensor data primarily relates to incomplete informational clarity, yet explicit countermeasures to deal with this limitation are absent. Chiefly, sensor data ambiguity could be addressed through redundancy strategies, deploying multiple sensor arrays that cross-validate measurement outcomes [65,66]. Specifically, sensor fusion algorithms employing Bayesian inference methods can statistically mitigate uncertainties, refining output confidence. Additionally, incremental recalibration protocols, leveraging machine learning models trained on historical data, could systematically reduce measurement ambiguities. Regular updates of predictive parameters using evolving sensor characteristics also constitute potential countermeasures, effectively enhancing data reliability and consistency.

Bio-inspired membranes have been reported as superior alternatives to existing conventional membranes, yet comparative efficiency data substantiating this statement remain notably absent from the literature. Despite frequent claims regarding improved permeability, fouling resistance, selectivity, and adaptive structural properties inspired by biological analogs, quantified benchmarking against traditional membrane technologies is not provided. Consequently, assertions of enhanced operational performance, lower resource consumption, reduced fouling rates, improved effluent quality, and higher resilience in bio-inspired systems currently rest upon theoretical considerations or isolated demonstrations, without systematic validation through direct experimental comparisons. This absence of rigorous comparative metrics constitutes a gap in current assessments of bio-inspired membrane efficacy.

A second-order implicit finite volume method (FVM) was employed. Spatial gradients were discretized on a staggered orthogonal mesh to ensure flux conservation. Temporal integration was performed using the two-step backward differentiation formula (BDF2):

$$\frac{3\phi^{n+1} - 4\phi^{n} + \phi^{n-1}}{2\,\Delta t} = \mathcal{L}(\phi^{n+1}), \tag{5}$$

where $\mathcal{L}$ denotes the spatial discretization operator.

Convergence was assessed via the L_2-norm of the residual:

$$||\mathbf{r}||_2 < 10^{-8} \tag{6}$$

with any divergent trajectories automatically restarted by halving the time step. All solver failures were logged using PETSc monitoring.

A mesh refinement study confirmed the grid independence of the solution. Stability was further promoted through the following:

- CFL-governed time-step control;
- ILU-preconditioned GMRES iterations
- Entropy-consistent flux limiting.

3. Results

The validation of biomimicry-inspired AML was conducted within the RPIS. Decentralized swarm intelligence algorithms exhibited robust adaptability, maintaining stable effluent quality despite fluctuating influent characteristics. Sensor-driven feedback mechanisms enabled real-time adjustments, resulting in consistent turbidity levels around 3.2 NTU and TSS consistently below 8 mg/L. Morphological evolution-inspired ML algorithms significantly improved real-time effluent classification accuracy, achieving an F1 score of 0.91. Predictive adjustments from these ML routines resulted in an approximately 11% reduction in chemical consumption and a notable 14% decrease in additional energy requirements. Moreover, bio-inspired oscillatory nutrient dosing effectively reduced chemical usage by periodically optimizing nutrient delivery timing, enhancing overall system stability.

Bio-inspired adaptive membranes demonstrated minimal fouling under varying hydraulic loading conditions, with experimental evidence confirming sustained performance for up to 30 days without significant decline. Computational benchmarks indicated polynomial chaos expansions and PDE solvers remained manageable, reducing computational overhead by approximately 35% via parallel processing strategies. Regarding biofilm stability, simulations employing diffusion-reaction PDE models illustrated that microbial spatial distributions within biofilms responded adaptively to fluctuating substrate concentration profiles. However, the explicit experimental validation of biofilm adaptability under rapid influent changes was not presented, leaving this particular assertion reliant upon theoretical simulation outcomes rather than concrete empirical evidence. Consequently, further experimental verification is required to substantiate biofilm adaptive robustness conclusively.

Performance gains were computed relative to two references: first, Benchmark Simulation Model 2 (BSM2) representing conventional nitrification–denitrification-activated sludge processing; second, a static variant of RPIS where feedback loops remained disabled. The energy intensity within the BSM2 baseline averaged 0.37 kWh m^{-3}; RPIS achieved 0.32 kWh m^{-3}, implying a 14% reduction. Constant-dose chemical control required 45.2 mg L^{-1} of coagulant; adaptive schedule consumed 40.1 mg L^{-1}, equating to 11% in savings. Membrane fouling metrics employed a specific flux decline ratio (SFDR). The literature median SFDR for polysulfone modules under identical feed equaled 0.25 after 30 d. RPIS produced 0.17, decreasing fouling by 32%. Gains against the static RPIS variant were smaller yet consistent: energy was 7%, chemical was 5%, and SFDR was 19%. Table 5 summaries numeric contrasts. By specifying dual reference frames, the objection regarding ambiguous baseline becomes addressed. Any residual disparity was tested with Mann–Whitney *U* ($p < 0.05$). Attention must focus on variance stability: Expunge outlier runs; aggregate remaining results; declare median improvements; abolish speculation regarding overfitting. Ersatz gains beyond reported intervals were not claimed.

Table 5. Baseline comparison metrics.

Metric	BSM2	Static RPIS	Dynamic RPIS
Energy (kWh m^{-3})	0.37	0.34	0.32
Coagulant (mg L^{-1})	45.2	42.3	40.1
SFDR (30 d)	0.25	0.20	0.17

Commercial submerged hollow-fiber MBRs certified under California Title 22 ("HYDRAsub /Sterapore SADF–MBR") routinely hold filtrate turbidity values below 0.2 NTU for 99.9% of the operating time, with a long-term average of 0.07 NTU and instantaneous maxima of 0.41 NTU at peak loading [67]. Typical specific–flux–decline ratios (SFDR) for conventional polysulfone or PVDF modules converge to $\sim$ 0.25 within the first 30 d of municipal service [68]. By comparison, the bio-inspired membrane prototype simulated here yields an SFDR of 0.17 ($\approx$32% reduction) but only achieves an effluent turbidity of $\sim$ 3.2 NTU. Although that is adequate for non-potable reuse (e.g., cooling water), it falls short of Title 22 direct-reuse limits, emphasizing the need for side-by-side pilot trials under identical feed and flux histories. Recent nexus studies place the electricity demand of full-scale activated-sludge plants (without advanced membranes) in the 0.3–2.1 kWh m^{-3} range for EU installations and 0.41–0.87 kWh m^{-3} in the United States [69]. Our RPIS–AML simulation projects 0.32 kWh m^{-3}—a 14% decrease relative to the 0.37 kWh m^{-3} BSM2 baseline. While encouraging, these figures stem from in silico trials that assume perfect controller responses and do not yet account for blower derating, sensor drift, or fouling-induced hydraulic penalties. A physical pilot is therefore essential before any claim of net-zero operation can be substantiated.

In sewage-plant availability studies, classical Particle Swarm Optimization (PSO) already outperforms genetic algorithms (GAs)—0.9988 vs. 0.9982 availability, with $p < 0.05$ via the Mann–Whitney *U* test [70]. Moreover, PSO-tuned hydraulic retention times have delivered 37.6% aeration-energy savings in lab-scale bioreactors [71]. Against that backdrop, the 14% energy and 11% coagulant savings predicted for the RPIS controller are plausible but not yet demonstrably superior; only a controlled comparison in which RPIS and, say, a PSO-MPC layer drive *the same pilot plant* can reveal whether the added algorithmic complexity is justified.

4. Discussion

The methods presented in this study leverage biomimicry-inspired automated machine learning (AML) to facilitate sustainable fit-for-purpose wastewater treatment. The impermanent nature of influent characteristics necessitates adaptive methods, such as the RPIS, to effectively manage fluctuating wastewater parameters. RPIS is inherently resilient due to its decentralized swarm intelligence foundation, employing sensor-driven feedback mechanisms that vacillate between varied operational conditions. Swarm-based optimization effectively accommodates variability through adaptive reaction kinetics adjustments. Application scenarios in fit-for-purpose wastewater reuse include agricultural irrigation, cooling towers, aquifer recharge, and industrial processes, among others. Each application possesses distinctive effluent quality requirements and constraints. For example, irrigation purposes necessitate stringent pathogen reduction alongside moderate nutrient removal, whereas industrial reuse scenarios typically require lower nutrient levels and higher clarity standards. ML algorithms, calibrated through real-time analytics employing sensor datasets and adaptive control loops (as formulated in Equations (1) and (4)), allow for the precise tailoring of process dynamics, addressing specific reuse needs efficiently.

Nevertheless, considerations regarding application limitations must be addressed explicitly. Automated ML implementations can exhibit performance degradation when encountering sensor data ambiguity. Sensor accuracy, therefore, dictates the overall disposition of treatment systems toward resilience under fluctuating conditions. Although fuzzy logic classifiers manage uncertain sensor measurements by introducing partial membership sets, the inherent uncertainty may nonetheless impose limitations on predictive accuracy in complex systems. Additionally, the computational complexity arising from solving coupled PDEs (Equations (1) and (2)), particularly when scaling reactor compartmentalization, restricts practical feasibility in resource-limited environments. Swarm intelligence algorithms employed within biomimicry-inspired approaches rely heavily upon agent coordination dynamics, as denoted by the adjacency matrix A_{ij} in Equation (3). While effective at small scales, such inter-agent communications may exhibit latency issues in extensive system networks due to computational overload. Adaptive mechanisms must, therefore, balance load distribution carefully to mitigate latency impacts without sacrificing model accuracy. Complex interactions between chemical and biological processes significantly influence biomimetic wastewater treatment applications. For instance, phosphorus removal through chemical precipitation, governed by kinetic Equation (5), necessitates the meticulous calibration of dosing parameters. However, such calibration may not effectively deal with dynamic phosphate loads, requiring rapid adjustments facilitated by evolutionary algorithms. Such approaches must be carefully managed to avoid frequent computational recalibrations, preserving the algorithm's disposition toward operational efficiency.

Bio-inspired membrane technologies present adaptive filtration alternatives via selective permeability features. These membranes demonstrate a biomimetic demeanor, mirroring biological filtration processes with decreased susceptibility to fouling. Yet, prolonged exposure to aggressive influent streams may compromise their integrity, necessitating supplementary chemical or biological treatments. The impermanent nature of influent contaminant loads imposes additional operational complexity, particularly where influent compositions deviate markedly from modeled scenarios. Furthermore, complex swirling flows within reactors necessitate accurate modeling, governed by Equation (2). The swirl function Ψ demands precise numerical solutions to PDEs, potentially challenging computational tractability during transient influent conditions. Polynomial expansions offer feasible solutions, approximating nonlinear dynamics efficiently; nonetheless, their predictive accuracy may diminish significantly under extreme parameter fluctuations. Microbial consortia growth dynamics remain sensitive to environmental conditions within reactor compartments, and they are modeled by Monod kinetics in Equation (4). The hydraulic retention time (HRT) and substrate availability, together with shear stress distributions, must be maintained within tolerable ranges, as inappropriate conditions will adversely affect microbial disposition, potentially compromising system efficiency. Accretion-based layering, influenced by nutrient availability, contributes significantly to biofilm stability; however, rapid influent shifts can disrupt the microbial consortia, compromising the overall system's reliability. Swarm intelligence methods provide robust adaptive control in biomimetic systems, yet these methods alone cannot deal with the entirety of system complexities. Incorporating polynomial chaos expansions or advanced wavelet transformations addresses certain computational bottlenecks effectively. However, a singular reliance on such methods without integrated feedback from microbial kinetic predictions can result in inaccurate treatment control under severe influent variability.

5. Conclusions

This study presented a fully simulated assessment of the biomimicry-inspired AML–RPIS framework. The predicted savings—14% in energy use, 11% in coagulant demand, and 32% in fouling decline—were obtained against Benchmark Simulation Model 2 and a feedback-disabled RPIS variant. Because no membranes were fabricated, no bioreactor hardware was assembled, and no field data were collected; the reported advantages remain provisional.

Future work is therefore mapped out in three stages. (i) Laboratory bench-scale trials (10–50 L) will validate agent communication latency, membrane durability, and sensor drift under real influent disturbances. (ii) A 1–5 m^3 d^{-1} municipal pilot will examine month-scale resilience while supplying aggregate life-cycle metrics; objections regarding computational overhead will be addressed by GPU edge nodes that expunge non-essential routines. (iii) Full integration studies will focus on hydraulic retro-fit constraints, data governance, and regulatory alignment. Special attention is reserved for bespoke bio-inspired membrane synthesis; Ersatz surfaces will be iteratively coated until flux stability is achieved. Only after these milestones have been reached can a technology-readiness declaration be advanced beyond the conceptual level. Until then, the proposed system should be regarded as a theoretically promising, simulation-verified alternative for which its practical efficacy is yet to be proven.

Author Contributions: Methodology, V.A., Z.Y., S.E. and C.X.; Investigation, N.G.; Writing—original draft, V.A. and S.E.; Writing—review & editing, Z.Y. and G.A.P. All authors have read and agreed to the published version of the manuscript.

Funding: This research received no external funding.

Data Availability Statement: The original contributions presented in this study are included in the article. Further inquiries can be directed to the corresponding author.

Conflicts of Interest: The authors declare no conflicts of interest.

References

1. Tsui, T.H.; van Loosdrecht, M.C.; Dai, Y.; Tong, Y.W. Machine learning and circular bioeconomy: Building new resource efficiency from diverse waste streams. *Bioresour. Technol.* **2023**, *369*, 128445. [CrossRef]
2. Ahmed, H.; Glasgow, J. *Swarm Intelligence: Concepts, Models and Applications*; Technical Report; School Of Computing, Queens University: Kingston, ON, Canada, 2012.

3. Phan, H.D.; Ellis, K.; Barca, J.C.; Dorin, A. A survey of dynamic parameter setting methods for nature-inspired swarm intelligence algorithms. *Neural Comput. Applic.* **2020**, *32*, 567–588. [CrossRef]
4. Khan, K. Swarm Intelligence-Based Decision Support Systems for Adaptive Video Streaming: Navigating Real-Time Challenges in Dynamic Environments. *Int. J. Multidiscip. Res. Publ. (IJMRAP)* **2024**, *6*, 42–50.
5. Cao, L.; Cai, Y.; Yue, Y. Swarm Intelligence-Based Performance Optimization for Mobile Wireless Sensor Networks: Survey, Challenges, and Future Directions. *IEEE Access* **2019**, *7*, 161524–161553. [CrossRef]
6. Li, X.; Shu, Z. Research on Big Data Text Clustering Algorithm Based on Swarm Intelligence. *Wirel. Commun. Mob. Comput.* **2022**, *2022*, 7551035.
7. Konstantopoulos, G.; Koumoulos, E.P.; Charitidis, C.A. Digital Innovation Enabled Nanomaterial Manufacturing; Machine Learning Strategies and Green Perspectives. *Nanomaterials* **2022**, *12*, 2646. [CrossRef]
8. Ellery, A. Are There Biomimetic Lessons from Genetic Regulatory Networks for Developing a Lunar Industrial Ecology? *Biomimetics* **2021**, *6*, 50. [CrossRef]
9. Capodaglio, A.G. Fit-for-purpose urban wastewater reuse: Analysis of issues and available technologies for sustainable multiple barrier approaches. *Crit. Rev. Environ. Sci. Technol.* **2021**, *51*, 1619–1666.
10. Conrad, C.L.; Ben Yin, Y.; Hanna, T.; Atkinson, A.J.; Alvarez, P.J.; Tekavec, T.N.; Reynolds, M.A.; Wong, M.S. Fit-for-purpose treatment goals for produced waters in shale oil and gas fields. *Water Res.* **2020**, *173*, 115467. [CrossRef]
11. Shivani Reddy, S.; Shravani, B.; Sreeja, K.; Mourya, M. A comparison study of MABC, ACO, ABC and PSO load balancing algorithms. In Proceedings of the 4th Smart Cities Symposium (SCS 2021), Manama, Bahrain, 21–23 November 2021; Volume 2021, pp. 223–228. [CrossRef]
12. Rao, R.; Rai, D.; Balic, J. Surface Grinding Process Optimization Using Jaya Algorithm. In *Computational Intelligence in Data Mining—Volume 2. Advances in Intelligent Systems and Computing*; Behera, H., Mohapatra, D., Eds.; Springer: New Delhi, Indian, 2016; Volume 411. [CrossRef]
13. Mittal, M.; Kaur, G. BBO Comparison with other Nature Inspired Algorithms to Resolve Mixels. *Int. J. Adv. Res. Comput. Eng. Technol. (IJARCET)* **2013**, *2* .
14. Chaudhari, K.; Thakkar, A. Travelling Salesman Problem: An Empirical Comparison Between ACO, PSO, ABC, FA and GA. In *Emerging Research in Computing, Information, Communication and Applications*; Shetty, N., Patnaik, L., Nagaraj, H., Hamsavath, P., Nalini, N., Eds.; Advances in Intelligent Systems and Computing; Springer: Singapore, 2019; Volume 906. [CrossRef]
15. Shashikumar, U.; Tsai, P.C.; Wang, C.T.; Lay, C.H.; Ponnusamy, V.K. Beyond biomimicry: Innovative bioinspired materials strategies and perspectives for high-performance energy storage devices. *Process Saf. Environ. Prot.* **2024**, *191*, 1193–1217. [CrossRef]
16. Melendez-Alvarez, J.; He, C.; Zhang, R.; Kuang, Y.; Tian, X. Emergent Damped Oscillation Induced by Nutrient-Modulating Growth Feedback. *ACS Synth. Biol.* **2021**, *10*, 1227–1236. [CrossRef] [PubMed]
17. Stocchi, F.; Vacca, L.; Ruggieri, S.; Olanow, C. Intermittent vs continuous levodopa administration in patients with advanced Parkinson disease: A clinical and pharmacokinetic study. *Arch. Neurol.* **2005**, *62*, 905–910. [CrossRef] [PubMed]
18. Haisenleder, D.; Yasin, M.; Marshall, J. Regulation of gonadotropin, thyrotropin subunit, and prolactin messenger ribonucleic acid expression by pulsatile or continuous protein kinase-C stimulation. *Endocrinology* **1995**, *136*, 13–19. [CrossRef]
19. Sulaiman, H.; Yusof, A.A.; Mohamed Nor, M.K. Automated Hydroponic Nutrient Dosing System: A Scoping Review of pH and Electrical Conductivity Dosing Frameworks. *AgriEngineering* **2025**, *7*, 43. [CrossRef]
20. Kunwar, S.R.; Bogati, S.; Fonsah, E.G.; Amgain, L.P. Economic assessment of adopting nutrient Expert® wheat model vs conventional wheat fertilizer application management in Morang, Nepal. *J. Agric. Stud.* **2019**, *7*, 38–48. [CrossRef]
21. Parmar, B.; Vishwakarma, A.; Gupta, V.; Jha, P.; Biswas, A.; Hati, K.; Malgaya, G.; Nagwanshi, A.; Singh, S.; Yadav, R. Influence of Different Residue Retention and Nutrient Levels on Crop Productivity of Soybean (Glycine max L.) under Conservation Agriculture. *Int. J. Plant Soil Sci* **2023**, *35*, 1192–1203. [CrossRef]
22. Gandhi, H.; Tandon, K.; Gite, S.; Pradhan, B.; Alamri, A. Navigating the complexity of money laundering: Anti–money laundering advancements with AI/ML insights. *Int. J. Smart Sens. Intell. Syst.* **2024**, *17*, 1–29. [CrossRef]
23. Song, Y.; Wang, Y.; Xu, T.; Shi, X.; Wang, A.J.; Wang, H.C. Chapter 32—Data-driven management strategies for carbon emissions and emerging contaminants control in wastewater treatment plants. In *Water Security: Big Data-Driven Risk Identification, Assessment and Control of Emerging Contaminants*; Liang, B., Gao, S.H., Wang, H.C., Wang, A.J., Eds.; Elsevier: Dordrecht, The Netherlands, 2024; pp. 537–549. [CrossRef]
24. Garg, M.; Kumari, S.; Agarwal, S. The Integration of Artificial Intelligence in Advanced Wastewater Treatment Systems. In *The AI Cleanse: Transforming Wastewater Treatment Through Artificial Intelligence*; Garg, M., Ed.; Springer Water, Springer: Cham, Switzerland, 2024. [CrossRef]
25. Xu, R.; Li, Y.; Luo, Y.; Fang, F.; Feng, Q.; Cao, J.; Luo, J. Prediction and evaluation of indirect carbon emission from electrical consumption in multiple full-scale wastewater treatment plants via automated machine learning-based analysis. *Acs Es&T Eng.* **2022**, *3*, 360–372.

26. Cairone, S.; Hasan, S.W.; Choo, K.H.; Lekkas, D.F.; Fortunato, L.; Zorpas, A.A.; Korshin, G.; Zarra, T.; Belgiorno, V.; Naddeo, V. Revolutionizing wastewater treatment toward circular economy and carbon neutrality goals: Pioneering sustainable and efficient solutions for automation and advanced process control with smart and cutting-edge technologies. *J. Water Process Eng.* **2024**, *63*, 105486. [CrossRef]
27. Chukwunweike, J.; Anang, A.N.; Adeniran, A.A.; Dike, J. Enhancing manufacturing efficiency and quality through automation and deep learning: Addressing redundancy, defects, vibration analysis, and material strength optimization. *World J. Adv. Res. Rev.* **2024**, *23*, 1272–1295. [CrossRef]
28. Soo, A.; Gao, L.; Shon, H.K. Machine learning framework for wastewater circular economy—Towards smarter nutrient recoveries. *Desalination* **2024**, *592*, 118092. [CrossRef]
29. Gulshin, I.; Kuzina, O. Optimization of Wastewater Treatment Through Machine Learning-Enhanced Supervisory Control and Data Acquisition: A Case Study of Granular Sludge Process Stability and Predictive Control. *Automation* **2025**, *6*, 2. [CrossRef]
30. Nti, I.K.; Teimeh, M.; Nyarko-Boateng, O.; Adekoya, A.F. Electricity load forecasting: A systematic review. *J. Electr. Syst. Inf. Technol.* **2020**, *7*, 13. [CrossRef]
31. Ayasrah, F.T.M.; Alsharafa, N.S.; Sivaprakash, S.; Srinivasa Rao, B.; Sengan, S.; Kumaran N. Strategizing Low Carbon Urban Planning Through Environmental Impact Assessment by Artificial Intelligence Driven Carbon Footprint Forecasting. *J. Mach. Comput.* **2024**, *4*, 1140–1151. [CrossRef]
32. Lin, D.; Zhang, L.; Xia, X. Hierarchical model predictive control of greenhouse climate to reduce energy cost. In Proceedings of the 2020 2nd International Conference on Industrial Artificial Intelligence (IAI), Shenyang, China, 23–25 October 2020; pp. 1–6. [CrossRef]
33. Gong, J.Y.; Zhou, X.L.; Faraj, Y.; Zou, L.B.; Zhou, C.H.; Xie, R.; Wang, W.; Liu, Z.; Pan, D.W.; Ju, X.J.; et al. Coumarin-Based Light-Responsive Composite Nanochannel Membranes for Precise Controlled Release of Pesticides. *Adv. Funct. Mater.* **2024**, *34*, 2314642.
34. Lowe, M.; Qin, R.; Mao, X. A Review on Machine Learning, Artificial Intelligence, and Smart Technology in Water Treatment and Monitoring. *Water* **2022**, *14*, 1384. [CrossRef]
35. Nikhar, C.K.; Vyas, G.S.; Dalvi, R.S.; Bhoye, D.Y. A critical review on applications of machine learning in wastewater treatment: Insights and implications for distillery wastewater. *Water Qual. Res. J.* **2024**, *60*, 1–18.
36. Zari, M.P. Biomimetic design for climate change adaptation and mitigation. *Archit. Sci. Rev.* **2010**, *53*, 172–183. [CrossRef]
37. López, M.; Rubio, R.; Martín, S.; Croxford, B. How plants inspire façades. From plants to architecture: Biomimetic principles for the development of adaptive architectural envelopes. *Renew. Sustain. Energy Rev.* **2017**, *67*, 692–703. [CrossRef]
38. Zhang, W.; Zeng, X.; Huang, Y.; Liang, J.; Wang, X.; Guo, J.; Li, Z.; Yang, K.; Zhang, J. Based on machine learning: Energy consumption optimization and energy efficiency evaluation for photovoltaic electro-fenton technology in wastewater treatment plant. *Renew. Energy* **2025**, *243*, 122626. [CrossRef]
39. Huang, R.; Wang, H.; Mąkinia, J.; Jin, S.; Zhou, Z.; Zhang, M.; Yu, C.; Xie, L. Optimizing the chemical removal of phosphorus for wastewater treatment: Insights from interpretable machine learning modeling with binary classification of elasticity and productivity. *Resour. Conserv. Recycl.* **2025**, *215*, 108147. [CrossRef]
40. Ma, Y.; Qiao, Y.; Chen, M.; Rui, D.; Zhang, X.; Liu, W.; Ye, L. How small is big enough? Big data-driven machine learning predictions for a full-scale wastewater treatment plant. *Water Res.* **2025**, *274*, 123041. [CrossRef] [PubMed]
41. Hou, P.; Liu, S.; Hu, D.; Zhang, J.; Liang, J.; Liu, H.; Zhang, J.; Zhang, G. Predicting biomass conversion and COD removal in wastewater treatment by phototrophic bacteria with interpretable machine learning. *J. Environ. Manag.* **2025**, *375*, 124282. [CrossRef]
42. Fabris, M. Distributed Optimization Strategies for Mobile Multi-Agent Systems. Ph.D. Thesis, University of Padova, Padua, Italy, 2019.
43. Besseris, G.J. Screening dense and noisy DOX-datasets with NN-blending and "dizzy" swarm intelligence: Profiling a water quality process. *Swarm Evol. Comput.* **2017**, *32*, 219–233. [CrossRef]
44. Calzolari, G.; Sumathy, V.; Kanellakis, C.; Nikolakopoulos, G. Decentralized Multi-Agent Reinforcement Learning Exploration with Inter-Agent Communication-Based Action Space. In Proceedings of the 2024 24th International Conference on Control, Automation and Systems (ICCAS), Jeju, Republic of Korea, 29 October–1 November 2024; pp. 319–324. [CrossRef]
45. Hu, S.; Shen, L.; Zhang, Y.; Tao, D. Learning multi-agent communication from graph modeling perspective. *arXiv* **2024**, arXiv:2405.08550.
46. Mulyani, H.; Budianto, G.P.I.; Margono, M.; Kaavessina, M. The influence of pH adjustment on kinetics parameters in tapioca wastewater treatment using aerobic sequencing batch reactor system. In *Proceedings of the AIP Conference Proceedings*; AIP Publishing: Jatinangor, Indonesia, 2018; Volume 1931.
47. Lai, J.; Then, Y.; Hwang, S.; Tam, Y.C.; Chua, C.C.N. Modelling Temperature Profiles in Food Waste Composting: Monod Kinetics Under Varied Aeration Conditions. *Process Integr. Optim. Sustain.* **2025**. [CrossRef]

48. Lv, Y.; Wang, L.; Liu, X.; Chen, B.; Zhang, M. Degradation kinetics of aromatic VOCs polluted wastewater by functional bacteria at laboratory scale. *Sci. Rep.* **2022**, *12*, 19053. [CrossRef]
49. West, A.J. Improving Optimization Capabilities of the Swarm Algorithm Through Augmenting Stigmergy with Other Means of Inter-Ant (Inter-Agent) Communication. Ph.D. Thesis, California State University Channel Islands, Camarillo, CA, USA, 2016.
50. He, H.J.; Koppel, A.; Bedi, A.S.; Stilwell, D.J.; Farhood, M.; Biggs, B. Decentralized Multi-agent Exploration with Limited Inter-agent Communications. In Proceedings of the 2023 IEEE International Conference on Robotics and Automation (ICRA), London, UK, 29 May–2 June 2023; pp. 5530–5536. [CrossRef]
51. Reyes-Vidal, Y.; López-Maldonado, J.; Casta neda, F.; Orozco, G.; Rivera, F. Evaluation of Inlet Design and Flow Rate Effect on Current Density Distribution in a Microbial Electrolysis Cell Using Computational Simulation Techniques, Coupling Hydrodynamics and Bioanode Kinetics. *Int. J. Chem. React. Eng.* **2018**, *16*, 309–323. [CrossRef]
52. Labuza, T.P.; Fu, B. Growth kinetics for shelf-life prediction: Theory and practice. *J. Ind. Microbiol.* **1993**, *12*, 309–323. [CrossRef]
53. Luo, X.; Tian, A.; Chen, Y.; Zhou, Y.; Tang, Q. Theoretical one-dimensional porous media model for microbial growth on pore plugging and permeability evolution and its verification. *J. Air Waste Manag. Assoc.* **2023**, *73*, 777–785. [CrossRef] [PubMed]
54. Gaillard, J.F.; Rabouille, C. Using Monod Kinetics in Geochemical Models of Organic Carbon Mineralization in Deep-Sea Surficial Sediments. In *Deep-Sea Food Chains and the Global Carbon Cycle*; Rowe, G.T., Pariente, V., Eds.; Springer: Dordrecht, The Netherlands, 1992; pp. 309–324. [CrossRef]
55. Saunshi, N.; Karp, S.; Krishnan, S.; Miryoosefi, S.; Jakkam Reddi, S.; Kumar, S. On the inductive bias of stacking towards improving reasoning. *Adv. Neural Inf. Process. Syst.* **2024**, *37*, 71437–71464.
56. Wu, Y.; Gadsden, S.A. Machine learning algorithms in microbial classification: A comparative analysis. *Front. Artif. Intell.* **2023**, *6*, 1200994. [CrossRef] [PubMed]
57. Wang, H.C.; Wang, Y.Q.; Wang, X.; Yin, W.X.; Yu, T.C.; Xue, C.H.; Wang, A.J. Multimodal Machine Learning Guides Low Carbon Aeration Strategies in Urban Wastewater Treatment. *Engineering* **2024**, *36*, 51–62. [CrossRef]
58. Sun, J.; Xu, Y.; Nairat, S.; Zhou, J.; He, Z. Prediction of biogas production in anaerobic digestion of a full-scale wastewater treatment plant using ensembled machine learning models. *Water Environ. Res. A Res. Publ. Water Environ. Fed.* **2023**, *95*, e10893. [CrossRef]
59. Lee, Z.; Wu, Y.C.; Wang, X. Automated Machine Learning in Waste Classification: A Revolutionary Approach to Efficiency and Accuracy. In Proceedings of the 2023 12th International Conference on Computing and Pattern Recognition, Qingdao, China, 27–29 October 2023; pp. 299–303.
60. Salles, R.; Mendes, J.; Araújo, R.; Melo, C.; Moura, P. Prediction of Key Variables in Wastewater Treatment Plants Using Machine Learning Models. In Proceedings of the 2022 International Joint Conference on Neural Networks (IJCNN), Padua, Italy, 18–23 July 2022; pp. 1–9. [CrossRef]
61. Freitas, D.; Lopes, L.G.; Morgado-Dias, F. Particle Swarm Optimisation: A Historical Review Up to the Current Developments. *Entropy* **2020**, *22*, 362. [CrossRef]
62. Abualigah, L. Particle Swarm Optimization: Advances, Applications, and Experimental Insights. *Comput. Mater. Contin.* **2025**, *82*, 1539–1592. [CrossRef]
63. Wang, J.; Lin, Y.; Liu, R.; Fu, J. Odor source localization of multi-robots with swarm intelligence algorithms: A review. *Front. Neurorobotics* **2022**, *16*, 949888. [CrossRef]
64. Saba, T.; Rehman, A.; Haseeb, K.; Alam, T.; Jeon, G. Cloud-edge load balancing distributed protocol for IoE services using swarm intelligence. *Clust. Comput.* **2023**, *26*, 2921–2931. [CrossRef]
65. Khamis, K.; Bradley, C.; Gunter, H.J.; Basevi, G.; Stevens, R.; Hannah, D.M. Calibration of an in-situ fluorescence-based sensor platform for reliable BOD5 measurement in wastewater. *Water Sci. Technol. A J. Int. Assoc. Water Pollut. Res.* **2021**, *83*, 3075–3091. [CrossRef]
66. Guz, Ł.; Łagód, G.; Jaromin-Glen, K.; Suchorab, Z.; Sobczuk, H.; Bieganowski, A. Application of Gas Sensor Arrays in Assessment of Wastewater Purification Effects. *Sensors* **2015**, *15*, 1–21. [CrossRef] [PubMed]
67. Freeman, B.; Jones, N.; Bartels, C.; Sunano, S.; Itonaga, T. Operation of a Submerged Hollow Fiber Membrane Bioreactor for Wastewater Treatment on Meeting Reclamation Criteria. Available online: https://membranes.com/wp-content/uploads/Documents/Technical-Papers/Application/Waste/Operation-of-a-Submerged-Hollow-Fiber-Membrane-Bioreactor-for-Wastewater-Treatment-on-Meeting-Reclamation-Criteria.pdf (accessed on 4 May 2025)
68. Bernardes, A.F.; Meng, Z.; Campos, L.C.; Coppens, M.O. Bio-inspired anti-fouling strategies for membrane-based separations. *Chem. Commun.* **2025**, *61*, 5064–5071. [CrossRef] [PubMed]
69. Radini, S.; Marinelli, E.; Akyol, Ç.; Eusebi, A.L.; Vasilaki, V.; Mancini, A.; Frontoni, E.; Bischetti, G.B.; Gandolfi, C.; Katsou, E.; et al. Urban water-energy-food-climate nexus in integrated wastewater and reuse systems: Cyber-physical framework and innovations. *Appl. Energy* **2021**, *298*, 117268. [CrossRef]

70. Saini, M.; Kumar, A.; Saini, D.K.; Gupta, P. Availability optimization of power generating units used in sewage treatment plants using metaheuristic techniques. *PLoS ONE* **2023**, *18*, e0284848. [CrossRef]
71. Muloiwa, M.; Dinka, M.; Nyende-Byakika, S. Modelling and optimizing hydraulic retention time in the biological aeration unit: Application of artificial neural network and particle swarm optimization. *S. Afr. J. Chem. Eng.* **2024**, *48*, 292–305. [CrossRef]

water

MDPI

Article

Integrating Spatial Concerns into Water Reuse Regulations: Insights from the European Union and the Iberian Peninsula

Teresa Fidélis [1,*], Arsham Afyouni [2] and Fayaz Riazi [1]

[1] GOVCOPP and Department of Environment and Planning, University of Aveiro, 3810-193 Aveiro, Portugal; fayaz.riazi@ua.pt
[2] Department of Environment and Planning, University of Aveiro, 3810-193 Aveiro, Portugal; arshamafyouni@ua.pt
* Correspondence: teresafidelis@ua.pt

Abstract: Water scarcity in Southern Europe, driven by climate conditions and water-intensive land use, is promoting water reuse adoption. Water reuse regulations are emerging, but little is known about integrating spatial concerns into their contents. This study examines how spatial issues are addressed within water reuse regulations adopted by the European Union (EU), Portugal, and Spain. Through a comparative content analysis, this research explores the inclusion and distribution of key terms related to water drivers, spatial concepts, and land use types within key sections, preamble, objectives, permitting, risk assessment, monitoring, and governance. The findings show that Portugal and Spain exhibit poorer integration of water scarcity compared to the EU, and Portugal does not address it in its objectives. In contrast, broad spatial terms are more prominent in Portugal, while Spain emphasises conservation and environmental areas more. Spatial terms are distributed differently across sections, reflecting different regulatory approaches. Surprisingly, none of the regulations link to plans. They mention risk management plans and, occasionally, circular economy and river basin management plans. Agriculture and urban activities dominate, although Portugal emphasises industry and green areas. This study highlights the need for more spatially informed water reuse regulations.

Keywords: integration; land use; land use planning; spatial planning; water reuse; water scarcity

1. Introduction

Increasing water scarcity events in various locations, alongside changes in population and climate, have intensified the need for coordinated efforts between land use planning and water resource management [1]. Water scarcity is often understood as an imbalance between supply and demand, but it can also be a matter of access [2]. Even when water is physically available, some regions or communities may experience scarcity due to governance failures, unequal distribution, or inadequate infrastructure [2]. Moreover, certain land use types or practices can also exacerbate water scarcity [3]. Water scarcity is therefore a constraint interrelated with spatial planning and the implementation of climate adaptation actions [4]. Integrating land use, water, and other relevant policies (e.g., energy and agriculture) is necessary to adopt a more holistic approach to managing water resources. Land use planning helps prevent water scarcity and fosters adequate locations for water reuse initiatives [5]. Land use planning may support the development of wastewater reuse infrastructure and initiatives, which can help meet growing water demand in a sustainable manner [4]. In addition, it can facilitate early and broad engagement with the public, key stakeholders, and organisations in the planning process for water reuse initiatives [6]. Despite the role of land use planning in preventing water scarcity, it often operates in isolation from water management, leading to missed opportunities for integrated solutions [1].

Water management considerations should be integrated with land use strategies in water reuse policies and regulations, which must ensure not only the sustainable supply of water but also equitable access to it across different territories. This approach helps to avoid the "take, make, waste" model and design new infrastructure with water efficiency and reuse in mind [5]. However, questions remain about what must be integrated and how this should occur [7]. Integration can be understood in various ways, such as governance integration, which involves bringing stakeholders together at different levels and incorporating local knowledge [8]. It can also refer to the alignment of policies and procedures. In the specific context of this study, integration focuses on embedding water resources and water initiatives as core principles within land use plans and strategies [9]. Relevant to water scarcity and reuse, this integration can be understood as the need to coordinate and align different aspects of water management, land use planning, and environmental protection to address water scarcity and promote sustainable water reuse [10]. Within this scope, two primary approaches have emerged: a technical approach, which emphasises innovations for water reuse, recycling, and reduced consumption, and a governance approach, which focuses on institutional and organisational dimensions, including policy and regulations [5], as well as spatial and water-related considerations. Governance is a complex concept, particularly in the context of sustainability, encompassing a wide range of definitions, frameworks, and approaches [11,12]. Although it has been the focus of extensive scholarly debate, a detailed exploration falls beyond the scope of this article. Governance approaches for water reuse involve, among other aspects, an integrated water resources lens; adapted institutional arrangements and decision-making processes; social, economic and territorial power relations balances; stakeholder participation mechanisms; co-creation processes; and adaptive strategies [13–16].

Water reuse regulations need to consider land use and territorial features and be tailored to specific local contexts [4,17]. Water reuse poses risks that necessitate effective monitoring and control. This involves evaluating the total volume of water available, including its quantity, quality, and intended purpose, to ensure appropriate treatment and means for transport or distribution. Additionally, potential impacts on soil, human health, and other environmental factors must be considered in relation to local land use pressures and territorial features [5]. Moreover, water reuse must be managed through a robust risk management plan that identifies relevant actors and their responsibilities. This plan should consider land use and territorial features, recognising, for example, that water reuse for agricultural irrigation may not be suitable in all areas [18].

This article aims to understand how water scarcity and territorial concerns are incorporated into water reuse regulations in the European Union (EU) and the two Iberian countries, Portugal (PT) and Spain (ESP), where climate-related pressures on water availability have increasingly intensified. In the scope of this study, the broad concept of "land use planning" is often used as an umbrella term to encompass land use, spatial, location, and territorial concerns. This research addresses the following questions: Are spatial concerns integrated into water reuse regulations? If yes, how do the patterns of this integration compare between different jurisdictions? For this purpose, the article develops a content analysis to examine how water scarcity and spatial concerns are embedded in the referred water reuse regulations. It analyses the EU (Regulation (EU) 2020/741 of the European Parliament and of the Council), Portuguese (Decree-Law No. 119/2019), and Spanish (Royal Decree 1085/2024) regulations. Although studies on regulations and policies are frequent in the literature, this study offers a novel approach by analysing regulations and categorising specific features of their content through a systematic content analysis approach, providing straightforward and comparable insights about real-world regulatory frameworks.

This article is structured into five sections. The second section outlines brief insights identified in the literature on the research topic. The third section presents the research design, materials, and analytical approaches. The fourth section presents the findings derived from the analysis. The fifth section presents the discussion where the findings are discussed, considering the methodology, data, literature, and practical implications. The conclusions summarise the main contributions and potential directions for future research.

2. Insights from the Literature

The scientific literature conceptualises the term "integration" regarding several critical concerns. First, integration is conceptualised as the need for strong articulation between land use planning and water resources planning systems [9,19,20]. This approach may integrate water scarcity and reuse into planning instruments and consider approval processes for treatment systems [19]. This integration is critical to maintaining and restoring surface water quality, implementing sustainable management measures, protecting ecosystem services, and reinforcing adaptation strategies to climate change. The integration could be achieved by establishing measures and rules in integrated plans that prevent land use impacts on water quality and quantity, creating synergies between water protection and land use control [20]. In addition, integration refers to the institutional arrangement integrating planning and regulatory approaches across multiple agencies and levels of government [9,21,22]. Faragher and Carden, 2023 [21], argue that the complexity of water systems spans various levels of government and involves coordination within different departments and directorates. Furthermore, diverse stakeholder perspectives and knowledge are crucial in decision-making, particularly in land use planning and water governance [23,24].

This section provides key insights from the scientific literature about the integration between water scarcity, territorial concerns, and water reuse. It aims to clarify the emerging concerns raised by the scientific community and to support the analysis of how water reuse regulations address these concerns, which are developed further in this article. It is based on a set of sixty-three scientific articles or reviews published in journals in English in the last five years, identified in the Scopus platform using the keywords "water scarcity" and "land use planning" or "spatial planning", and "water reuse" and "land use planning" or "spatial planning" in the fields of environmental and social sciences, selected in November 2024.

The analysis of the identified articles crossing water scarcity and land use planning or spatial planning reveals that climate change is emerging as a common and critical challenge [25,26]. The interconnectedness between water scarcity and land use is recognised [19,21] but still poorly explored by specific studies. In contexts of water scarcity, land use is often closely tied to water management, watersheds, water allocation, and relevant legislation. In contrast, land use planning is most often associated with broad development control and climate change [9,20]. The timeline of publications indicates that, although land use is not a new topic in the context of water scarcity and reuse, land use planning has only recently gained attention.

The analysis of the location of publications shows that while the United States and China lead in the number of publications, European countries such as the Netherlands, Spain, and Germany appear to have stronger collaborative links with other nations. In the European context, contributions are prominent in Germany, the United Kingdom, Austria, Romania, Spain, and the Netherlands. Interestingly, the Netherlands, Spain, and Portugal publications act as bridges, linking Europe to countries on other continents, such as South Africa, Australia, and China. Additionally, publications from the United States show a strong research collaboration with European countries, facilitating connections between Canada and Europe. Despite China's significant contributions as a leading publisher in the field, its collaboration with other countries remains relatively limited.

A further analysis of key research themes revealed four distinct groups of authors, each focusing on interconnected topics within water resources, land use, and climate change. One group centred their research on the Soil and Water Assessment Tool (SWAT), hydrological modelling, and climate change (e.g., [27,28]). Another group focused on water resource management and climate adaptation, particularly in China (e.g., [29–31]). A third group examined broader environmental impacts, including life cycle analysis, water footprints, and ecosystem sustainability [32,33]. Lastly, another group of authors focused on land and water use in drylands, including remote sensing, irrigation practices, and vegetation management in Central Asia (e.g., [34–36])

Land use planning offers diverse contributions to addressing water scarcity and fostering water reuse initiatives. One contribution referred to in the literature relates to the alignment of goals between water management and land use planning policy instruments and the creation of synergies between them, especially in locations with high water consumptive uses, complex water conflicts, or constraining water-related territorial features such as areas prone to scarcity [9,19,37]. Sustainable land use practices, such as reforestation and agroforestry, permeable urban land uses and runoff reduction, groundwater recharges, and restraining water uses to water availability, are among the suggestions found in the literature (see, for example, [25,26,38,39]). These measures may also facilitate adaptation to natural disasters like floods and soil erosion, particularly in wetter climates [40]. Land use planning strategies play a critical role in curbing excessive water use and supporting water reuse in high-density development settings [1,41] or facilitating the adoption of water reuse technologies by creating infrastructures for rainwater harvesting or distributing treated water for reuse [42]. Nevertheless, the need for a basin-level, holistic approach to balance socio-economic demands with ecological sustainability is stressed as a relevant criterion to consider when integrating water reuse technologies and approaches into land use plans [21,43]. According to the literature, specific interventions, such as limiting agricultural land conversion and preserving forests, are crucial for maintaining water flow and quality, as noted by Pouya and Turkoglu, 2020 [44], and Carse, 2012 [45]. Promoting drought-resistant crops and sustainable irrigation practices is also mentioned as relevant to support water reuse while reducing overall demand [46,47]. Adopting land use controls around reservoirs and implementing zoning regulations to prevent adverse impacts on water quality and availability are also referred to as valuable steps [20]. Together, these diverse strategies highlight the multifaceted role of land use planning in ensuring sustainable water resource management and reuse.

Water reuse regulations should account for the critical influence of location and land use on water production and availability and recognise the need to engage a broad range of stakeholders, including both land and water users, beyond solely private water rights holders [48]. The literature identifies three main concerns regarding the integration of water reuse regulations with land use planning. The first relates to the need for a holistic approach integrating water reuse planning and land use concerns [19,20,44]. Water reuse regulations and water resource management plans must be integrated with land use planning and zoning regulations to ensure that land use activities within water basins are compatible with and support the protection of water resources [37]. In addition, the distribution infrastructure on land must be integrated into planning strategies and instruments [19]. Second, the regulations must ensure institutional coordination across sectors and vertically at different levels [19,41]. Third, water reuse regulations must adopt more flexible regulations, increasing monitoring, control, and assessment. Accordingly, land use activities should be rigorously controlled, with their environmental impacts systematically assessed, potentially through permitting procedures [20,23,49].

From the brief insights drawn from the literature presented in the previous paragraphs, although the importance of integrating water management with land use planning and water-related territorial features is, in general, recognised, little is known about the specificities of water reuse regulations and how they should incorporate spatial concerns, especially in scarcity contexts. This article seeks to enrich this subject within the scientific community by adding new insights.

More specifically, from a regulatory standpoint, water reuse frameworks may attend to regional hydrological, climatic, ecological, and infrastructural variability, ensuring enforceability, equity, and sustainability. In areas identified as water-scarce or drought-prone, regulations attend to minimum levels of water reuse and establish enforceable financial or operational requirements, with reuse thresholds proportionate to the severity of local water stress. Regulatory approaches may consider reuse zones based on geographic and climatic parameters, including precipitation regimes, elevation, and coastal proximity, thereby ensuring spatial precision in regulatory applicability. They may also differentiate requirements according to land use categories and related environmental features, like specifying standards for greywater reuse in urban contexts, treated wastewater in agricultural applications, and industrial process water recovery in manufacturing zones. At the river basin scale, reuse regulations should align with integrated water reuse management principles, ensuring legal coherence across jurisdictions, protection of downstream rights, and basin-wide sustainability. In ecologically sensitive areas, regulations may consider constraints or require adaptive measures to prevent degradation, incorporating provisions for nature-based solutions such as constructed wetlands, where appropriate. Finally, regulatory frameworks should be tiered according to governance capacity and infrastructure availability, establishing enforceable minimum standards while allowing for flexible compliance pathways suited to local conditions and ensuring the due involvement of the various water users and stakeholders in the decision-making processes.

The following section presents the methodology adopted to explore further how water reuse regulations integrate these concerns.

3. Materials and Methods

3.1. Introduction to the Documents for Analysis

The emergence of water reuse in Europe has been driven by advancements in treatment technologies, escalating water scarcity, and the need for cost-effective solutions for non-potable applications [5,18]. The European Environment Agency has highlighted that water scarcity affects a significant portion of EU land and population annually, underscoring the urgency of implementing such measures [50]. The Water Reuse Regulation adopted by the European Union in 2020, establishing minimum standards for reclaimed water quality to ensure safety and public confidence, may contribute to addressing these challenges in member states. Before this, Portugal, facing increasing water scarcity, proactively adopted water reuse regulations in 2019, setting a precedent for other member states. Spain followed in 2024 with its regulations, reflecting a regional commitment to sustainable water management. Both countries, especially their southern regions, increasingly integrate water reuse and desalination strategies to adapt to climate change and secure water resources for current and future needs [4,6].

The EU Regulation 2020/741 establishes minimum standards for reclaimed water used in agricultural irrigation, addressing water scarcity caused by urbanisation, agriculture, and climate change. It promotes treated wastewater reuse as a safe, cost-efficient alternative to conventional water sources while harmonising standards across member states to encourage adoption. The regulation supports the circular economy by encouraging nutrient recovery, reducing dependency on synthetic fertilisers, and enhancing sustainability. Key provisions include mandatory risk management plans to ensure water safety and monitoring requirements for consistent water quality. The regulation also requires permitting water reuse projects, ensuring transparency, traceability, and public confidence. Unifying standards and promoting innovative practices enhance water security while protecting the environment and public health. In addition, it is also worth mentioning two more recent related documents which complement the EU water reuse regulatory landscape: the 2022 Commission Guidelines (2022/C 298/01), which focus on compliance and transparency, and the 2024 Delegated Regulation (EU) 2024/1765, which supplements EU Regulation 2020/741 by providing technical specifications for risk management. Their emphasis, however, is mainly on technical management and stakeholder coordination, providing no specific guidance on spatial concerns. This explains solely focusing on the 2020 regulations.

The Portuguese Water Reuse Regulations are established under the Decree-Law No. 119/2019. These regulations address water scarcity in Portugal, which is driven by increased consumption, urbanisation, and climate change. It promotes the reuse of treated wastewater for non-potable purposes such as irrigation, landscaping, and industrial use, offering a cost-effective and environmentally sustainable alternative to conventional water sources. By integrating water reuse into national strategies, the law supports circular economy principles, reduces greenhouse gas emissions, and enhances resilience to climate change. The law establishes a framework for safe water reuse, requiring risk-based evaluations to ensure compliance with health and environmental standards. It sets out licensing procedures for producing and using reclaimed water and outlines strict monitoring requirements to maintain water quality. These measures aim to protect public health, safeguard natural ecosystems, and reduce pressure on traditional water supplies. The Spanish Water Reuse Regulations are established under Royal Decree 1085/2024, reinforcing Spain's commitment to sustainable water management in response to increasing water scarcity, particularly in the southern regions. These regulations promote the reuse of treated wastewater for non-potable applications such as agricultural irrigation, landscape irrigation, industrial processes, and urban uses, aligning with circular economy principles. By integrating water reuse into national policies, the decree reduces dependence on freshwater resources, lowers water treatment costs, and enhances climate resilience. It establishes a risk-based framework to ensure the safe use of reclaimed water, setting quality standards, monitoring protocols, and licensing requirements for its production and application. These measures aim to protect public health, safeguard ecosystems, and support sustainable economic activities, aligning with European Union regulations on water reuse and ensuring harmonised and controlled implementation across Spain.

3.2. Analytical Steps

To address the research questions outlined in Section 1, an analytical approach based on a content analysis, a method widely tested in similar studies within environmental policy [20,51,52], is used to examine the three water reuse regulations. It assumes that the stronger the embeddedness of spatial concerns, the stronger their capacity to adopt water reuse regulations in a spatial manner. The analytical steps consisted of (i) conducting a preliminary reading of the three documents to identify relevant terms related to water scarcity and spatial concerns; (ii) developing a coding table with the selected terms for analysis and their rationale (see Table 1); (iii) counting the frequency of each term's occurrence across the entire document to gain an overall perspective of embeddedness of the spatial and land use concerns; and iv) counting the frequency of each term within main sections of the document, such as the preamble; scope and objectives; permitting, compliance and enforcement; risk assessment; monitoring and quality standards; and governance and public engagement (as detailed in Table 2), to display how the terms are incorporated in each regulatory frameworks. To reduce language differences, the English versions of the documents were analysed. Terms were searched both in the text and the tables of the documents. To ensure reliability and avoid biases, the coding table was first applied independently by each author, and then, the results were discussed and merged. The agreement measurement between the three authors, using Fleiss's kappa [53], was 0.82, which is considered almost perfect. Since the documents vary in length, word frequency was normalised using the formulas below to ensure a fair comparison. For this purpose, the following equations were used:

$$The\ frequency\ of\ land\ use\ planning\ related\ word = \frac{\begin{array}{c}Number\ of\ times\ a\ specific\ word\ is\ mentioned\\ in\ a\ document\end{array}}{Total\ number\ of\ words\ in\ a\ document} * 100$$

$$The\ frequency\ of\ land\ use\ planning\ related\ word = \frac{\begin{array}{c}Number\ of\ times\ a\ specific\ word\ is\ mentioned\\ in\ a\ part\ of\ the\ document\end{array}}{Total\ number\ of\ words\ in\ the\ document} * 100$$

Table 1. Coding table with the list of terms searched in the regulations.

Themes	Keywords	Query Words	Rational
Water Driver	Scarcity	water scarcity, drought	While drought refers to a temporary natural event caused by reduced rainfall, scarcity represents a long-term imbalance where water demand exceeds sustainable supply. Scarcity often reflects structural pressures from overuse or pollution, while drought highlights short-term climatic extremes. Identifying these terms helps assess whether regulations adopt a proactive, strategic framing or a reactive, emergency-based approach to water management. Their use signals how urgently and structurally water reuse is positioned within national or EU-level legal frameworks.
Spatial Terms	Land/Territory	territory, territorial, spatial, land, soil	These terms are relevant because they reflect how regulations consider the physical and territorial context of water reuse practices. For instance, zoning refers to spatial regulations that determine permissible land uses and their compatibility with water reuse logistics. For instance, restricting reuse infrastructure near sensitive land uses (e.g., residential zones or nature reserves) is a zoning-based concern. Infrastructure refers to the physical systems required for transporting, storing, and distributing treated wastewater, which must be spatially integrated within land use and planning instruments. References to river basins and territory indicate an ecological or hydrological approach, while terms like local and land show sensitivity to land use and planning boundaries. The inclusion of transport or transfer is especially relevant, as treated wastewater often needs to be conveyed across jurisdictions or land uses, raising logistical, environmental, and governance considerations. These keywords highlight the extent to which regulations incorporate spatially explicit spatial planning and infrastructure provisions.
	Local	place, local, location, space	
	Area	area, zone	
	Transport	transfer, transport, transportation	
	River basin	river, river basin, hydrographic basin	
	Plan	plan, planning	

Table 1. *Cont.*

Themes	Keywords	Query Words	Rational
Activities and land use types	Agriculture	agriculture, agricultural	These terms are used to identify sectors where water reuse is applied or encouraged. Their frequency helps map regulatory priorities and potential mismatches between legal intentions and real-world practices. For instance, strong mentions of agriculture or urban indicate alignment with water-intensive land uses, while terms like wetland, ecosystem, forest, or green area point to stricter environmental concerns. Understanding which land uses are emphasised reveals how comprehensive the regulatory vision is in promoting water reuse across different territorial functions.
	Urban	urban, public space	
	Industry	industry, industrial	
	Green areas	wetland, ecosystem, landscape, green area, garden	
	Forest	forest, forestry	

Table 2. Regulations' major sections and their relevance for the analysis.

Sections	Relevance to Spatial and Activities Keywords
Preamble	Acknowledges the importance of the concept. Spatial words, if they appear, may set the context for addressing resource constraints through spatial planning [19,20].
Scope and Objectives	Defines overarching concerns and provides a high-level acknowledgment of land use planning in reuse strategies [19], including the objectives, scopes, and obligations.
Permitting, Compliance, and Enforcement	Defines procedures for obtaining official permission to operate (licensing) and how parties ensure they follow the rules and the ability to effectively control land use activities essential for water reuse. Spatial terms may reflect how land use planning supports compliance (e.g., site selection for reuse facilities) [23,49].
Risk Assessment	Acknowledges the risk associated with water reuse and its mitigation, and highlights potential risks in water reuse systems that may be tied to land use concerns and the threats posed by land use changes to water resources recharge, especially in drought-prone zones [21].
Monitoring and Quality Standards	Covers monitoring mechanisms, and quality and quantity indicators to ensure safe water reuse. The inclusion of spatial terms may indicate the land use requirements for reuse, highlighting how different land contexts may affect the implementation of water reuse systems [20].
Governance and Public Engagement	Covers the cooperation, delegation, institutions, and public information. Spatial terms like local or territorial could highlight how governance structures support land use planning for water reuse, and their crucial role in institutional arrangements that integrate planning and regulatory approaches across multiple agencies and levels of government [9,21].

The following section presents the results of the analysis, derived from the application of the methodological framework outlined earlier.

4. Findings

4.1. Overall Embeddedness of the Spatial and Land Use Concerns in the Regulations

The analysis of the frequency of the terms referred to in the three documents (European Union (EU), Portugal (PT), and Spain (ESP)), as represented in Figure 1, reveals that their patterns differ slightly.

The EU regulations use the terms "scarcity" and "drought". However, "scarcity" is predominant. "Water availability" is also present, though less frequently than the previous terms. Though present, spatial terms like "area", "local", "space", and "territory" are not frequently used. The use of "plan" is mainly related to risk management plans, with only a single mention tied to the EU action plan for the circular economy and the river basin management plan. The references regarding land use types indicate that "agriculture" and "urban" are the most frequently referenced types of land uses and activities, followed by "industry" and marginally by "forest". Terms such as wetlands, coastal zones, or green are also absent from the EU regulation, despite their relevance to ecosystem-based reuse strategies and environmental planning.

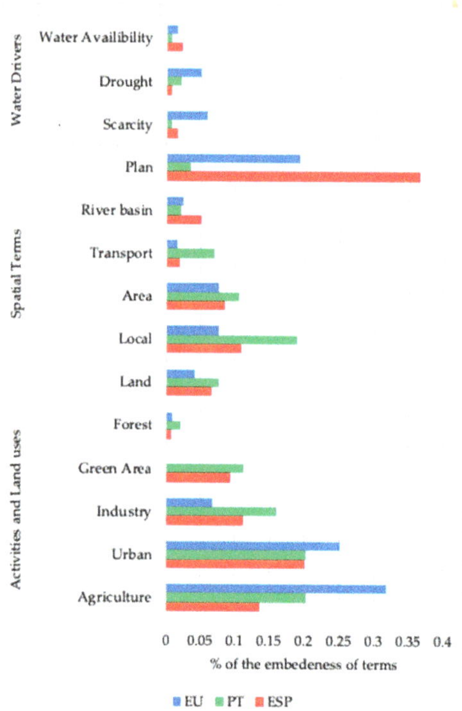

Figure 1. Frequency of the terms in the regulations (%).

Compared with the EU, the Portuguese regulation refers to "scarcity" or "drought" much less frequently, though "drought" is slightly more referred to than "scarcity". References to "water availability" are minimal and appear even less frequently than in the EU regulation. Spatial terms, like "area" and "land", are only marginally more referred to than in the EU regulations. However, terms like "local" or "territory" have higher occurrences. Although "plan" is mentioned in the Portuguese regulation, it is scarcely referred to in comparison to the other documents. Unlike the EU Regulations, where references to "plan" are mainly related to management plans, in Portugal, they are related to landscape and river basin management plans. For land use-associated terms, despite following the pattern of the EU regulations, showing a stronger embeddedness of terms regarding "agriculture" and "urban", though on a smaller scale, it shows a stronger emphasis on "industry" and "green areas".

The Spanish regulations follow the pattern of the EU regulations, giving more substantial use of the term "scarcity" than "drought". Moreover, "water availability" is mentioned more frequently than in the EU and even surpasses the use of "scarcity" and "drought" within the same regulation. On the contrary, the use of terms related to areas or zones is poorer. It refers to the "river basin" in the context of water reuse, with a stronger emphasis than the EU regulations. The use of the word "plan" is the most frequent in comparison with the other two documents. It is mainly related to risk management plans. Regarding land use-associated terms, the Spanish regulation differs from the EU regulation by including many more references to "forest", followed by "agriculture", "urban", and "industry". "Green areas", though enriched with terms like "wetland", are less referred to in these regulations.

When comparing the Portuguese and Spanish regulations, the Portuguese place slightly more emphasis on "drought", while the Spanish regulation places more emphasis on "scarcity". For spatial terms, the Portuguese regulation more frequently uses terms like "local" and "territory", while the Spanish regulation highlights "river basin" and "wetland". In terms of activities and land uses, the former gives more weight to "urban" and "industry" while the latter, on the other hand, places more emphasis on "forest". The most notable difference is the use of "plan", the most frequently used word in the Spanish regulation, mainly referring to risk management plans. In contrast, the Portuguese regulation uses "plan" far less, and when it appears, it is associated with landscape planning and river basin management plans. Table 3 presents a few examples of sentences in the documents using the terms.

Table 3. Examples of sentences of the regulations using the analysed terms present in the documents.

Scopes	Extracted Sentences Using Terms Under Analysis
Water Drivers	"The water resources of the Union are increasingly coming under pressure, leading to **water scarcity** and a deterioration in water quality. In particular, climate change, unpredictable weather patterns, and **drought** are contributing significantly to the strain on the availability of freshwater, arising from urban development and agriculture." (EU, Preamble) "The frequency and intensity of **droughts** and their environmental and economic damage have increased dramatically over the past 30 years. Thus, situations of high needs, combined with low or even absence of rainfall and high evapotranspiration, may cause situations of imbalance and **scarcity** in the availability of water." (PT, Preamble) "Ultimately, water reuse is currently seen as an essential element within integrated water management that allows it to contribute to the fulfilment of both the protection objectives of the public hydric domain, and the environmental objectives (...); confront water **scarcity** and **drought;** promote the circular economy; and support adaptation to climate change." (ESP Preamble)
Spatial Terms	"Hazard means a biological, chemical, physical or radiological agent that has the potential to cause harm to people, animals, crops or plants, other **terrestrial** biota, aquatic biota, **soils** or the environment in general." (EU, Article 3) "For the purposes of this Regulation, it should be possible for treatment operations and urban wastewater reclamation operations to take place in the same physical **location**, using the same facility, or different, separate facilities." (EU, Preamble) "Sprinkler irrigation at distances greater than 70 m from **residential areas** or **spaces** for public use." (PT, Annex II, Table 1) "In the **places** of production and use of water for reuse, information and signage must be placed, in the terms defined by decree of the member of the Government responsible for the area of the environment." (PT, Article 28) "These **plans** will recognize the obligation of, at least, the progressive replacement of the use of water for human consumption for the bucket of streets, in watertight and ornamental flows or the **river** of large **urban green areas**, for the use of regenerated water or other supply sources, all depending on the requirements that are detailed for each **territorial** scope in hydrological **planning**." (ESP, Article 26-3) "The use of regenerated waters for the artificial recharge of water bodies, as well as to satisfy the water needs of **wetlands** and other aquatic ecosystems will be considered, for the purposes of this Regulation, not as uses but as environmental destinations, and its process will be carried out in accordance with that established in the RDPH, being subject to the quality requirements specified in annexes I and II of this regulation." (ESP Article 3-3)
Activities and land use types	"Water reuse could contribute to the recovery of the nutrients contained in treated **urban** wastewater, and the use of reclaimed water for irrigation purposes in agriculture or forestry could be a way of restoring nutrients, such as nitrogen, phosphorus and potassium, to natural biogeochemical cycles." (EU, Preamble) "Water quality standards for reuse for irrigation to protect **agricultural**, **forestry**, and soil crops." (PT, Annex I, A) "Likewise, as a means to promote the circular economy and reinforce adaptation to climate change, it is necessary to regulate the use of regenerated water for certain uses, encouraging reuse within the **urban** water cycle itself." (ESP, Preamble)

4.2. Embeddedness of the Terms in the Major Sections of the Regulations

The analysis of the frequency of the terms referred to in the three documents based on their major structural sections, namely the preamble, objectives, permitting, risk assessment, monitoring, and governance sections, as represented in Figure 2, reveals interesting results.

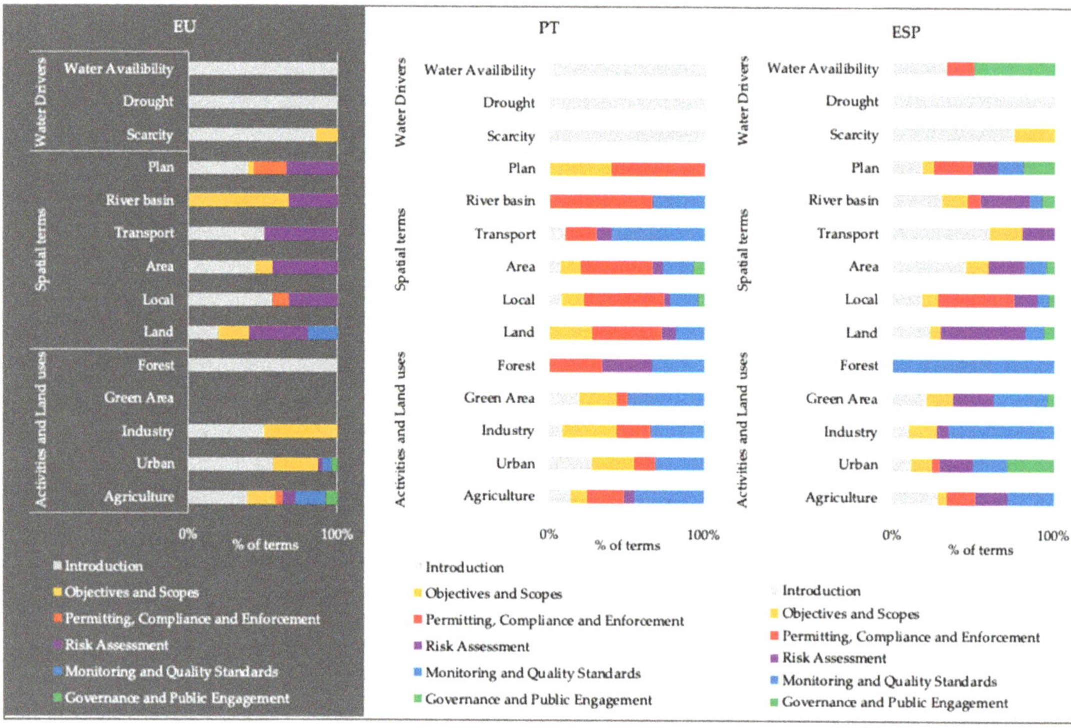

Figure 2. Embeddedness of spatial and land use-related terms in the content of the regulations (%).

Most of the terms related to water drivers, "drought" and "scarcity", are mainly referred to in the preamble of the three regulations, which have no force of law. Still, in the EU and Spanish regulations, "scarcity" also appears in the objectives section. In the EU regulation, the embeddedness of spatial terms is strong in the preamble, followed by the objectives and risk assessment sections. In contrast, in the Portuguese regulation, the spatial terms are mostly referred to in the scope and objectives, permitting, and monitoring sections. This document has no references to spatial concerns in the governance and public engagement section. The Spanish regulation shows spatial terms primarily in the objectives, permitting, and risk assessment sections. It also includes a few references in the governance and public engagement section, in contrast with the EU and the Portuguese regulations, in which they are scarcely mentioned. Regarding activities and land use types, the EU regulation mainly includes these terms in the preamble and objectives, with "urban" and "agriculture" appearing across all sections. In the Portuguese regulation, these terms are distributed among the different sections of the document. The Spanish regulation also refers to these terms in most sections, except for "forest", which is only mentioned in the monitoring section. The references to activities in the governance and public engagement section are higher than those in the other regulations. Table 4 summarises the major key regulatory contrasts.

Table 4. Main key regulatory contrasts among the regulations.

	EU	Portugal	Spain
Water Driver	Scarcity and drought	Drought	Water availability, scarcity
Spatial Concerns	Plan, area, local	Local, transport	Plan, river basin, wetland
Land Uses	Agriculture, urban	Industry, green areas, forest	Urban, green areas
Section Focus	Introduction Objectives and scope Risk assessment	Permitting, compliance and enforcement Monitoring and quality standards	Monitoring and quality standards Risk assessment Governance and public engagement

5. Discussion

This section discusses the key findings considering the research question (Are spatial concerns integrated into the water reuse regulations? If yes, how do the patterns of this integration compare between different country regulations?) and the quality of the findings, considering the methodology and their added value to the broader scientific community and regulators.

5.1. The Data, the Method, and the Findings

Water reuse regulations should account for the critical influence of location and land use on water production and availability and recognise the need to engage a broad range of stakeholders, including land and water users. At the EU level, water reuse regulations establish minimum standards for reclaimed water; the land uses and activities involved; and guidance for permitting, risk management and monitoring. At the member state levels, regulations derive from the general requirements and translate them into national and territorial specificities. Their narrative is influenced by institutional, administrative, and political traditions and cultural contexts, which were not considered in this article. The analysis of the regulations focused on substantive terms, which were searched through a systematic quantitative and qualitative method to assess the embeddedness of water scarcity, spatial terms, and land use types. This analysis of the overall embeddedness was performed mainly by counting the frequency of specific words or content to understand their contextual use [54]. The assessment of the distribution in the different sections of the regulation is complex and prone to error since the structure of the documents varies. To avoid misinterpretation of the texts, the different authors who cross-read the documents checked the content analysis. Also, the analysis was undertaken on regulations translated to English, facilitating the comparison yet missing the particularities of each mother language, likely to be lost after the translations. The analysis covered the EU and the two countries of the Iberian Peninsula, a territory vulnerable to drought events and facing water scarcity problems. It offered interesting insights into the different regulatory approaches regarding the inclusion of spatial concerns. Extending this study to other countries could offer new and valuable insights into spatial-wise water reuse regulations.

While the integration between land use and water resource planning is acknowledged as essential in the literature [8,12], the findings show that the integration of spatial concerns is present, though in tiny proportions, in the narrative and with different approaches within each regulatory analysed document. The comparative analysis of the EU, Portuguese, and Spanish regulations highlights notable differences in regulatory approaches regarding the inclusion of spatial concerns. The EU and Spanish regulations emphasise a long-term imbalance between water supply and demand. In contrast, Portugal uses more the perspective of a temporary phenomenon caused by reduced rainfall. This distinction indicates differences in how each regulation perceives and responds to these challenges. The Portuguese regulations, dating from 2019, precede the EU regulations, emerging as a regional response to severe drought events and the need to reduce freshwater consumption.

While it aligns with earlier EU directives like Directive 91/271/EEC on urban wastewater treatment, it lacks the harmonised criteria and cross-border coherence established by the 2020 EU regulation. This temporal difference also explains the limited emphasis on integrated risk management and the need for future updates to enhance alignment. The Spanish regulations, approved after that of the EU, mirror the perspectives of the EU on this long-term planning perspective. The stronger integration of broad spatial concerns in the Portuguese regulations suggests a stronger connection to land use planning. The Spanish regulations, however, stress more the need to consider the river basin in the context of water reuse, reinforcing a hydrological approach. While agriculture and urban activity uses are the most frequently referenced across all regulations, showing their importance in water reuse strategies, Portugal also places a stronger emphasis on industry and green areas than the EU and Spain. Also, Spain strongly focuses on ecological and environmental concerns, referring more to forests and wetlands. While spatial concerns are mainly addressed in the EU regulations' objectives and risk management sections, they are more widely distributed in the various sections of Portugal's and Spain's regulations. Portugal has a prominent presence in the permitting and monitoring sections, while Spain is also prominent in risk assessment. Similar results are evident in the references to land uses and activities. Moreover, spatial concerns are mainly addressed in the EU regulations' objectives and risk management sections, but they are more widely distributed in the various sections of Portugal's and Spain's regulations. Yet, Portugal's regulations are prominent in the permitting and monitoring sections, while Spain is also prominent in risk assessment. Similar results are evident in the references to land uses and activities. No specific references to spatial planning or spatial plans were found in any of the analysed water reuse regulations.

5.2. The Findings and the Literature

This study advances the discourse on land-water governance by bridging the gap between water reuse regulatory analysis and spatial planning. While prior research has predominantly focused on water reuse's technical and hydrological dimensions, this study highlights the spatial and regulatory dimensions, addressing a key gap in the literature [49,55]. It emphasises the need for enhanced policy coordination, explicit spatial frameworks, and the adoption of integrated tools (e.g., territorial zoning) that align water reuse with water scarcity, ecological, and spatial planning priorities [21,41]. Given that water availability and production are influenced by land use and spatial concerns [41], the incorporation of these spatial considerations into water reuse regulations necessitates integrating water scarcity and reuse into planning instruments, ensuring that land use activities enhance both water quality and quantity, and fostering synergies between water protection and spatial planning measures [9,19,20]. To this end, it is essential to tailor water reuse policies to regional hydrological, climatic, and infrastructural conditions while integrating zoning and permitting procedures for effective land-water governance [19,41,44]. However, this requires a thorough understanding of the local contexts and regulatory frameworks, calling for guidance about context-based solutions that account for territorial specificities. Effective policies must adapt to regional water stress levels, existing governance structures, and land use patterns to ensure the enforceability and sustainability of water reuse initiatives [20,23,49]. Furthermore, the findings emphasise that effective integration requires coordination across multiple government levels and agencies [19,41]. The findings can also be discussed through the lens of IWRM and adaptive governance, two key frameworks for environmental governance [13]. The EU regulations (EU 2020/741) reflect IWRM by promoting coherence across water use, land use, and sustainability through standardised risk management and cross-sectoral coordination. Portugal's regulations, developed earlier, follow a more sectoral approach, focusing on specific reuse, disregarding an IWRM approach. In contrast, Spain's regulations are more aligned with adaptive governance principles, emphasising public participation, multi-level coordination, and stakeholder engagement, making it more flexible and responsive to diverse interests and changing conditions. Moreover, Portugal's regulations do not explicitly reference public engagement mechanisms, which may limit transparency, reduce community involvement, and weaken the legitimacy of water reuse initiatives. In contrast, Spain's regulations include provisions for public information, awareness campaigns, and stakeholder participation, reflecting a more participatory governance approach [13,15,16]. This difference highlights how governance design can directly influence the acceptance and sustainability of water reuse practices.

Based on our findings, water reuse regulations may be strengthened by requiring zoning overlay checks for the location of water reuse projects, with due consideration for the functioning of the river basin, related hydrological units, and ecological water flows. Incorporating spatial concerns into permitting processes, as seen in the EU and Portugal, may also ensure alignment with land use and water quality protection. Additionally, technical guidance for infrastructure location, inspired by Portugal's planning approach, may enhance regulatory clarity. Strengthening monitoring protocols and stakeholder engagement, particularly in drought-prone areas, mirrors Spain's commitment to public participation. Finally, regulations should incorporate ecosystem-sensitive planning, as exemplified by the EU and Spain, to protect critical areas like wetlands and river basins.

Ensuring adequate inclusion of spatial concerns into water reuse regulations requires policy efforts that align land use and water reuse strategies, particularly through risk-based permitting, spatial planning tools, and governance coordination. Future research should broaden the scope of regulatory comparisons to include additional countries and examine real-world implementation while also exploring the factors that influence the integration of spatial concerns and land use planning in water reuse regulations to generate deeper insights for more effective and climate-resilient water governance.

6. Conclusions

Water reuse regulations are emerging, but little is known about the integration of scarcity and spatial concerns into their contents. Through a systematic content analysis, this article reveals significant differences in the embeddedness of the concerns in the EU, Portugal, and Spain water reuse regulations. The findings of this article offer valuable contributions to both the scientific community and regulatory bodies by revealing the varying approaches to integrating spatial concerns within water reuse regulations across different jurisdictions. By emphasising the disparities between the European Union, Portugal, and Spain, this research not only highlights how these regulations address the fundamental challenge of water scarcity but also underscores the importance of contextual considerations in policymaking. The identified differences, such as Portugal's orientation towards land use planning and Spain's ecological focus, unveil how national options and characteristics influence water management regulatory strategies. This insight can help foster more effective regulatory frameworks that are responsive to local water scarcity issues and environmental contexts. For regulators, these findings serve as a valuable resource for enhancing policy efficacy related to water reuse. By revealing the prominence of spatial concerns in specific regulatory sections, such as permitting and risk assessment, this study encourages regulators to adopt a more holistic approach when crafting laws. The emphasis on integrating land use considerations into water management can guide the development of more sustainable and adaptive policies that balance socio-economic demands with ecological preservation. Furthermore, the comparative analysis highlights best practices that can be shared and adapted among jurisdictions, promoting collaboration and knowledge exchange aimed at achieving water resource sustainability in the face of growing climate challenges.

Author Contributions: Conceptualisation, T.F.; Literature review, A.A. and F.R.; Methodology, T.F.; Data collection and analysis: T.F., A.A. and F.R.; Validation, T.F., A.A. and F.R.; Graphic design, A.A. and F.R.; Writing—original draft preparation, A.A. and F.R.; Writing—review and editing, T.F. and F.R.; Supervision, T.F. All authors have read and agreed to the published version of the manuscript.

Funding: This work was financially supported by the UID Research Unit on Governance, Competitiveness and Public Policies (GOVCOPP), funded by national funds through FCT—Fundação para a Ciência e a Tecnologia.

Data Availability Statement: The data are collected in an Excel database and are available on request to the authors.

Acknowledgments: The authors are grateful to the members of their university department for the support and incentive to develop and publish this research work and to the Spanish colleagues for facilitating access to their regulations.

Conflicts of Interest: The authors declare no conflicts of interest.

References

1. Shandas, V.; Lehman, R.; Larson, K.; Bunn, J.; Chang, H. Stressors and Strategies for Managing Urban Water Scarcity: Perspectives from the Field. *Water* **2015**, *7*, 6775–6787. [CrossRef]
2. Sen, A. *Poverty and Famines: An Essay on Entitlement and Deprivation*; Oxford University Press: Oxford, UK, 1981. Available online: https://books.google.com/books?hl=en&lr=&id=FVC9eqGkMr8C&oi=fnd&pg=PA11&dq=Sen,+A.+(1981).+Poverty+and+Famines:+An+Essay+on+Entitlement+and+Deprivation.+Oxford+University+Press&ots=hvZ4Ghc1Hn&sig=XaQ5zAfuQNGWAFp4dgMV-8qEnZY (accessed on 5 May 2025).
3. Peiris, M.T.O.V.; Gowshitharan, N. Impact assessment of land use on water scarcity in urban areas: Case of Jaffna Peninsula, Sri Lanka. *Water Policy* **2022**, *24*, 1146–1165. [CrossRef]
4. UN-Water. *UN-Water Analytical Brief on Unconventional Water Resources*; United Nations: Manhattan, NY, USA, 2020; pp. 1–59.
5. UNESCO i-WSSM. *Water Reuse Within a Circular Economy Contex*; UNESCO i-WSSM: Hwaseong, Republic of Korea, 2020.
6. EC–European Commission. Guidelines on Integrating Water Reuse into Water Planning and Management in the context of the WFD. In *Common Implementation Strategy for the Water Framework Directive*; EC–European Commission: Brussels, Belgium, 2016; pp. 1–95.
7. Biswas, A.K. Integrated Water Resources Management: A Reassessment. *Water Int.* **2004**, *29*, 248–256. [CrossRef]
8. Lerner, A.M.; Eakin, H.C.; Tellman, E.; Bausch, J.C.; Aguilar, B.H. Governing the gaps in water governance and land-use planning in a megacity: The example of hydrological risk in Mexico City. *Cities* **2018**, *83*, 61–70. [CrossRef]
9. You, J. Connecting Land and Water Planning in Colorado. *J. Plan. Educ. Res.* **2024**, *44*, 1970–1987. [CrossRef]
10. EMF–Ellen MacArthur Foundation. *Delivering the Circular Economy: A Toolkit for Policymakers*; Ellen MacArthur Foundation Publishing: Isle of Wight, UK, 2015.
11. Treib, O.; Bähr, H.; Falkner, G. Modes of governance: Towards a conceptual clarification. *J. Eur. Public Policy* **2007**, *14*, 1–20. [CrossRef]
12. Massuga, F.; Doliveira, S.L.D.; Kuasoski, M.; Mangoni, S.S. Modes of governance for sustainability transitions: Conceptual definitions. *Environ. Policy Gov.* **2024**, *34*, 663–678. [CrossRef]
13. Lemos, M.C.; Agrawal, A. Environmental Governance. *Annu. Rev. Environ. Resour.* **2006**, *31*, 297–325. [CrossRef]
14. Cleaver, F.; Hamada, K. Good' water governance and gender equity: A troubled relationship. *Gend. Dev.* **2010**, *18*, 27–41. [CrossRef]
15. Riazi, F.; Fidélis, T.; Matos, M.V.; Sousa, M.C.; Teles, F.; Roebeling, P. Institutional arrangements for water reuse: Assessing challenges for the transition to water circularity. *Water Policy* **2023**, *25*, 218–236. [CrossRef]
16. Matos, M.V.; Fidélis, T.; Sousa, M.C.; Riazi, F.; Miranda, A.C.; Teles, F. Institutional challenges in water reuse and circularity: Insights from co-creation processes in Southern Europe and Middle East. *Water Policy* **2025**, *27*, 429–446. [CrossRef]
17. Brears, R.C. Common Implementation Strategy for the Water Framework Directive and the Floods Directive. In *Regional Water Security*; Wiley: Hoboken, NJ, USA, 2016.

18. Maffettone, R.; Gawlik, B.M. *Technical Guidance—Water Reuse Risk Management for Agriculture Irrigation Schemes in Europe*; Publications Office of the European Union: Luxembourg, 2022. [CrossRef]
19. Bauer, S.; Wagner, M. Possibilities and Challenges of Wastewater Reuse—Planning Aspects and Realized Examples. *Water* **2022**, *14*, 1619. [CrossRef]
20. Rodrigues, C.; Fidélis, T. The integration of land use in public water reservoirs plans—A critical analysis of the regulatory approaches used for the protection of banks. *Land Use Policy* **2019**, *81*, 762–775. [CrossRef]
21. Faragher, T.P.T.; Carden, K. Groundwater governance for improving city water resilience in Cape Town, South Africa. *Front. Sustain. Cities* **2023**, *5*, 1062661. [CrossRef]
22. Godden, L.; Kung, A. Water Law and Planning Frameworks Under Climate Change Variability: Systemic and Adaptive Management of Flood Risk. *Water Resour. Manag.* **2011**, *25*, 4051–4068. [CrossRef]
23. Juntti, M.; Downward, S.D. Interrogating sustainable productivism: Lessons from the Almerían miracle. *Land Use Policy* **2017**, *66*, 1–9. [CrossRef]
24. Peiris, M.T.O.V.; Gowshitharan, N. Impact of urbanisation on ground water resources: Experience from developing countries. *Int. J. Agric. Resour. Gov. Ecol.* **2023**, *19*, 105–122. [CrossRef]
25. Hidayat, Y.; Rachman, L.M.; Wahjunie, E.D.; Baskoro, D.P.T.; Purwakusuma, W.; Yusuf, S.M.; Araswati, F.D. Water Balance Prediction by Simulating Land Use Planning and Water Retention Infrastructure in Upper Cisadane Sub-Watershed, West Java, Indonesia. *J. Pengelolaan Sumberd. Alam Dan Lingkung. (J. Nat. Resour. Environ. Manag.)* **2024**, *14*, 415. [CrossRef]
26. Yang, S.; Yang, D.; Zhao, B.; Ma, T.; Lu, W.; Santisirisomboon, J. Future Changes in High and Low Flows under the Impacts of Climate and Land Use Changes in the Jiulong River Basin of Southeast China. *Atmosphere* **2022**, *13*, 150. [CrossRef]
27. Arnold, N.; Richards, K.; Willis, I.; Sharp, M. Initial results from a distributed, physically based model of glacier hydrology. *Hydrol. Process.* **1998**, *12*, 191–219. [CrossRef]
28. Yang, W.; Andréasson, J.; Graham, L.P.; Olsson, J.; Rosberg, J.; Wetterhall, F. Distribution-based scaling to improve usability of regional climate model projections for hydrological climate change impacts studies. *Hydrol. Res.* **2010**, *41*, 211–229. [CrossRef]
29. Wang, Y.; Quan, Q.; Shen, B. Spatio-temporal variability of drought and effect of large scale climate in the source region of Yellow River. *Geomat. Nat. Hazards Risk* **2019**, *10*, 678–698. [CrossRef]
30. Wang, Z.; Ficklin, D.L.; Zhang, Y.; Zhang, M. Impact of climate change on streamflow in the arid Shiyang River Basin of northwest China. *Hydrol. Process.* **2012**, *26*, 2733–2744. [CrossRef]
31. Zhang, H.; Wang, B.; Lan, T.; Shi, J.; Lu, S. Change-point detection and variation assessment of the hydrologic regime of the Wenyu River. *Toxicol. Environ. Chem.* **2016**, *98*, 358–375. [CrossRef]
32. Pfister, S.; Koehler, A.; Hellweg, S. Assessing the Environmental Impacts of Freshwater Consumption in LCA. *Environ. Sci. Technol.* **2009**, *43*, 4098–4104. [CrossRef]
33. Rockström, J.; Falkenmark, M.; Karlberg, L.; Hoff, H.; Rost, S.; Gerten, D. Future water availability for global food production: The potential of green water for increasing resilience to global change. *Water Resour. Res.* **2009**, *45*. [CrossRef]
34. Conrad, S.A.; Hall, M.; Cook, S.; Geisenhoff, J. Key decisions for sustainable utility energy management. *Water Supply* **2010**, *10*, 721–729. [CrossRef]
35. Lamers, L.P.M.; Dolle, G.E.T.; Van Den Berg, S.T.G.; Van Delft, S.P.J.; Roelofs, J.G.M. Differential responses of freshwater wetland soils to sulphate pollution. *Biogeochemistry* **2001**, *55*, 87–101. [CrossRef]
36. Khamzina, A.; Lamers, J.P.A.; Vlek, P.L.G. Nitrogen fixation by Elaeagnus angustifolia in the reclamation of degraded croplands of Central Asia. *Tree Physiol.* **2009**, *29*, 799–808. [CrossRef]
37. Fidelis, T.; Roebeling, P. Water resources and land use planning systems in Portugal—Exploring better synergies through Ria de Aveiro. *Land Use Policy* **2014**, *39*, 84–95. [CrossRef]
38. Scherr, S.J.; Sthapit, S. *Mitigating Climate Change Through Food and Land Use*; Worldwatch Institute: Washington, DC, USA, 2009; Volume 179.
39. Narany, T.S.; Ramli, M.F.; Fakharian, K.; Aris, A.Z. A GIS-index integration approach to groundwater suitability zoning for irrigation purposes. *Arab. J. Geosci.* **2016**, *9*, 502. [CrossRef]
40. An, L.; Zhong, S.; Shen, L. Dynamic Effects of Climate and Land Use Policies on Water Yield in Drylands—A Case Study in the Northwest of China. *Water* **2022**, *14*, 3940. [CrossRef]
41. Cremades, R.; Sanchez-Plaza, A.; Hewitt, R.J.; Mitter, H.; Baggio, J.A.; Olazabal, M.; Broekman, A.; Kropf, B.; Tudose, N.C. Guiding cities under increased droughts: The limits to sustainable urban futures. *Ecol. Econ.* **2021**, *189*, 107140. [CrossRef]
42. Ekmekcioğlu, Ö.; Koc, K.; Dabanli, I.; Deniz, A. Prioritizing urban water scarcity mitigation strategies based on hybrid multi-criteria decision approach under fuzzy environment. *Sustain. Cities Soc.* **2022**, *87*, 104195. [CrossRef]
43. Fovargue, R.E.; Rezapour, S.; Rosendahl, D.; Wootten, A.M.; Sabzi, H.Z.; Moreno, H.A.; Neeson, T.M. Spatial planning for water sustainability projects under climate uncertainty: Balancing human and environmental water needs. *Environ. Res. Lett.* **2021**, *16*, 034050. [CrossRef]

44. Pouya, S.; Turkoglu, H. Evaluation of the water resource plans in Turkey based on sustainable water management principles. *Sustain. Water Resour. Manag.* **2020**, *6*, 91. [CrossRef]
45. Carse, A. Nature as infrastructure: Making and managing the Panama Canal watershed. *Soc. Stud. Sci.* **2012**, *42*, 539–563. [CrossRef]
46. Lei, Y.; Zhang, H.; Chen, F.; Zhang, L. How rural land use management facilitates drought risk adaptation in a changing climate—A case study in arid northern China. *Sci. Total Environ.* **2016**, *550*, 192–199. [CrossRef]
47. Mabhaudhi, T.; Chimonyo, V.; Modi, A. Status of Underutilised Crops in South Africa: Opportunities for Developing Research Capacity. *Sustainability* **2017**, *9*, 1569. [CrossRef]
48. Álvez, A.; Espinosa, P.; Castillo, R.; Iglesias, K.; Bañales-Seguel, C. An Urgent Dialogue between Urban Design and Regulatory Framework for Urban Rivers: The Case of the Andalién River in Chile. *Water* **2022**, *14*, 3444. [CrossRef]
49. Zhao, Y.; Wang, C.; Cao, X.; Song, S.; Wei, P.; Zhu, G. Integrated environmental assessment of a diversion-project-type urban water source considering the risks of novel and legacy contaminants. *Sci. Total Environ.* **2024**, *951*, 175380. [CrossRef] [PubMed]
50. EEA–European Environment Agency. *Drivers of and Pressures Arising from Selected Key Water Management Challenges: A European Overview*; EEA–European Environment Agency: Copenhagen, Denmark, 2021.
51. Mendes, R.; Fidélis, T.; Roebling, P.; Teles, F.; Farrelly, M. What is spatial planning saying? A conceptual and methodological framework to assess the institutionalization of nature using critical discourse analysis. *Crit. Discourse Stud.* **2024**, *21*, 274–292. [CrossRef]
52. Fidélis, T.; Cardoso, A.S.; Riazi, F.; Miranda, A.C.; Abrantes, J.; Teles, F.; Roebeling, P.C. Policy narratives of circular economy in the EU—Assessing the embeddedness of water and land in national action plans. *J. Clean. Prod.* **2021**, *288*, 125685. [CrossRef]
53. Fleiss, J.L. Measuring nominal scale agreement among many raters. *Psychol. Bull.* **1971**, *76*, 378–382. [CrossRef]
54. Phillips, N.; Lawrence, T.B.; Hardy, C. Discourse and Institutions. *Acad. Manag. Rev.* **2004**, *29*, 635. [CrossRef]
55. Fidelis, T.; Rodrigues, C. The integration of land use and climate change risks in the Programmes of Measures of River Basin Plans—Assessing the influence of the Water Framework Directive in Portugal. *Environ. Sci. Policy* **2019**, *100*, 158–171. [CrossRef]

MDPI AG
Grosspeteranlage 5
4052 Basel
Switzerland
Tel.: +41 61 683 77 34

Water Editorial Office
E-mail: water@mdpi.com
www.mdpi.com/journal/water

www.ingramcontent.com/pod-product-compliance
Lightning Source LLC
Chambersburg PA
CBHW040825250926
51850CB00007BA/1261
* 9 7 8 3 7 2 5 8 6 0 7 5 3 *